Sets and Ordered Structures

Sergiu Rudeanu
University of Bucharest, Romania

Contents

Foreword

Professor Sergiu Rudeanu, the author of this eBook, is a well-known research mathematician, working in the foundations of mathematics and mathematical logic.

In the 1960s he pioneered with P. L. Ivănescu (Hammer) the use of Boolean methods in operational research (their book *Boolean Methods in Operations Research and Related Areas*, Springer, Berlin, 1968 was published in French by Dunod, Paris in 1970; it is still cited after more than 40 years since publication). His many papers in Boolean algebras, their generalisations and applications—undoubtedly, Professor Rudeanu's main area of research—are complemented by some important books, including S. Rudeanu, *Boolean Functions and Equations*, North-Holland and Elsevier, Amsterdam, New York, 1974, V. Boicescu, A. Filipoiu, G. Georgescu, S. Rudeanu, *Łukasiewicz-Moisil Algebras*, North-Holland, Amsterdam, 1991, S. Rudeanu, *Lattice Functions and Equations*, Springer-Verlag, London, 2001, and R. Padmanabhan, S. Rudeanu, *Axioms for Lattices and Boolean Algebras*, World Scientific, NJ, 2008.

Professor Rudeanu is also a highly appreciated lecturer who taught many generations of undergraduate and graduate students at the University of Bucharest. His lectures are crystal clear, well-organised and rigorous, but not intimidating; many of them have been collected in textbooks published, in Romanian, by the University of Bucharest Publishing Company. My wife Elena and myself fondly remember his lectures from decades ago: they influenced our careers in no single way.

The eBook *Sets and Ordered Structures* presents, in Rudeanu's classical style, basic methods and results of the theory of ordered sets as tools for various branches of mathematics. The important notions and concepts are first presented in a gentle, informal manner; then, more abstract or axiomatic versions follow. The author is enthusiastic about his subject and his passion is well transmitted to the reader.

The eBook's itinerary starts with elements of naive set theory followed by a sketch of axiomatic set theory and categories; they fix the "mathematical language" for the remainder of the eBook. The next chapter presents partially and totally ordered sets, including a discussion of the axiom of choice and well-ordered sets. The next chapter deals with transfinite algebras, cardinal and ordinal numbers; the more abstract part presents von Neumann's construction of ordinals. The chapter on lattice theory includes modular, distributive, Boolean and complete lattices, as well as closure operators and Galois connections; these tools will be used in the study of universal algebras. The next chapter discusses meet and join representations in a lattice, and then use them for set-theoretical embeddings and the topological duality for distributive lattices. The last chapter presents a few applications in algebra, topology, universal algebra, formal concept analysis, and mathematical logic.

This eBook is an essential companion for mathematicians—students or professionals alike—computer scientists, and information-engineers. Its right place is on the iPad of the reader, who can read, browse, or annotate it. I hope that the reader will enjoy this eBook as much as I do.

Cristian S. Calude
The University of Auckland
New Zealand

Preface

This eBook presents several basic methods and results of the theory of ordered sets which are currently used in various branches of mathematics. The first chapter includes, besides the usual introductory elements of set theory, a sketch of axiomatic set theory and of the theory of categories, which provides a useful language. Then the fundamentals of partially and totally ordered sets are presented, including a few properties equivalent to the axiom of choice and culminating with well-ordered sets. The next chapter deals with transfinite numbers and includes von Neumann's construction of ordinals. An introduction to lattice theory presents modular, distributive, Boolean and complete lattices, as well as closure operators and Galois connections; these include tools which serve as prerequisites to universal algebra. Chapter 5 first deals with meet and join representations in a lattice, then comes back to the set-theoretical framework: set-theoretical embeddings and topological duality for distributive lattices. Chapter 6 sketches a few applications.

It is hoped that this volume may serve as a textbook for undergraduate and graduate mathematics students, as well as a reference book for mathematicians working in fields different from set theory or algebra. Perhaps a more appropriate title would have been "Sets and Ordered Structures for the Working Mathematician". The interest of the eBook for an algebraist consists in implicit teaching suggestions.

Chapter 1 begins with a quite informal sketch of the Zermelo-Fraenkel axiomatic system for set theory, having in mind the idea that although the axiomatic line will not be followed in the sequel, the mathematician working within the framework of naive set theory cannot be allowed to ignore the existence of foundations. The next section gathers in a systematic presentation the most frequently used properties of correspondences, relations and functions; thus e.g. each of the concepts of injection, surjection and bijection is characterized by 6-8 equivalent conditions. Categories and functors are then introduced; although the book is not directed towards category theory, a few categorical ideas and facts will be explicitly illustrated within the concrete categories which constitute the object of this eBook. On the other hand, the eBook will sometimes gain in rigour and conciseness by making use of the categorical language, e.g. in the sections devoted to duality for posets and to the algebraic and order-theoretical aspects of lattices and representation theory.

Chapters 2 and 3 are devoted to those concepts and properties related to ordered sets which are of current use throughout the whole mathematics. The systematization offered here includes the complete existential theory of the distinguished elements which a partially ordered set may possess: least/greatest element, minimal/maximal elements, g.l.b./l.u.b. (*i.e.*, the determination of all logical implications between these concepts). The general study of partially ordered sets, or posets for short, is then successively specialized to chains, or totally ordered sets, and well-ordered sets. Ordinals are introduced after as many of their properties as possible have been proved within the more elementary framework of well-ordered sets. The theory of ordinals and cardinals is based on the axiom stating that an abstract entity can be associated with each class of isomorphic well-ordered sets, such that to distinct classes correspond distinct entities. However in the end of chapter 3 we present von Neumann's construction, in which every ordinal is obtained as the set of all lower ordinals.

Chapters 4 and 5 study a few topics from lattice theory, which is another specialization of the idea of order. The selection has been guided by the following three aims: to construct lattice-theoretical tools broadly used in various branches of mathematics, to sketch the lattice-theoretical background necessary to universal algebra, and to build up the representation theory for certain classes of lattices (and even of posets), thus returning, in a certain sense, to the purely set-theoretical framework. The presentation starts with semilattices, according to a general tendency of this eBook, namely to establish the results under no more hypotheses than are actually needed, whenever this does not complicate the proofs. Then lattices, modular and distributive lattices, Boolean algebras and complete lattices (in particular, compactly generated lattices) are successively studied. Closure operators and Galois connections are also investigated, in view of their numerous applications. Filters and ideals are studied both for their own interest and due to their rôle in the representation theory of various types of posets and lattices. Under the general heading of representation are included, on the one hand, meet and join decompositions of elements in certain types of lattices, and on the other hand, set-theoretical embeddings for various types of posets and lattices, as well as the duality theory for distributive lattices with 0 and 1, in the general setting of H.A. Priestley.

Chapters 1-5 represent our selection of the fundamentals of ordered sets used in mathematics, while Chapter 6 actually sketches several applications.

Many examples and exercises are included. The eBook is self-contained with the slight exception of certain examples taken from other fields of mathematics. The bibliography contains items which we recommend to the reader desirous to have a deeper insight of the various topics presented in this eBook.

Acknowledgements

I began to write a draft of this eBook while I was a visiting professor at the University of Oran, Algeria, where I found excellent working conditions. Then I experimented in my courses at the University of Bucharest, with good results, several teaching ideas now included in the eBook.

I am grateful to Professor Cristian S. Calude for his invaluable help in the realization of this eBook.

Sergiu Rudeanu

University of Bucharest, Romania
E-mail: srudeanu@yahoo.com

Chapter 1

Sets

As stated in the Preface, the aim of this eBook is to study several order-theoretical tools which are currently used in various fields of mathematics, particularly in set theory, which, in its turn, provides a tool for the whole mathematics. Thus we begin quite naturally with the construction of the essentials of the set-theoretical framework, with special emphasis on order-theoretical aspects.

The first chapter could be more precisely titled "Set theory before the introduction of the concept of order" and comprises three sections. The first one sketches an axiomatic construction of set theory. This is motivated by our belief that although most mathematicians work within the framework of naive set theory, no mathematician could ignore the existence of foundations. However, as the axiomatic line will not be followed in the sequel, the presentation is quite informal. The next section gathers in a systematic presentation the most frequently used properties of correspondences, relations and functions; thus e.g. each of the concepts of injection, surjection and bijection is characterized by 6-8 equivalent conditions. The last section sketches a few categorical prerequisites which will enable us to use the language of categories whenever it will be convenient in the subsequent sections.

The last two sections have also the aim of specifying terminology and notation.

Keywords: Zermelor-Fraenkel system, axiom, paradox, correspondence, function, category, morphism, bimorphism, isomorphism, functor, natural transformation.

1.1 A Sketch of Axiomatic Set Theory

The aim of this section is an informal and very brief presentation of the Zermelo-Fraenkel system of axioms for set theory, together with some of its

Sergiu Rudeanu

easiest and most important consequences; for more details, the reader is referred to Halmos [1960]. Our presentation has been also much influenced by Fraenkel and Bar Hillel [1958] and Barnes and Mack [1975]. We conclude with a brief discussion of classes and the concept of universe.

The Zermelo-Fraenkel axiomatization of set theory (ZF) handles certain objects called *sets* and a relationship $\in$ between them, called *membership*; if A and B are sets, $A \in B$ is read "(the set) A belongs to (the set) B". So in ZF members of sets are sets themselves, whereas in everyday mathematics this is usually not the case. However this discrepancy is not really troublesome, because we can identify the objects x of any mathematical theory with the sets $\{x\}$. Besides, many fundamental mathematical objects are realized as sets: an ordered pair is a set (cf. the definition $(X, Y) = \{\{X\}, \{X, Y\}\}$ below); a binary relation and in particular a function is a triplet, hence a set, or, if identified with its graph, it is a set of ordered pairs; a quotient set is a set of cosets, hence of sets; natural numbers are sets and, more generally, ordinal numbers can be realized as sets (cf. the construction after ZF6 below and Ch.3§3); integer numbers are cosets; rational numbers are cosets; etc.

We assume that logic is prior to ZF, which enables us to infer theorems from the axioms. In so doing we use several definitions which introduce many new mathematical signs, and abbreviations which allow us to shorten reasonings. The first definition is that of *inclusion*: it is an abbreviation for the property "for every Z, the relation $Z \in X$ implies $Z \in Y$". This situation is stenographed as $X \subseteq Y$ and is read "X is included in Y (warning: not "contained", as one sometimes hears!), or "X is a *subset* of Y; we may also write $Y \supseteq X$ instead of $X \subseteq Y$.

It is easily seen that this definition implies the *reflexivity* and *transitivity* of the inclusion relation:

$$X \subseteq X ,$$

$$X \subseteq Y \,\&\, Y \subseteq Z \Longrightarrow X \subseteq Z ,$$

respectively. The *equality* of sets is defined by

$$X = Y \Longleftrightarrow X \subseteq Y \,\&\, Y \subseteq X ;$$

the implication $\Longleftarrow$ from the above definition is known as the *antisymmetry* of the inclusion relation. Now we see that the equality is a reflexive, symmetric and transitive relation:

$$X = X ,$$

$$X = Y \Longrightarrow Y = X ,$$

$$X = Y \,\&\, Y = X \Longrightarrow X = Z ;$$

it is important to note that $X = Y$ means $Z \in X \Longleftrightarrow Z \in Y$, which is expressed by saying that a set is uniquely determined by its elements. We can also introduce the relation of *strict inclusion*:

$$X \subset Y \Longleftrightarrow X \subseteq Y \,\&\, X \neq Y ,$$

where $X \neq Y$ means $\neg(X = Y)$; we may also write $Y \supset X$ instead of $X \subset Y$. We see that for every X, Y, Z we have, with a similar convention for $\not\subset$,

$$X \not\subset X ,$$

$$X \subset Y \Longrightarrow Y \not\subset X ,$$

$$X \subset Y \,\&\, Y \subset Z \Longrightarrow X \subset Z ,$$

$$X \subset Y \,\&\, Y \subseteq Z \Longrightarrow X \subset Z ,$$

$$X \subseteq Y \,\&\, Y \subset Z \Longrightarrow X \subset Z ,$$

$$X \subseteq Y \Longleftrightarrow X \subset Y \text{ or } X = Y ;$$

the verification of all the above properties is routine.

Note that the above results have been established solely on a logical base, with no supplementary assumption. Now we introduce axioms.

ZF1 (axiom of extensionality). *For every X, Y, Z, we have $X \in Y \,\&\, X = Z \Longrightarrow Z \in Y$.*

Note that the following alteration of ZF1 is valid without assuming any axiom: $X \in Y \,\&\, X = Z \Longrightarrow X \in Z$. So this property and axiom ZF1 show that each term of a membership relation $X \in Y$ can be replaced by an equal set.

ZF2 (axiom scheme of subsets). *For every set X and every property $\pi(Y)$ defined for sets, there is a set Z such that $Y \in Z$ if and only if $Y \in X$ and Y has property π.*

The term "property" is taken informally in our description; in a formal presentation of ZF, $\pi(Y)$ is a formula of the predicate calculus of first order, containing the free variable Y. Roughly speaking, a formula is defined inductively: free individual variables and individual constants are terms; any functor applied to terms yields a term; any predicate applied to terms yields a formula; propositional constants are formulas; propositional connectives applied to formulas yield formulas; quantifiers applied to to formulas yield formulas in which the quantified variables become bound individual variables.

The set constructed in ZF2 is denoted $\{Y \in X \mid \pi(Y)\}$ and is uniquely determined by Z and π in view of the remark stating that a set is uniquely determined by its elements.

ZF2 implies in particular the existence and uniqueness of the *empty set* $\varnothing$, which has no elements. For take an arbitrary set Z and a property $\pi(Y)$ false for every Y, e.g. $Y \neq Y$; then $\varnothing = \{Y \in Z \mid Y \neq Y\}$. If $\varnothing'$ is another empty set, then since property $Y \in \varnothing$ is false for every Y, the implication $Y \in \varnothing \Longrightarrow Y \in \varnothing'$ is true, showing that $\varnothing \subseteq \varnothing'$; one proves similarly that $\varnothing' \subseteq \varnothing$, hence $\varnothing = \varnothing'$. Note that $\varnothing \subseteq X$ for every set X.

Also, if A is a subset of a set X, there is a unique set $\{Y \in X \mid \neg(Y \in A)\}$, known as the *complement* of A (with respect to X) and which is denoted by $X \setminus A$ or by $\mathbf{C}_X A$ or simply by $\mathbf{C}A$ when no confusion is possible.

ZF3 (axiom of pairing). *For every X, Y with $X \neq Y$, there exists a set Z having X and Y as members.*

Moreover, we can prove the existence of a set having X and Y as members and only them. For if Z denotes a set such that $X \in Z$ and $Y \in Z$, then the set $\{S \in Z \mid S = X \text{ or } S = Y\}$ exists by ZF2 and its sole members are X, Y. Moreover, it is immediately seen from the definition that this set is equal to any other set having X and Y as its sole members. The unique set consisting of X and Y is denoted $\{X, Y\}$.

ZF4 (axiom of union). *For every set X containing at least two elements* (i.e., $\exists A \, \exists B \; A \in X \;\&\; B \in X \;\&\; A \neq B$) *there exists a set V such that if $Z \in Y$ and $Y \in X$ then $Z \in V$.*

Now ZF2 implies the existence of the set

$$U = \{X \in V \mid \exists Y \, (Z \in Y \;\&\; Y \in X)\}$$

and in fact U does not depend on V. Indeed, let $\pi(Z)$ be the property $\exists Y \, (Z \in Y \;\&\; Y \in X)$. Then $U = \{Z \in V \mid \pi(Z)\}$ and if V' is any set such that $\pi(Z) \Longrightarrow Z \in V'$, define $U' = \{Z \in V' \mid \pi(Z)\}$. Then

$$Z \in U \Longrightarrow \pi(Z) \Longrightarrow \pi(Z) \,\&\, Z \in V' \Longrightarrow Z \in U' \;,$$

therefore $U \subseteq U'$ and similarly $U' \subseteq U$, hence $U = U'$. Thus the *union* U of the set X exists and is unique; it is denoted by $\bigcup_{Y \in X} Y$ or by $\bigcup\{Y \mid Y \in X\}$ or by $\bigcup\{Y \mid p(Y)\}$ if $p(Y)$ is any property characterizing the elements of X (in this notation the letter Y can be replaced by any letter other than X).

ZF5 (axiom of the power set). *For every set X there exists a set Q such that $Z \subseteq X \Longrightarrow Z \in Q$*

Moreover, ZF2 implies the existence of the set $P = \{Z \in Q \mid Z \subseteq X\}$, which is called the *power set* of the set Q. One proves as above that P is independent of the set Q. Thus the power set exists and is unique; it is denoted by 2^X.

At this point we can prove that the set $\{X, Y\}$ exists even if $X = Y$, that is, we can prove the existence of the *singleton* $\{X\}$ for every X. If $X = \varnothing$ remark that $2^\varnothing$ exists by ZF5 and $2^\varnothing = \{\varnothing\}$ because $Y \in 2^\varnothing \Longleftrightarrow Y \subseteq \varnothing \Longleftrightarrow Y = \varnothing$ (as $\varnothing \subseteq Y$ for every Y). If $X \neq \varnothing$ then the set $Y = \{X, \varnothing\}$ exists by ZF3 and also $\{Z \in Y \mid Z = X\}$ by ZF2; the latter set coincides with $\{X\}$.

Now we can define the concept of an *ordered pair* (X, Y) as the set

$$(X, Y) = \{\{X\}, \{X, Y\}\} \;. \tag{1.1.1}$$

It can be proved by a careful discussion of all possible cases (exercise!) that

$$(X, Y) = (T, V) \Longleftrightarrow X = T \,\&\, Y = V \;. \tag{1.1.2}$$

Furthermore we define *ordered triplets* by $(X, Y, Z) = (X, (Y, Z))$, which immediately implies

$$(X, Y, Z) = (S, T, V) \Longleftrightarrow X = S \,\&\, Y = T \,\&\, Z = V \;. \tag{1.1.3}$$

Let us also prove that $\bigcup\{Y \mid Y \in X\}$ exists even if X does not contain two elements: namely $\bigcup\{Y \mid Y \in \{A\}\} = A$ and $\bigcup\{Y \mid Y \in \varnothing\} = \varnothing$. Indeed,

$$Z \in \bigcup\{Y \mid Y \in \{A\}\} \iff \exists Y\,(Z \in Y \;\&\; Y \in \{A\})$$

$$\iff \exists Y\,(Z \in Y \;\&\; Y = A) \iff Z \in A$$

(the implication $\Longrightarrow$ has been noted after ZF1, while the converse follows by taking $Y = A$). Also,

$$Z \in \bigcup\{Y \mid Y \in \varnothing\} \iff \exists Y\,(Z \in Y \;\&\; Y \in \varnothing)$$

and the latter property is false; so, $\bigcup\{Y \mid Y \in \varnothing\}$ contains no element.

We have thus proved that the union of the members of a set X, denoted

(1.1.4) $$\textstyle\bigcup_{Y\in X} Y = \bigcup\{Y \mid Y \in X\} = \{Z \mid \exists Y\,(Y \in X \;\&\; Z \in Y\})\,,$$

which is the set of all Z that belong to some member of X, exists and is unique for every set X. We are now going to prove that the *intersection* of all members of a set X, denoted by

(1.1.5) $$\textstyle\bigcap_{Y\in X} Y = \bigcap\{Y \mid Y \in X\} = \{Z \mid \forall Y\,(Y \in X \Longrightarrow Z \in Y)\}\,,$$

which is the set of all Z that belong to all the members of X, exists and is unique for every non-empty set X. Indeed, as the set $\bigcup_{Y\in X} Y$ exists, we can apply ZF2 to construct the set

$$I = \{Z \in \bigcup_{Y\in X} Y \mid \forall Y\,(Y \in X \Longrightarrow Z \in Y)\}\,,$$

whose elements Z belong to every member Y of X. It thus remains to prove the converse: every Z which belongs to all the members of X , belongs to I. Let Z be such an element; as $X \neq \varnothing$, there exists an element $A \in X$, hence $Z \in A$, therefore $Z \in \bigcup_{Y\in X} Y$; so $Z \in I$, as desired.

It now follows that $\bigcup_{Y\in X}$ (that $\bigcap_{Y\in X}$) is the least (greatest) set which includes (is included in) all the sets $Y \in X$. In particular $\bigcap_{Y\in X} \subseteq \bigcup_{Y\in X}$.

If $p(Z)$ is a property which characterizes the members of X, then the intersection can be written $\bigcap\{Y \mid p(Z)\}$. When $X = \{A, B\}$, the union and intersection are denoted by $A \cup B$ and $A \cap B$, respectively. When $X = \{A_1, \ldots, A_n\}$, the union an intersection may be denoted by $\bigcup_{i=1}^{n} A_i$ and $\bigcap_{i=1}^{n} A_i$, respectively.

We can also prove that the *Cartesian product*

(1.1.6) $$A \times B = \{(X, Y) \mid X \in A \;\&\; Y \in B\}$$

of two sets exists for every A, B. The technique is the same as for the previous proofs. Note first that if $X \in A$ and $Y \in B$ then X and Y belong to $A \cup B$, hence $\{X\}$ and $\{X, Y\}$ are elements of $2^{A\cup B}$, therefore (1.1.1) shows that $(X, Y) \in C$, where $C = 2^{2^{A\cup B}}$. Now the idea is to construct the set $\Pi = \{Z \in C \mid \pi(Z)\}$, where $\pi(Z)$ is the property $\exists X\, \exists Y\,(X \in A \;\&\; Y \in B \;\&\; Z = (X, Y))$. It only remains to show that the property $Z = (X, Y)$ can be expressed in our object language. In view of (1.1.1), the latter equality means $\forall T\,(T \in Z \Longleftrightarrow T = \{X\}$ or $T = \{X, Y\})$ and we have already seen how to express the last two predicates.

Now we define a *correspondence* between two sets A and B as a triplet (A, R, B), where R is a subset of $A \times B$; if $A = B$, the correspondence is said to be a (*binary*) *relation* on the set A. The basic concept of a *function* also becomes available: it is a correspondence (A, F, B) such that the following properties hold:

$$\forall X\,(X \in A \Longrightarrow \exists Y\,(Y \in B\ \&\ (X, Y) \in F))\ ,$$

$$\forall X\,\forall Y\,\forall Z\,((X, Y) \in F\ \&\ (X, Z) \in F \Longrightarrow Y = Z)\ .$$

The proof that $(X, Y) \in F$ can be expressed within our object language is left to the reader. The set F is called the *graph* of the function and is often identified with the function itself.

Exercise 1.1.1 Prove that if $A \neq \varnothing$ there is no function $(A, F, \varnothing)$. For a given B, the unique function of the form $(\varnothing, F, B)$ is the *empty function* $(\varnothing, \varnothing, B)$.

Exercise 1.1.2 Prove that $\bigcup\{Y \mid Y \subseteq X$ and Y is a singleton$\} = X$.

We now introduce

ZF6 (axiom of infinity) *There exists a set* W *such that* $\varnothing \in W$ *and if* $X \in W$ *then* $X \cup \{X\} \in W$.

So far we have tacitly admitted that sets actually exist (e.g., the first step in proving the existence of $\varnothing$ was to take an arbitrary Z). This quite natural assumption can be dropped, in which case all previous results must be stated conditionally ("if sets exist, then ...") until the introduction of ZF6; at that moment the existence of sets is guaranteed and all the results become unconditionally valid.

The rôle of ZF6 is, however, more important than that of ensuring the existence of sets in general. Let us call any set having the properties in ZF6, a *successor set*. Then the intersection of all successor sets exist and is itself a successor set. For let $\pi(X)$ be the property "if W is a successor set then $X \in W$". Clearly the property of being a successor set can be expressed by a formula in our object language, which transforms $\pi(X)$ into such a formula. Now take an arbitrary successor set W and construct the set $\mathbb{N} = \{X \in W \mid \pi(X)\}$. It is easily seen that $\mathbb{N}$ consists in fact of all the sets having the property $\pi(X)$, hence $\mathbb{N}$ is independent of W and unique. As a matter of fact, $\mathbb{N}$ is the smallest successor set and consists of the elements

$$\varnothing,\ \varnothing \cup \{\varnothing\} = \{\varnothing\},\ \{\varnothing\} \cup \{\{\varnothing\}\} = \{\varnothing, \{\varnothing\}\}, \ldots\ .$$

$\mathbb{N}$ is just the set of natural numbers, for which the usual notation $0, 1, 2, \ldots$ can be introduced. See also Section 3.3.

ZF7 (axiom scheme of replacement) *For every set* A *and every property* $\rho(X, Y)$ *such that for every* $X \in A$ *there is at most one set* Y *satisfying* $\rho(X, Y)$, *there is a set* B *such that* $Y \in B \Longleftrightarrow \exists X\,(X \in A\ \&\ \rho(X, Y))$.

Given the situation described in ZF7, we introduce a function defined as follows. Take the set $D = \{X \in A \mid \exists Y\ \rho(X,Y)\}$, which exists by ZF2. Then $D \times B$ and $F = \{(X,Y) \in D \times B \mid \rho(X,Y)\}$ also exist by a previous result and again ZF2; moreover, (D, F, B) is a function. Roughly – and not quite correctly – speaking, this means that if the *domain* D of a function F is a set, so is its *range* $F(B) = \{F(X) \mid X \in A\}$. Here $F(X)$ denotes the unique element such that $(X, F(X)) \in F$.

Exercise 1.1.3 If $A = \varnothing$ or $A = \{X\}$, the existence of the set B and of the function (A, F, B) can be established without using ZF7.

Now we introduce

ZF8 (von Neumann's axiom). *For every* $A \neq \varnothing$ *there exists a set* $X \in A$ *such that* $A \cap X = \varnothing$.

The rôle of ZF8 is to exclude the possibility of the existence of a set A such that $A \in A$, of a set A such that $A \in B$ and $B \in A$ for some B,..., of a set A such that $A \in A_1$, $A_1 \in A_2$, $\ldots$, $A_{n-1} \in A_n$, $A_n \in A$, for some $A_1, A_2, \ldots, A_n$; etc. For assume $A \in A$ for some A. Then $A \in A \cap \{A\} \neq \varnothing$ and $X \in \{A\}$ implies $\{A\} \cap X = \{A\} \cap A \neq \varnothing$ although $\{A\} \neq \varnothing$, thus contradicting ZF8; etc.

Let us prove the existence of the *direct product* or *unordered Cartesian product* or simply *product* $\prod_{Y \in X} Y$ of the sets Y of a set X: this is the set of all functions $(X, F, \bigcup_{Y \in X} Y)$ such that

$$\forall Y\, \forall X\, (Y \in X\ \&\ (Y,Z) \in F \Longrightarrow Z \in Y)\,.$$

Using the notation $F(Y)$ for the element Z, the above condition can be written in the form

$$\forall Y\, (Y \in X \Longrightarrow F(Y) \in Y)\,.$$

Given X, the latter property is actually a formula $\pi(F)$, so that it only remains to prove the existence of a set containing all the functions F from X to $\bigcup_{Y \in X} Y$. More generally, for every two sets A and B, the set $\{A\} \times (2^{A \times B} \times \{B\})$ exists and is nothing but the set of all correspondences $(A, R, B) = (A, (R, B))$; now obviously ZF2 implies the existence of the set of all functions from A to B.

The direct product $\prod_{Y \in X} Y$ is uniquely determined by X and can also be denoted by $\prod\{Y \mid Y \in X\}$ or by $\prod\{Y \mid p(Y)\}$ if $p(Y)$ is a property characterizing the members of X. In particular if $X = \{Y_1, \ldots, Y_n\}$, we may use the notation $\prod_{i=1}^n Y_i$.

Exercise 1.1.4 Prove that:

a. $\varnothing \in X \Longrightarrow \prod\{Y \mid Y \in X\} = \varnothing$.

b. $\prod\{Y \mid Y \in \varnothing\} = \{(0,0,0)\}$. (Hint: $(0,0,0)$ is the unique correspondence between $\varnothing$ and $\bigcup_{Y \in \varnothing} Y$).

c. $\prod\{Y \mid Y \in \{A\}\}$ has "as many" elements as A.

d. $\prod\{Y \mid Y \in \{A, B\}\}$ has "as many" elements as $A \times B$ and as $B \times A$.

If $X \neq \varnothing$, the elements of $\prod\{Y \mid Y \in X\}$ are called *choice functions* on X; the reason is that such a function "chooses" an element $F(Y) \in Y$ for each $Y \in X$. If $X = \varnothing$ there is an improper "choice function"; cf. Exercise 1.1.4.b.

Exercise 1.1.4.a can be restated to the effect that if X contains the empty set, there is no choice function on X. It is quite natural to ask whether the converse also holds; Exercise 1.1.4.c,d provides affirmative answers in very particular cases. The general case is also solved affirmatively with the aid of a new axiom.

ZF9 (axiom of choice). *For every $X \neq \varnothing$ such that $\varnothing \notin X$ and $Y \cap Z = \varnothing$ for every $Y, Z \in X$ with $Y \neq Z$, there is a choice function defined on X.*

Let us prove that ZF9 implies the following property: *for every non-empty set X there is a choice function defined on X.* It will follow that the latter property, sometimes called the *general principle of choice*, is in fact equivalent to ZF9; for this reason many authors disregard the above form of ZF9 and reserve the term "axiom of choice" for the general principle of choice.

Let $\varnothing \notin X \neq \varnothing$. Denote by $\rho(Y, V)$ the property $V = Y \times \{Y\}$. Clearly for every $Y \in X$ there is at most one V such that $\rho(Y, V)$; therefore ZF7 ensures the existence of an element Z such that

$$V \in Z \Longleftrightarrow \exists Y \,(Y \in X \;\&\; Y \times \{Y\} = V)\,.$$

Then $X \neq \varnothing$ implies $Z \neq \varnothing$. Furthermore, as $Y \neq \varnothing$ for all $Y \in X$, it follows that $V \neq \varnothing$ for all $V \in Z$. Finally, $V, V' \in Z$ and $V \neq V'$ imply $V = Y \times \{Y\}$, $V' = Y' \times \{Y'\}$ for some $Y, Y' \in X$ with $Y \neq Y'$, therefore $V \cap V' = \varnothing$ (because $S \in V \cap V'$ would imply $S = (T, Y) = (T', Y')$, hence $Y = Y'$). Thus the set Z fulfills the hypotheses of ZF9, hence there is a choice function $(Z, F, \bigcup_{Y \in X})$ on Z. Let now $\sigma(Y, T)$ be the property $(Y \times \{Y\}, (T, Y)) \in F$, where $T \in Y$. For every $Y \in X$, if T, T' are such that $\sigma(Y, T)$ and $\sigma(Y, T')$ hold, then $(T, Y) = (T', Y')$, hence $T = T'$. Thus σ satisfies the hypotheses of ZF7 and there is a function G associated with σ as we have seen before. The domain D of G is X, because for every $Y \in X$ we have $Y \times \{Y\} \in Z$, hence there exists (T, Y) such that $(Y \times \{Y\}, (T, Y)) \in F$, which means $\sigma(Y, T)$. Moreover, if $(Y, T) \in G$, this means $\sigma(Y, T)$, that is, $(Y \times \{Y\}, (T, Y)) \in F$, hence $(T, Y) \in Y \times \{Y\}$, finally $T \in Y$. Thus G is a choice function on X. As a corollary note that *for every non-empty set X there is a choice function defined for the non-empty sets belonging to X.*

The whole of set theory, as is currently used in various branches of mathematics, can be developed from ZF1–ZF9 in the same manner as we have done before. Thus e.g.:

Exercise 1.1.4 Prove that for every X, Y, Z:

(1.1.7) $$X \cup Y = Y \cup X,\; X \cap Y = Y \cap X\,,$$

(1.1.8) $$X \cup X = X,\; X \cap X = X,$$

(1.1.9) $$X \cup (X \cap Y) = X,\; X \cap (X \cup Y) = X\,,$$

(1.1.10) $$(X \cup Y) \cup Z = X \cup (Y \cup Z),\; X \cap (X \cup Y) = X \cap (Y \cap Z)\,,$$

(1.1.11) $$X \cup (Y \cap Z) = (X \cup Y) \cap (X \cup Z),\; X \cap (Y \cup Z) = (X \cap Y) \cup (X \cap Z),$$

and if A, B are subsets of X, then

$$A \cup \mathbf{C}A = X,\ A \cap \mathbf{C}A = \varnothing\ , \tag{1.1.12}$$

$$\mathbf{CC}A = A\ , \tag{1.1.13}$$

$$\mathbf{C}(A \cup B) = \mathbf{C}A \cap \mathbf{C}B,\ \mathbf{C}(A \cap B) = \mathbf{C}A \cup \mathbf{C}B\ . \tag{1.1.14}$$

However we stop here the development of Zermelo-Fraenkel set theory and conclude with a few words about paradoxes. One of the conditions to be fulfilled by a good axiomatic system for set theory is, of course, to avoid the set-theoretical paradoxes constructed so far. This is the case with the ZF system.

Thus, for instance, the Richard paradox concerns "the least natural number which cannot be defined by a sentence of less than a hundred words", the existence of which is contradictory because the sentence between the quotation marks is precisely a sentence of less than a hundred words which defines that number. But this paradox is based on the existence of the set of those natural numbers which can be defined by a sentence of less than a hundred words and such a set cannot be constructed within ZF (we cannot apply ZF2 because the property in discussion cannot be expressed as a formula of the ZF formal language).

As another example, consider the Russell paradox, which asks whether the set S of all sets X with the property $X \notin X$ satisfies this property or not. If the answer is yes, this means $S \notin S$, thus implying $S \in S$ by the very definition of S; if the answer is no, this means $S \in S$, hence $S \notin S$, again by the definition of S. So both possible answers yield a contradiction. However we cannot construct the set $\{X \mid X \notin X\}$ within the ZF system and this fact eliminates the paradox. We can only state:

Exercise 1.1.5 Let A be an arbitrary set and put $B = \{X \in A \mid X \notin X\}$. Prove that $B \notin A$ (Hint: $B \notin B$).

The point in the solution of the Russell paradox is that, unlike Cantor's unrestricted concept of a set, the existence of the set of all X satisfying $\pi(X)$ is <u>not</u> guaranteed for every π, even if π can be expressed by a formula; we can only state that for every A the set $\{X \in A \mid \pi(X)\}$ exists by ZF2. Thus in particular there is no set of all sets, no set of all groups, no set of all topological spaces, etc. There are other axiomatic systems for set theory, e.g., the Bernays-Gödel system, which associates with each formula $\pi(X)$ an entity called the *class* of all objects X satisfying $\pi(X)$. A membership relation holds between certain pairs of classes and every class X for which there exists a class Y such that $X \in Y$ is called a *set*; classes which are not sets, i.e., which do not appear as left sides of certain membership relations, are said to be *proper classes*. Thus the sets which occur in everyday mathematics are indeed sets of this axiomatic system, while such paradoxical sets as the set of all sets, the set of all groups, the set of all topological spaces, etc., are rejected, but there exist proper classes such as the class of all sets, the class of all groups, the class of all topological spaces, etc. Classes turn out to be a natural and convenient tool in the development of contemporary mathematics. In this eBook we will use the class language whenever necessary. In particular thee following property will b needed, which is established in class-and-set- axiomatic systems: *every subclass of a set is a set.*

There is also an attempt to avoid classes even in contexts in which they appear in a natural way, e.g. in category theory. This approach is as follows: define a *universe* as a set U with the following properties:

(i) $X \in Y \in U$ imply $X \in U$;

(ii) $X \in U$ and $Y \in U$ imply $\{X, Y\} \in U$, $(X, Y) \in U$ and $X \times Y \in U$;

(iii) $X \in U$ implies $2^X \in U$ and $\bigcup_{Y \in X} Y \in U$;

(iv) the set $\mathbb{N}$ of natural numbers belongs to U;

(v) if $F : A \longrightarrow B$ is a surjective function with $A \in U$ and $B \subseteq U$, then $B \in U$.

It turns out that the above properties (i)-(v) are sufficient in order to preserve all set-theoretical constructions involved in a given mathematical theory within a given universe, thus eliminating classes. (However it seems to the present author that the concept of a class is a more natural and more reliable concept than that of a universe).

1.2 Correspondences, Relations and Functions

This section collects in a systematic presentation several basic definitions and results concerning correspondences, relations and functions. Both functions and relations are presented as particular cases of correspondences; the rôle of the properties of coinjectivity and cosurjectivity is emphasized. (The important particular concept of order relation will be introduced in the next chapter and studied throughout the entire eBook.) For the sake of convenience we will recall some basic concepts from the first section, the more so as the notation must be changed, because in the first section everything was denoted by capitals, while in the remainder of this eBook we adopt the usual convention of denoting sets by capitals and their elements by small letters.

The *union* and *intersection* of a family of sets $\{A_i \mid i \in I\}$ are defined by $\bigcup_{i \in I} A_i = \{x \mid \exists i \in I \ x \in A_i\}$ and $\bigcap_{i \in I} A_i = \{x \mid x \in A_i \ \forall i \in I\}$, respectively. In particular $A \cup B = \{x \mid x \in A \text{ or } x \in B\}$ and $A \cap B = \{x \mid x \in A \ \& \ x \in B\}$.

Definition 1.2.1 Given two sets X and Y, a *correspondence* from X to Y is a triplet (X, ρ, Y) where $\rho \subseteq X \times Y$. The set X is called the *domain* of the correspondence, Y is the *codomain*, while ρ is said to be the *graph* of the correspondence. For every $x \in X$, $y \in Y$, the property $(x, y) \in \rho$ is denoted $x \rho y$.

Example 1.2.1 If X is a plane, Y is the set of all straight lines of X (X and the elements of Y are thought as sets of points), and $\rho = \{(x, y) \mid x \in X \ \& \ y \in$

Y & $x \in y\}$, then (X,ρ,Y) is a correspondence from X to Y. Also, if X is the set of straight lines in the three-dimensional Euclidean space, Y is the set of planes of that space and $\rho = \{(x,y) \mid x \in X \;\&\; y \in Y \;\&\; x \perp y\}$, we can construct the correspondence (X,ρ,Y).

As a matter of fact, mathematics is full of correspondences (including functions) which play a fundamental rôle. In the above example and in any correspondence occurring in mathematics we have an intuitive idea of a certain kind of correspondence between two sets, to which we can associate a correspondence in the sense of Definition 1.2.1. Whereas this is the "philogenetic" way corresponding to the historical development of mathematics, from the strictly logical point of view Definition 1.2.1 belongs to the set-theoretical foundation of mathematics, therefore it can be regarded as prior to other branches of mathematics, e.g. to geometry.

A correspondence (X,ρ,Y) is often identified with its graph ρ. This is possible whenever the domain X and the codomain Y result unambiguously from the context, otherwise there are infinitely many correspondences with the same graph ρ: if (X,ρ,Y) is a correspondence , so are $(\overline{X},\rho,\overline{Y})$ for every $\overline{X} \supseteq X$ and every $\overline{Y} \supseteq Y$. In this respect we introduce the following:

Definition 1.2.2 If (X,ρ,Y) and $(\overline{X},\rho,\overline{Y})$ are correspondences such that $X \subseteq \overline{X}$ and $Y \subseteq \overline{Y}$, we say that the former (the latter) is a *restriction* (an *extension*) of the latter (of the former).

Exercise 1.2.1 Given a correspondence (X,ρ,Y), find the smallest correspondence having the graph ρ (i.e., a correspondence which is a restriction of any correspondence having the graph ρ).

With each correspondence (X,ρ,Y) we associate a correspondence from 2^X to Y and a correspondence from X to 2^Y, the graphs of which are also denoted by ρ, as follows:

Definition 1.2.3 If (X,ρ,Y) is a correspondence, $A \subseteq X$, $B \subseteq Y$, $x \in X$, $y \in Y$, we set

$$A\rho y \iff a\rho y \,\forall\, a \in A \;, \tag{1.2.1'}$$

$$x\rho B \iff x\rho b \,\forall\, b \in B \;. \tag{1.2.1''}$$

Exercise 1.2.2 Prove that for every $y \in Y, x \in X$ we have $\varnothing\rho y$ and $x\rho\varnothing$.

Another construction which turns out to be very important is the following:

Definition 1.2.4 The *inverse* of a correspondence (X,ρ,Y) is the correspondence (Y,ρ^{-1},X), where for every $X \in X, y \in Y$,

$$y\rho^{-1}x \iff x\rho y \;. \tag{1.2.2}$$

It follows immediately that for every $A \subseteq X$, $y \in Y$ and every $B \subseteq Y$, $x \in X$,

$$A\rho y \iff y\rho^{-1}A \;, \tag{1.2.3'}$$

(1.2.3″) $$x\rho B \Longleftrightarrow B\rho^{-1}x\ .$$

Definition 1.2.5 If $(X,\rho\,,Y)$ is a correspondence and $A \subseteq X$, set

(1.2.4′) $$\rho(A) = \{y \in Y \mid \exists\, a\,(a \in A\ \&\ a\rho y)\}\ ;$$

in particular if $x \in X$, set $\rho(x) = \rho(\{x\})$, that is,

(1.2.5′) $$\rho(x) = \{y \in Y \mid x\rho y\}\ .$$

Definitions 1.2.4 and 1.2.5 immediately imply that for every $B \subseteq Y$, $y \in Y$,

(1.2.4″) $$\rho^{-1}(B) = \{x \in X \mid \exists\, b\,(b \in B\ \&\ x\rho b)\}\ ,$$

(1.2.5″) $$\rho^{-1}(y) = \{x \in X \mid x\rho y\}\ .$$

Warning The inverse correspondence ρ^{-1} in Definition 1.2.4 should not be confused with the sets $\rho^{-1}(B)$ and $\rho^{-1}(y)$ in Definition 1.2.5.

Now we can extend to correspondences a terminology which is usually applied to functions:

Definition 1.2.6 A correspondence (X,ρ,Y) is said to be:

(1) *injective*, if $\rho^{-1}(y)$ has at most one element, for every $y \in Y$;
(2) *surjective*, if $\rho(X) = Y$;
(3) *coinjective*, , if $\rho(x)$ has at most one element, for every $x \in X$;
(4) *cosurjective*, if $\rho^{-1}(Y) = X$.

In other words, the correspondence is :
(1) injective if $x\rho y\ \&\ x'\rho y \Longrightarrow x = x'$;
(2) surjective if for every $y \in Y$ there exists $x \in X$ such that $x\rho y$;
(3) coinjective if $x\rho y\ \&\ x\rho y' \Longrightarrow y = y'$;
(4) cosurjective if for every $x \in X$ there exists $y \in Y$ such that $x\rho y$.

Definitions 1.2.5 and 1.2.6 can be stated in a more suggestive language. Given a correspondence $X,\rho\,,Y)$,if $x\rho y$ let us refer to y as an *image* of x and to x as a *pre-image* of y. Then if $A \subseteq X$, $\rho(A)$ is the set of images of the elements of A, and if $B \subseteq Y$, $\rho^{-1}(B)$ is the set of pre-images of the elements of B. The correspondence is

(1) injective if every $y \in Y$ has at most one pre-image;
(2) surjective if every $y \in Y$ has at least a pre-image;
(3) coinjective if every $x \in X$ has at most one image;
(4) cosurjective if every $x \in X$ has at least an image.

We see that *functions* are characterized as coinjective and cosurjective correspondences. Every cosurjective correspondence $(X,\rho\,,Y)$ can be identified with the *multiple-valued function* $f : X \longrightarrow 2^Y$ which associates with each $x \in X$ the set of its images. Multiple-valued functions are used in graph theory (see e.g. Berge [1970]) and they are the basic tool in hyperalgebra (see e.g. Corsini and Leoreanu [2003], Davvaz and Leoreanu-Fotea [2008]). Every coinjective correspondence (X,ρ,Y) can be identified with the *partially defined function* $f : X \overset{\circ}{\longrightarrow} Y$ which is in fact the function $f : D \longrightarrow Y$ where D is the set of pre-images of the elements of Y and $f(x) = y \Longleftrightarrow x\rho y$. Partially defined functions are important in theoretical computer science (see e.g. Golan [1992]).

Remark 1.2.1 The correspondence (X,ρ,Y) is injective (surjective, coinjective, cosurjective) if and only if the inverse correspondence (Y,ρ^{-1},X) is coinjective (cosurjective, injective, surjective).

Exercise 1.2.3 Prove that for every correspondence (X,ρ,Y) : (i) $\rho(\varnothing) = \varnothing$; (ii) $A \subseteq A' \Longrightarrow \rho(A) \subseteq \rho(A')$; (iii) $\rho(A \cup A') = \rho(A) \cup \rho(A')$; (iv) $\rho(A \cap A') \subseteq \rho(A) \cap \rho(A')$; (v) $\rho(A \cap A') = \rho(A) \cap \rho(A')$ for all $A, A' \subseteq X \Longleftrightarrow \rho$ is injective; (vi) $\rho^{-1}(B \cap B') = \rho^{-1}(B) \cap \rho^{-1}(B')$ for all $B, B' \subseteq Y \Longleftrightarrow \rho$ is coinjective; (vii) $\rho(X \setminus A) \subseteq Y \setminus \rho(A)$ for all $A \subseteq X \Longleftrightarrow \rho$ is injective; (viii) $Y \setminus \rho(A) \subseteq \rho(X \setminus A) \Longleftrightarrow \rho$ is surjective.

In the following we will denote correspondences by their graphs.

Definition 1.2.7 Given two correspondences $\rho \subseteq X \times Y$ and $\sigma \subseteq Y \times Z$, their *composite* is the correspondence $\sigma \circ \rho \subseteq X \times Z$ defined by

$$x(\sigma \circ \rho)z \Longleftrightarrow \exists\, y\ (x\rho y\ \&\ y\sigma z)\ . \tag{1.2.6}$$

Proposition 1.2.1 *If* $\rho \subseteq X \times Y, \sigma \subseteq Y \times Z, \tau \subseteq Z \times T$, *then*

$$\tau \circ (\sigma \circ \rho) = (\tau \circ \sigma) \circ \rho\ , \tag{1.2.7}$$

$$(\sigma \circ \rho)^{-1} = \rho^{-1} \circ \sigma^{-1}\ , \tag{1.2.8}$$

$$(\rho^{-1})^{-1} = \rho\ . \tag{1.2.9}$$

PROOF: Clearly

$$(x,y) \in (\rho^{-1})^{-1} \Longleftrightarrow (y,x) \in \rho^{-1} \Longleftrightarrow (x,y) \in \rho\ ,$$

$$(z,x) \in \rho^{-1} \circ \sigma^{-1} \Longleftrightarrow \exists\, y\ (z,y) \in \sigma^{-1}\ \&\ (y,x) \in \rho^{-1}$$

$$\Longleftrightarrow \exists\, y\, (x,y) \in \rho\ \&\ (y,z) \in \sigma \Longleftrightarrow (x,z) \in (\sigma \circ \rho) \Longleftrightarrow (z,x) \in (\sigma \circ \rho)^{-1}.$$

To prove (1.2.7) we use the properties $\exists\, X\, \exists\, Y\, p(X,Y) \Longleftrightarrow \exists\, Y\, \exists\, X\, p(X,Y)$ and $\exists\, X\, (p\ \&\ q(X)) \Longleftrightarrow p\ \&\ \exists\, X\, q(X)$. So,

$$(x,t) \in \tau \circ (\sigma \circ \rho) \Longleftrightarrow \exists\, z\, (x,z) \in \sigma \circ \rho\ \&\ (z,t) \in \tau$$

$$\Longleftrightarrow \exists\, z\, \exists\, y\ (x,y) \in \rho\ \&\ (y,z) \in \sigma\ \&\ (z,t) \in \tau$$

$$\Longleftrightarrow \exists\, y\ (x,y) \in \rho\ \&\ \exists\, z\ (y,z) \in \sigma\ \&\ (z,t) \in \tau$$

$$\Longleftrightarrow \exists\, y\ (x,y) \in \rho\ \&\ (y,t) \in \tau \circ \sigma \Longleftrightarrow (x,y) \in (\tau \circ \sigma) \circ \rho\ .$$

□

Exercise 1.2.4 Find sufficient conditions for $\rho \circ \sigma = \sigma \circ \rho$.

Example 1.2.2 For every set X, the correspondence $(X, 1_X, X)$, where $1_X = \{(x,x) \mid x \in X\} = X \times X$, is called the *identity* of X, or the *diagonal* of X. So $x\, 1_X\, y$ means simply $x = y$.

Proposition 1.2.2 *For every correspondence $\rho \subseteq X \times Y$):*

$$\rho \circ 1_Y = 1_X \circ \rho = \rho\,. \tag{1.2.10}$$

PROOF: Routine verification left to the reader. □

Proposition 1.2.3 *For every correspondence (X,ρ,Y):*

(i) *If ρ is injective, then $(\rho^{-1} \circ \rho)(A) \subseteq A$ for every $A \subseteq X$.*

(ii) *If ρ is surjective, then $B \subseteq (\rho \circ \rho^{-1})(B)$ for every $B \subseteq Y$.*

(iii) *If ρ is coinjective, then $(\rho \circ \rho^{-1})(B) \subseteq B$ for every $B \subseteq Y$.*

(iv) *If ρ is cosurjective, then $A \subseteq (\rho^{-1} \circ \rho)(A)$ for every $A \subseteq X$.*

PROOF: Let ρ be injective and $x \in (\rho^{-1} \circ \rho)(A)$. Then $x\rho y$ for some $y \in \rho(A)$, which means $a\rho y$ for some $a \in A$; now the injectivity implies $x = a \in A$.

Let ρ be surjective and $b \in B$. Then $x\rho b$ for some $x \in X$; but this implies $x \in \rho^{-1}(B)$, hence $b \in \rho(\rho^{-1}(B))$.

Properties (iii) and (iv) are established similarly, or using (i), (ii) and Remark 1.2.1. □

Proposition 1.2.4 *For every two correspondences (X,ρ,Y) and (Y,σ,Z):*

(o) *If ρ and σ are injective (surjective, coinjective, cosurjective), then so is $\sigma \circ \rho$.*

(i) *If $\sigma \circ \rho$ is injective and σ is cosurjective, then ρ is injective.*

(ii) *If $\sigma \circ \rho$ is surjective, then σ is surjective.*

(iii) *If $\sigma \circ \rho$ is coinjective and ρ is surjective, then σ is coinjective.*

(iv) *If $\sigma \circ \rho$ is cosurjective, then ρ is cosurjective.*

PROOF: Routine verification left to the reader. □

Definition 1.2.8 By a *function* or *mapping* $f : X \longrightarrow Y$ from X to Y is meant a coinjective and cosurjective correspondence (X, f, Y). If the *domain* X and the *codomain* Y are understood, the function may be denoted simply by f. The notation $f(x) = y$ means that $y \in Y$ is unique with the property xfy[1]. An injective (a surjective) function is also called an *injection* (a *surjection*).

This definition coincides with the usual concept of a function, with which the reader is supposed to be familiar. The comment after Example 1.2.1 applies in particular to functions.

In the case of functions, Definition 1.2.5, its consequences and Definition 1.2.7 can be written in the form

$$f(A) = \{y \in Y \mid \exists x\,(x \in A \;\&\; y = f(x)\} = \{f(x) \mid x \in A\}\ ,$$

[1] This slightly disobeys Definition 1.2.5, which prescribes $f(x) = \{y\}$.

$$f^{-1}(B) = \{x \in X \mid \exists b\,(b \in B \;\&\; f(x) = b)\} = \{x \in X \mid f(x) \in B\}\ ,$$

$$f^{-1}(y) = \{x \in X \mid f(x) = y\}\ ,$$

if $f : X \longrightarrow Y$ and $g : Y \longrightarrow Z$, then $(g \circ f)(x) = g(f(x))$ for every $x \in X$.

Remark 1.2.2 1) A correspondence f is a function if and only if the correspondence f^{-1} is injective and surjective. 2) If the domain or the codomain of a correspondence is empty, so is its graph. 3) For every X the correspondence $(X, \varnothing, \varnothing)$ is injective, surjective and coinjective, while for every Y the correspondence $(\varnothing, \varnothing, Y)$ is an injective function. 4) The correspondence $(X, \varnothing, \varnothing)$ is cosurjective, or equivalently, a function, if and only if $X = \varnothing$. 5) An *empty mapping* $\perp_Y = (\varnothing, \varnothing, Y)$ is surjective if and only if $Y = \varnothing$. 6) We set $\perp = \perp_\varnothing$, that is, $\perp$ is the empty bijection (injection and surjection) $\perp = (\varnothing, \varnothing, \varnothing)$.

Hint. When dealing with the empty set we must use the fact that if p is a false sentence, then the sentence $p \to q$ is true for any sentence q.

Exercise 1.2.5 Prove that $f^{-1}(Y \setminus B) = X \setminus f^{-1}(B)$ for every $B \subseteq Y$.

Proposition 1.2.5 *For every two functions $f : X \longrightarrow Y$ and $g : Y \longrightarrow Z$, their composite $g \circ f$ is a function. Moreover:*

(o) *If f and g are injective (surjective), then so is $g \circ f$.*

(i) *If $g \circ f$ is injective, then f is injective.*

(ii) *If $g \circ f$ is surjective, then g is surjective.*

PROOF: Immediate from Proposition 1.2.4 □

Theorem 1.2.1 *The following conditions are equivalent for a function $f : X \longrightarrow Y$:*

(i) *f is injective;*

(ii) *for every $x, x' \in X$, if $x \neq x'$ then $f(x) \neq f(x')$;*

(iii) *for every $x, x' \in X$, if $f(x) = f(x')$ then $x = x'$;*

(iv) *for every $y \in Y$, the set $f^{-1}(y)$ has at most one element;*

(v) *for every set Z and every two functions $h_1, h_2 : Z \longrightarrow X$, if $f \circ h_1 = f \circ h_2$ then $h_1 = h_2$;*

(vi) *$f^{-1}(f(A)) \subseteq A$ for every $A \subseteq X$;*

(vii) *$f^{-1}(f(A)) = A$ for every $A \subseteq X$.*

PROOF: If $X = \varnothing$, conditions (i)-(vii) are satisfied by Remark 1.2.2 (all the sets occurring in (v), (vi) and (vii) are empty). So, in the proof below we can suppose $X \neq \varnothing$, hence $Y \neq \varnothing$.

(i)$\Longleftrightarrow$(iv): By Definition 1.2.6.

(ii)$\Longleftrightarrow$(iii): By a well-known logical property.

(iii)$\Longleftrightarrow$(iv): Because (iv) says that $\forall\, y\, (x, x' \in f^{-1}(y) \Longrightarrow x = x')$.

(iii)$\Longrightarrow$(v): If $f(h_1(x)) = f(h_2(x))$ then $h_1(x) = h_2(x)$.

(v)$\Longrightarrow$(iii): If $f(x) = f(x')$ set $Z = \{x, x'\}$ and $h_1(z) = x$, $h_2(z) = x'$ for $z \in Z$; then $(f \circ h_1)(z) = (f \circ h_2)(z)$, hence $x = h_1(z) = h_2(z) = x'$.

(i)$\Longrightarrow$(vi) and (vi)$\Longrightarrow$(vii) follow by Proposition 1.2.3, (i) and (iv), respectively.

(vii)$\Longrightarrow$(iii): If $f(x) = f(x')$ then $x' \in f^{-1}(f(\{x\})) = \{x\}$, that is, $x' = x$. □

Corollary 1.2.1 *If $X \neq \varnothing$, then a function $f : X \longrightarrow Y$ is injective if and only if*

(viii) *there exists a function $g : Y \longrightarrow X$ such that $g \circ f = 1_X$.*

PROOF: (iii)$\Longrightarrow$(viii): Notice first that $Y \neq \varnothing$ by Remark 1.2.2.4). Fix an element $a \in X$ and for every $y \in Y$ define $g(y) = x$ if $y = f(x)$ for some $x \in X$, else $g(y) = a$. This is a good definition in view of the injectivity of f and clearly $g \circ f = 1_X$.

(viii)$\Longrightarrow$(v): If $f \circ h_1 = f \circ h_2$ then

$$h_1 = 1_X \circ h_1 = g \circ f \circ h_1 = g \circ f \circ h_2 = 1_X \circ h_2 = h_2 .$$

□

Example 1.2.3 For every set X and every subset $A \subseteq X$, the mapping $i : A \longrightarrow X$ defined by $i(a) = a$ for every $a \in A$, is an injection, called the *inclusion* mapping. For every $a_0 \in A$, the function $g : X \longrightarrow A$ defined by $g(x) = x$ if $x \in A$, else $g(x) = a_0$, satisfies $g \circ i = 1_A$.

Theorem 1.2.2 *The following conditions are equivalent for a function $f : X \longrightarrow Y$:*

(i) *f is surjective;*

(ii) *for every $y \in Y$ there exists $x \in X$ with $f(x) = y$;*

(iii) *$f^{-1}(y) \neq \varnothing$ for every $y \in Y$:*

(iv) *$f(X) = Y$;*

PROOF: If $Y = \varnothing$ then $X = \varnothing$ because f is a function, hence conditions (i)-(iv) are satisfied by Remark 1.2.2. So we can suppose $Y \neq \varnothing$. Now if $X = \varnothing$ then conditions (i)-(iv) are false, hence they are still equivalent. Therefore we can suppose also $X \neq \varnothing$.

(i)$\Longleftrightarrow$(iv): By Definition 1.2.6.

(ii)$\Longleftrightarrow$(iii): Trivial.

(ii)$\Longleftrightarrow$(iv): Because (ii) says that $Y \subseteq f(X)$.

(ii)$\Longrightarrow$(v): If $h_1 \circ f = h_2 \circ f$, then for every $y \in Y$ there is $x \in X$ such that $h_1(y) = h_1(f(x)) = h_2(f(x)) = h_2(y)$, therefore $h_1 = h_2$. □

Corollary 1.2.2 *If $X \neq \varnothing \neq Y$, then the following conditions are equivalent for a function $f : X \longrightarrow Y$:*

(i) *f is surjective;*

(v) *for every set Z and every two functions $h_1, h_2 : Y \longrightarrow Z$, if $h_1 \circ f = h_2 \circ f$ then $h_1 = h_2$;*

(vi) *$B \subseteq f(f^{-1}(B))$ for every subset $B \subseteq Y$;*

(vii) *$f(f^{-1}(B)) = B$ for every subset $B \subseteq Y$.*

(viii) *there exists a function $g : Y \longrightarrow X$ such that $f \circ g = 1_Y$.*

PROOF: (ii)$\Longleftrightarrow$(v): If $h_1 \circ f = h_2 \circ f$, then for every $y \in Y$ there is $x \in X$ such that $h_1(y) = h_1(f(x)) = h_2(f(x)) = h_2(y)$, therefore $h_1 = h_2$.

(v)$\Longrightarrow$(iv): Suppose there exists $y_0 \in Y \setminus f(X)$. Define $h_1, h_2 : Y \longrightarrow \{0, 1\}$ by setting $h_1(y) = 1$ for all y, $h_2(y) = 1$ for all $y \neq y_0$ and $h_2(y_0) = 0$. Then $h_1 \circ f = h_2 \circ f$ but $h_1 \neq h_2$.

(i)$\Longrightarrow$(vi) and (vi)$\Longrightarrow$(vii) follow by Proposition 1.2.3, (ii) and (iii), respectively.

(vii)$\Longrightarrow$(iv): For every $y \in Y$ we have $y \in \{y\} = f(f^{-1}(\{y\})) \subseteq f(X)$, therefore $Y \subseteq f(X)$.

(ii)$\Longrightarrow$(viii): For every $y \in Y$ choose as $g(y)$ an element x such that $f(x) = y$ (using the axiom of choice). Then $x = g(y) \Longrightarrow f(x) = y$, hence $f(g(y)) = y$.

(viii)$\Longrightarrow$ (ii): Since $f(g(y)) = y$ for all $y \in Y$, property (ii) holds: take $x = g(y)$. □

Now we recall a well-known notion:

Definition 1.2.9 A function $f : X \longrightarrow Y$ is called a *bijection* or a *bijective* mapping if it is both injective and surjective.

The standard model of bijections consists of the identity mappings $1_X : X \longrightarrow X$. An exotic example of a bijection it the empty bijection $\perp = (\varnothing, \varnothing, \varnothing)$, which, according to Remark 1.2.2, is the unique bijection with $X = \varnothing$ or $Y = \varnothing$. So, in studying bijections, *we will assume that $X \neq \varnothing \neq Y$.*

If f is a bijection, then the inverse correspondence f^{-1} is also a bijection, by Remark 1.2.1. In this case $f(x) = y \Longleftrightarrow x = f^{-1}(y)$, therefore $f^{-1}(f(x)) = x$ and $f(f^{-1}(y)) = y$ hold identically. The latter identities can be written in the form

$$f^{-1} \circ f = 1_X \quad f \circ f^{-1} = 1_Y \,. \tag{1.2.10}$$

The above properties are refined in the next theorem.

Theorem 1.2.3 *The following conditions are equivalent for a function $f : X \longrightarrow Y$:*

(i) f is a bijection;

(ii) for every $y \in Y$ there is exactly one element $x \in X$ such that $f(x) = y$;

(iii) the correspondence f^{-1} is a function;

(iv) the correspondence f^{-1} is a bijection;

(v) the correspondence f^{-1} satisfies (1.2.10);

(vi) there is a correspondence $g \subseteq Y \times X$ such that

$$g \circ f = 1_X \text{ and } f \circ g = 1_Y; \tag{1.2.11}$$

(vii) there is a function $g : Y \longrightarrow X$ satisfying (1.2.11).

PROOF: (i)$\Longleftrightarrow$(ii): Trivial.

(i)$\Longrightarrow$(iv): By Remark 1.2.1.

(iv)$\Longrightarrow$(iii): Trivial.

(iii)$\Longrightarrow$(i): The function $f = (f^{-1})^{-1}$ is injective and surjective by Remark 1.2.1 applied to the function f^{-1}.

(i)$\Longrightarrow$(v): Already noted.

(v)$\Longrightarrow$(vi): Trivial.

(vi)$\Longrightarrow$(vii): We apply Proposition 1.2.4 to prove that the very correspondence g occurring in (vi) is a function:

$g \circ f$ is coinjective and f is surjective, hence g is coinjective;

$f \circ g$ is cosurjective, hence g is cosurjective.

(vii)$\Longrightarrow$(i): Again by Proposition 1.2.4:

$g \circ f$ is injective and g is cosurjective, hence f is injective;

$f \circ g$ is surjective, hence f is surjective. □

Corollary 1.2.3 *If $f : X \longrightarrow Y$ and $g \subseteq Y \times Z$ satisfy* (1.2.11), *then f and g are bijections and $g = f^{-1}$.*

PROOF: The hypothesis is condition (vi) in the theorem, so f is a bijection and g is a function. Therefore condition (1.2.11) reads

$$g(f(x)) = x \;\forall x \in X \text{ and } f(g(y)) = y \;\forall y \in Y \;,$$

that is, $f(x) = y \Longleftrightarrow g(y) = x$, proving that $g = f^{-1}$. □

In this eBook we will often use Corollary 1.2.3 to prove the existence of bijections between certain sets.

Bijective mappings are related to a very useful idea which can be explained as follows. If X is a set and there is a bijection $f : I \longrightarrow X$ between a certain set I and X, then we can think of X as the set $X = \{f(i) \mid i \in I\}$, where to distinct elements i, j of I there correspond distinct elements $f(i), f(j)$ of X. This is usually written in the form

$$X = (x_i)_{i \in I} \;,$$

where each $f(i)$ has been re-denoted x_i, and X is then referred to as a *family* of elements indexed by the *index set I*. Notice that any set X can be indexed, taking X itself as index set. From a formal point of view, a family could be defined as a pair consisting of a set X and an *indexing*, that is, a bijection having X as codomain.

Definition 1.2.10 A *binary relation* or simply a *relation* on a set X is a correspondence (X, ρ, X). When no confusion is possible, the relation (X, ρ, X) is again identified with its graph ρ.

Example 1.2.4 The set of all straight lines of a plane, endowed with the relation "parallel to", or the same set endowed with the relation "orthogonal to", are examples of relations in the intuitive sense of the word: they can be associated with relations in the sense of Definition 1.2.10 (in the same way as for Example 1.2.1). Two types of relations are fundamental in mathematics: equivalence relations and partial orders. The most important fact about the former is probably Theorem 1.2.4 below; as to the latter, let us recall Bourbaki's opinion according to which the algebraic, topological and order structures are the three fundamental types of mathematical structures. This eBook is devoted to several order-theoretical concepts and results which are employed in various branches of mathematics.

Definition 1.2.11 A binary relation $\equiv$ on a set X is called an *equivalence*, if it is *reflexive, symmetric* and *transitive*, i.e., for all $x, y, z \in X$,

(1.2.12) $$x \equiv x \ ,$$

(1.2.13) $$x \equiv y \Longrightarrow y \equiv x \ ,$$

(1.2.14) $$x \equiv y \ \& \ y \equiv z \Longrightarrow x \equiv z \ ,$$

respectively. For every $x \in X$, the set

(1.2.15) $$\hat{x} = \{y \in X \mid x \equiv y\}$$

of all elements equivalent to x is called the *coset* of x modulo $\equiv$. The set

(1.2.16) $$\{\hat{x} \mid x \in X\} = X/\equiv$$

of all cosets is called the *quotient set* of X modulo $\equiv$. The mapping $p : X \longrightarrow X/\equiv$ defined by $p(x) = \hat{x}$ is called the *canonical surjection* associated with $\equiv$.

The standard example of an equivalence relation is the identity 1_X, which in fact is a function (cf. Example 1.2.2), in which case $x \equiv y$ means simply $x = y$ and each coset $\hat{x}$ is the singleton $\{x\}$. However the significance of Definition 1.2.11 follows precisely from the fact that there are many equivalence relations which are different from the identity and which play an important rôle in various mathematical constructions; cf. Definition 1.2.12 and Theorem 1.2.4 below.

Example 1.2.5 As typical examples of equivalence relations we quote the relation "$d_1 \parallel d_2$ or $d_1 = d_2$", which is a slight generalization of the parallelism of lines in a plane (in which case the cosets can be interpreted as directions of the plane), the similar extension of the parallelism of planes in space, the similarity of triangles in the set of triangles of a plane, the congruence modulo a given integer in the set $\mathbb{Z}$ of integers, etc. The reader is urged to supply further examples.

Example 1.2.6 A very important class of equivalence relations is the following. If $f : X \longrightarrow Y$ is a function, the relation ker f defined on X by

$$x \ker f x' \Longleftrightarrow f(x) = f(x') \tag{1.2.17}$$

is an equivalence called the *kernel* of f. It is easy to see that $\ker f = f^{-1} \circ f$.

Example 1.2.7 The *universal equivalence* ω_X on a set X is the equivalence $\omega_X = (X, X \times X, X)$, which, as usual, can be identified with its graph $X \times X$; or, it can be denoted simply by ω when no confusion is possible. Clearly an equivalence relation on X is ω_X if and only if its only coset is X.

Remark 1.2.3 The empty relation $\perp = (\varnothing, \varnothing, \varnothing)$ is the unique relation on the empty set and it is an equivalence relation.// Hint. As for Remark 1.2.2.

Proposition 1.2.6 *If $\equiv$ is an equivalence relation on a set $X \neq \varnothing$, then the following properties hold for every $x, y \in X$:*
(i) $x \in \hat{x}$,
(ii) $x \equiv y \Longleftrightarrow \hat{x} = \hat{y}$.

PROOF: (i): By (1.2.12).
(ii): If $\hat{x} = \hat{y}$ then $y \in \hat{x}$ by (i), hence $x \equiv y$. Conversely, if $x \equiv y$, take $z \in \hat{x}$. Then $x \equiv z$, hence $y \equiv z$ by (1.2.13) and (1.2.14), therefore $z \in \hat{y}$. Thus $\hat{x} \subseteq \hat{y}$ and similarly $\hat{y} \subseteq \hat{x}$. □

Definition 1.2.12 A *partition* of a set $X \neq \varnothing$ is a set of pairwise disjoint non-empty subsets of X, the union of which equals X. In other words, it is a family $(A_i)_{i \in I}$ of subsets of X such that

$$A_i \neq \varnothing \quad (\forall\, i \in I)\ , \tag{1.2.18}$$

$$A_i \cap A_j = \varnothing \quad (\forall\, i, j \in I,\ i \neq j)\ , \tag{1.2.19}$$

$$\textstyle\bigcup_{i \in I} A_i = X\ . \tag{1.2.20}$$

Theorem 1.2.4 *Let X be a non-empty set. Then the following hold:*
(i) *The cosets of an equivalence relation on X form a partition of X.*
(ii) *Conversely, if $(A_i)_{i \in I}$ is a partition of X, then the relation $\equiv$ defined by*

$$x \equiv y \Longleftrightarrow \exists\, i \in I \ \ x \in A_i \ \&\ y \in A_i \tag{1.2.21}$$

is an equivalence on X.
(iii) *The mapping described in* (i) *and* (ii) *establishes a bijection between the equivalence relations on X and the partitions of X.*

PROOF: (i): We have $\hat{x} \neq \varnothing$ for all x and $\bigcup_{x \in X} \hat{x} = X$ by Proposition 1.2.6. To prove (1.2.19), assume $\hat{x} \cap \hat{y} \neq \varnothing$. Take $z \in \hat{x} \cap \hat{y}$. Then $x \equiv z$ and $z \equiv y$, hence $x \equiv y$, therefore $\hat{x} = \hat{y}$.
(ii): If $x \in X$, then $x \in A_i$ for some $i \in I$, hence $x \equiv x$. Clearly the relation $\equiv$ is symmetric. If $x \equiv y$ and $y \equiv z$, then there exist $i, j \in I$ such that $x, y \in A_i$ and $y, z \in A_j$, hence $y \in A_i \cap A_j \neq \varnothing$, therefore $i = j$ and $x, z \in A_i$, proving that $x \equiv z$.

(iii) We apply the technique provided in Corollary 1.2.3.

If $\equiv$ is an equivalence relation and $\sim$ is the equivalence associated with the cosets of $\equiv$ by (i), we must prove that $\sim$ coincides with $\equiv$. But

$$x \sim y \Longleftrightarrow \exists\, z \;\; x, y \in \hat{z} \Longleftrightarrow \exists\, z \;\; x \equiv z \equiv y \Longleftrightarrow x \equiv y\,,$$

where the last implication $\Longleftarrow$ follows with $z = x$.

If $(A_i)_{i \in I}$ is a partition and $\equiv$ is the equivalence associated with it by (ii), we must prove that the cosets of $\equiv$ are the A_is. Indeed, take $x \in A_i$. Then for all $y \in X$ we have $y \in A_i \Longleftrightarrow x \equiv y \Longleftrightarrow y \in \hat{x}$, proving that $A_i = \hat{x}$. □

We conclude this section with a few results that will be useful in the sequel.

Exercise 1.2.6 Prove that every mapping $f : X \longrightarrow Y$ can be decomposed in the form

$$f = i \circ \widehat{f} \circ p\,, \tag{1.2.22}$$

where $p : X \longrightarrow \ker f$ is the canonical surjection and $i : f(X) \longrightarrow Y$ is the inclusion. Conclude that every function is the composite of a surjection and an injection. (Hint. Define $\widehat{f}(\widehat{x}) = f(x)$.)

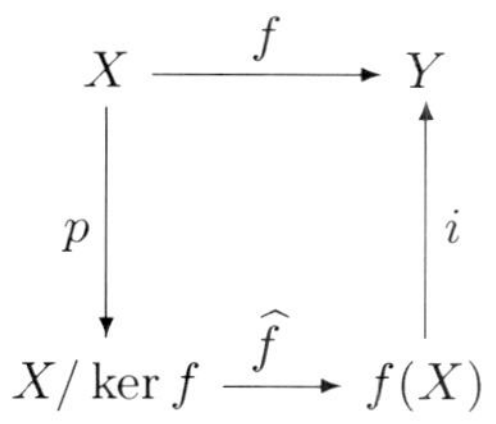

Fig. 1.1

Proposition 1.2.7 *Let $f : X \longrightarrow Y$ be a surjection and $g : X \longrightarrow Z$ a function. Then the following hold:*

(i) *There is a (necessarily unique) function $h : Y \longrightarrow Z$ such that $g = h \circ f$ if and only if $\ker f \subseteq \ker g$.*

(ii) *h is an injection if and only if $\ker f = \ker g$.*

(iii) *h is a surjection of and only if g is a surjection.*

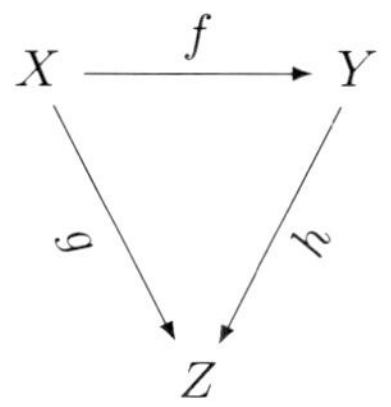

Fig. 1.2

PROOF: (i): Suppose $\ker f \subseteq \ker g$. If $y \in Y$, take $x \in X$ such that $y = f(x)$ and put $h(y) = g(x)$. The function h is well defined because if $y = f(x')$ then $x \ker f\ x'$, hence $x \ker g\ x'$, that is, $g(x) = g(x')$. Conversely, if $g = h \circ f$, then $g(x) = h(f(x))$ for every $x \in X$. This proves both the inclusion $\ker f \subseteq \ker g$ and the uniqueness of h.

(ii): If h is injective, then $g(x) = g(x') \Longrightarrow h(f(x)) = h(f(x')) \Longrightarrow f(x) = f(x')$; this means $\ker g \subseteq \ker f$. Conversely, assume $\ker g = \ker f$ and take $y, y' \in Y$ such that $h(y) = h(y')$. But $y = f(x)$ and $y' = f(x')$ for some $x, x' \in X$, hence $g(x) = h(f(x)) = h(f(x')) = g(x')$, that is, $(x, x') \in \ker g = \ker f$, therefore $y = f(x) = f(x') = y'$.

(iii): Assume h is surjective and take $z \in Z$. Then $z = h(y)$ for some $y \in Y$; but $y = f(x)$ for some $x \in X$, which yields $z = g(x)$. Conversely, assume g is surjective and take again $z \in Z$. Then $z = g(x)$ for some $x \in X$, therefore $z = h(f(x))$. □

The next exercise is a kind of dual to Proposition 1.3.7.

Exercise 1.2.7 Let $f : X \longrightarrow Y$ be an arbitrary function and $h : Z \longrightarrow Y$ an injection. Prove that: (i) there exists a (necessarily unique) function $g : X \longrightarrow Z$ such that $f = h \circ g$ if and only if $f(X) \subseteq h(Z)$; (ii) g is an injection if and only if f is an injection; (iii) g is a surjection if and only if $f(X) = h(Z)$. (Hint. Put $g(x) = z$ if $f(x) = h(z)$.)

Exercise 1.2.8 Put $2 = \{0, 1\}$ and $Y^X = \{f \mid f : X \longrightarrow Y\}$ for every two sets X and Y. Establish a bijection between $\{f \mid f : X \longrightarrow 2\}$ and the set $\{Z \mid Z \subseteq X\}$, just justifying the notation 2^X for the latter set (known as the set of all *parts* of X). Also, interpret Y^X as a direct product. This justifies the fact that whenever a direct product $\prod_{i \in I} Y_i$ is specialized by taking $Y_i = Y$ for all $i \in I$, we obtain the set Y^I of all mappings $f : I \longrightarrow Y$. (Hint. Map f to $\{x \in X \mid f(x) = 1\}$ and Z to the *characteristic function* of Z, that is, the function $f : I \longrightarrow X$ such that $f(x) = 1 \Longleftrightarrow x \in Z$. Further, establish a bijection between Y^X and $\prod_{x \in X}(Y \times \{x\})$).

1.3 Categories

We provide below several definitions and quite elementary results from category theory and illustrate them by examples. Our selection is strictly limited to what will be needed in the subsequent chapters in order to use the categorical language whenever this results in a gain of rigour and conciseness. The reader interested in more details and applications of category theory in various branches of mathematics is referred e.g. to Arbib and Manes [1975], Freyd [1966], Mac Lane [1971], Popescu [1971], Popescu and A. Radu [1971], Gh. Radu [1988].

We adopt the framework of sets and classes as explained in §1.2.

Definition 1.3.1 A *category* $\mathbf{C}$ consists of a class $\mid \mathbf{C} \mid$ of elements called the *objects* of $\mathbf{C}$, such that with each pair (A, B) of objects is associated a set

$\mathrm{Hom}_{\mathbf{C}}(A, B)$ of elements called the *morphisms* from A to B, such that axioms I-III below are satisfied. To state these conditions we introduce the notation $f : A \longrightarrow B$ or $A \xrightarrow{f} B$ for $f \in \mathrm{Hom}_{\mathbf{C}}(A, B)$. $f : A \longrightarrow B$

I. For every three objects A, B, C of $\mathbf{C}$ there is a mapping

$$\circ : \mathrm{Hom}_{\mathbf{C}}(A, B) \times \mathrm{Hom}_{\mathbf{C}}(B, C) \longrightarrow \mathrm{Hom}_{\mathbf{C}}(A, C) \;, \tag{1.3.1}$$

called *composition*, the image of a pair (f, g) being denoted by $g \circ f$ or simply by gf, such that the following *associativity law* is fulfilled: for every $A \xrightarrow{f} B \xrightarrow{g} C \xrightarrow{h} D$,

$$h \circ (g \circ f) = (h \circ g) \circ f \;. \tag{1.3.2}$$

II. For every object A of $\mathbf{C}$ there is a morphism $1_A : A \longrightarrow A$, called the *identity* of A, such that

$$f \circ 1_A = f \text{ and } 1_A \circ g = g \tag{1.3.3}$$

for every object B of $\mathbf{C}$ and every morphisms $f : A \longrightarrow B$ and $g : B \longrightarrow A$.

III. The sets $\mathrm{Hom}_{\mathbf{C}}$ are pairwise disjoint, i.e., for every $A, B, A', B' \in \mid \mathbf{C} \mid$,

$$(A, B) \neq (A', B') \Longrightarrow \mathrm{Hom}_{\mathbf{C}}(A, B) \cap \mathrm{Hom}_{\mathbf{C}}(A', B') = \varnothing \;. \tag{1.3.4}$$

The sets $\mathrm{Hom}_{\mathbf{C}}(A, B)$ may also be denoted by $\mathrm{Hom}(A, B)$ when no confusion is possible, or by $\mathbf{C}(A, B)$. In view of the associativity, the morphism (1.3.2) can unambiguously be written $h \circ g \circ f$ or hgf.

Mathematics are full of categories. For most of them, the objects are sets endowed with a certain structure, while the morphisms are those mappings between the sets which "preserve" the structure; as a matter of fact, this is the reason which explains and justifies the function-like notation in Definition 1.3.1. Yet, besides the categories which appear naturally throughout mathematics, many "pathological" categories can be constructed which have curious properties and in particular categories whose objects are not sets and/or whose morphisms are not mappings between sets.

The examples below try to illustrate each of the kinds of categories explained above. Each category will be described by simply indicating its objects and morphisms. Whenever the objects are sets, possibly equipped with a certain structure, and the morphisms are mappings between the support sets of the objects, it is tacitly meant that the composition of morphisms is the usual composition of functions; recall that the composite of two functions is again a function by Proposition 1.2.5 and composition is associative (even for correspondences) by Proposition 1.2.1, property (1.2.7). Therefore in these cases the verification of axiom I in Definition 1.3.1 reduces to the checking of the following facts: 1) each $\mathbf{C}(A, B)$ is actually a set (not a proper class!), and 2) if $f \in \mathbf{C}(A, B)$ and $g \in \mathbf{C}(B, C)$, then the usual composite $gf \in \mathbf{C}(A, C)$ (in other words, if f and g "preserve" the structure, so does gf). As to axiom II, unless otherwise stated, the identity morphism $1_A : A \longrightarrow A$ of an object A is simply the identity mapping of the support set of A, which is known to fulfill property (1.3.3) (exercise!). Therefore in these cases the verification of axiom II reduces to the verification of the fact that $1_A \in \mathbf{C}(A, B)$ (in other words, the identity mapping "preserves" the structure), which is usually a trivial property. Finally, axiom III can always be regarded as automatically verified. The idea is that if the sets $\mathbf{C}(A, B)$ are not pairwise disjoint from the beginning, one can replace each $f \in \mathbf{C}(A, B)$ by the triplet (f, A, B). Then there is a bijection between each old set $\mathbf{C}(A, B)$ and its new variant $\mathbf{C}(A, B) \times \{A\} \times \{B\}$, so that the new sets are pairwise disjoint although they essentially coincide with the old sets.

With these remarks in mind, let us pass to our list of examples.

Example 1.3.1 The category **Cor** of sets and correspondences. This means that $\mid \mathbf{Cor} \mid$ is the class of all sets and for every two sets A, B,

$$\mathbf{Cor}(A, B) = \{A\} \times 2^{A \times B} \times \{B\} \tag{1.3.5}$$

consists of all correspondences from A to B; so $\mathbf{Cor}(A, B)$ is actually a set (cf. §1, comments related to the direct product). The composition of morphisms is given by Definition 1.2.7 and it is associative by Proposition 1.2.1, property (1.2.7). The identity morphism is the correspondence $(A, 1_A, A)$ (as a matter of fact, it is a function). Property (1.3.3) is established in Proposition 1.2.2.

Example 1.3.2 The category **Set** of sets and functions, i.e., $\mid \mathbf{Set} \mid$ is the class of all sets, $\mathbf{Set}(A, B)$ is the set of all mappings from A to B (see again §1, comments on the direct product), and the operation of composition is the usual composition of functions.

In each of the subsequent examples of categories $\mathbf{C}$ consisting of certain sets and mappings, each $\mathbf{C}(A, B)$ is a subset of $\mathbf{Set}(A, B)$ (whose existence can be proved by ZF2 within ZF set theory).

The following three examples are based on Proposition 1.2.5.

Example 1.3.3 The category of all sets and all injective functions.

Example 1.3.4 The category of all sets and all surjective functions.

Example 1.3.5 The category of all sets and all bijections.

Example 1.3.6 The category of all finite sets and all mappings between them.

Example 1.3.7 The category **Mon** of all monoids and all homomorphisms of monoids.

Example 1.3.8 The category **Grp** of all groups and all homomorphisms of groups.

Example 1.3.9 The category **Ab** of all Abelian groups and all homomorphisms of groups between them.

Example 1.3.10 The category **Rng** of all rings and all homomorphisms of rings between them.

Example 1.3.11 The category **Rng1** of all rings with unit and all homomorphisms of rings with unit between them.

Warning We emphasize that in this eBook we reserve the term "morphism" for the meaning in Definition 1.3.1 and use the conventional term "homomorphism" for the morphisms of categories like those in Examples 1.3.8–1.3.12, although in the contemporary literature there is a strong tendency of using only the term "morphism".

Example 1.3.12 The category **Vec** of all vector spaces over a given field and all linear mappings.

Example 1.3.13 The category **Top** of all topological spaces and all continuous mappings.

Example 1.3.14 Let X be a non-empty set and ρ a reflexive and transitive relation on X. Define a category by taking X as class of objects and for every $a, b \in X$,

$$\mathrm{Hom}(a,b) = \begin{cases} \{(a,b)\} & \text{if } a\rho b, \\ \varnothing & \text{if } \neg(a\rho b), \end{cases}$$

and set $(b,c) \circ (a,b) = (a,c)$ if $\mathrm{Hom}(a,b) \neq \varnothing \neq \mathrm{Hom}(b,c)$, else $\circ$ is the empty function. The easy verifications are left to the reader.

Example 1.3.15 Let X be a non-empty class. Define a category by taking X as class of objects and for every $a, b \in X$,

$$\mathrm{Hom}(a,b) = \begin{cases} \{a\} & \text{if } a = b, \\ \varnothing & \text{if } a \neq b, \end{cases}$$

set $a \circ a = a$ and let (1.3.1) be the empty function if $a \neq b$ or $b \neq c$. The easy verifications are left to the reader.

Example 1.3.16 (the category of *pointed sets*). Let X be a class consisting of pairs of the form (A, a), where A is a set and $a \in A$. Define a category by taking X as class of objects and for every two objects $(A, a), (B, b)$ let $\mathrm{Hom}((A, a), (B, b))$ be reduced to the constant mapping $f : A \longrightarrow B$, $f(x) = b\ \forall x \in A$. Note that the identity morphism of (A, A) is not 1_A unless $A = \{a\}$.

Proposition 1.3.1 *In any category, the identity morphism 1_A of each object A is unique.*

PROOF: (same as in group theory). Let $1'_A$ be another identity. Then by applying (1.3.3) to the identities 1_A and $1'_A$ we get $1'_A \circ 1_A = 1'_A$ and $1'_A \circ 1_A = 1_A$, respectively, therefore $1_A = 1'_A$. □

There is another definition of the concept of a category, equivalent to Definition 1.3.1. Consider a class of elements called *arrows*, together with a class of distinguished pairs (f, g) of arrows, called *composable arrows*, and an operation which associates with each such pair (a, b) an arrow $g \circ f$, called their *composite*. We also say that "$g \circ f$ is defined"for "f, g are composable arrows". An arrow u is called an *identity* if $f \circ u = f$ whenever $f \circ u$ is defined and $u \circ g = g$ whenever $u \circ g$ is defined. Assume the following conditions are fulfilled. (i) For each arrow f there exist two identity arrows u and u', called the *source* and the *target*, such that $f \circ u$ and $u' \circ f$ are defined, respectively. (ii) The composite $h \circ (g \circ f)$ is defined if and only if $(h \circ g) \circ f$ is defined, in which case (1.3.2) holds. We say that the structure thus defined is a *category*.

Exercise 1.3.1 Prove the equivalence of the two definitions of categories. (Hint. Establish a "dictionary" which identifies morphisms with arrows and objects with identity arrows, the domain and the codomain of a morphism being the source and the target of the arrow, respectively).

Definition 1.3.2 The *dual* of a category $\mathbf{C}$ is the category $\mathbf{C}^0$ obtained as follows. Let

$$|\mathbf{C}^0| = |\mathbf{C}| \tag{1.3.6}$$

and for every objects A, B,

$$\mathrm{Hom}_{\mathbf{C}^0}(A, B) = \mathrm{Hom}_{\mathbf{C}}(B, A)\ , \tag{1.3.7}$$

the composition

$$\mathrm{Hom}_{\mathbf{C}^0}(A, B) \times \mathrm{Hom}_{\mathbf{C}^0}(B, C) \longrightarrow \mathrm{Hom}_{\mathbf{C}^0}(A, C) \tag{1.3.8}$$

in $\mathbf{C}_0$ being the same as the composition

$$\mathrm{Hom}_{\mathbf{C}}(C, B) \times \mathrm{Hom}_{\mathbf{C}}(B, A) \longrightarrow \mathrm{Hom}_{\mathbf{C}}(C, A) \tag{1.3.9}$$

in $\mathbf{C}$.

In other words, a morphism $f : A \longrightarrow B$ in $\mathbf{C}$ is regarded as a morphism $f : B \longrightarrow A$ in $\mathbf{C}^0$, and if $g : B \longrightarrow C$ is another morphism of $\mathbf{C}$, the composite $g \circ f : A \longrightarrow C$ in $\mathbf{C}$ is regarded as the composite $f \circ g : C \longrightarrow A$ in $\mathbf{C}^0$. It is immediately seen that $\mathbf{C}^0$ is actually a category; in particular the identity 1_A of an object A is the same.

Example 1.3.17 The category **Cor** in Example 1.3.1 is *self dual*, i.e., $\mathbf{Cor}^0 = \mathbf{Cor}$; the same is true for the categories in Examples 1.3.5 and 1.3.15. None of the categories in Examples 1.3.2–1.3.4, 1.3.6–1.3.14 and 1.3.17 coincides with its dual. Incidentally, the duals of the categories in Examples 1.3.2–1.3.4 and 1.3.6-1.3.13 are new examples of categories whose objects are sets but whose morphisms are not mappings.

Definition 1.3.2 enables us to associate with each concept **c** defined for an arbitrary category **C**, a *dual concept* $\mathbf{c}^0$: this is the concept **c** applied to the dual category $\mathbf{C}^0$, then translated into the language of the category **C**. Moreover, whenever we prove a theorem **t** valid for an arbitrary category **C**, a *dual theorem* $\mathbf{t}^0$ is automatically established: this is theorem **t** applied to the dual category $\mathbf{C}^0$, then translated into the language of the category **C**. This method of automatically proving new theorems is known as the *duality principle.*

Although the definitions and the reasoning in the above paragraph are rather rough, whenever we deal with specific concepts and theorems, the application of the duality principle becomes rigorous. In other words, we could say that the above duality principle is in fact a common scheme for a whole collection of rigorous proofs which are not explicitly stated but summarized by expressions like "by the (application of) the duality principle" or simply "by duality".

Thus, for instance, consider the following concepts:

Definition 1.3.3 A morphism $f : A \longrightarrow B$ is said to be:

(i) a *monomorphism,* if for every object X and every morphisms $u, v : X \longrightarrow A$, if $f \circ u = f \circ v$ then $u = v$;

(ii) an *epimorphism,* if for every object Y and every morphisms $u, v : B \longrightarrow Y$, if $u \circ f = v \circ f$ then $u = v$.

(iii) A morphism which is both a monomorphism and an epimorphism is called a *bimorphism.*

It follows immediately that the concepts of monomorphism and epimorphism are dual to each other. So are properties (i) and (ii) in Proposition 1.3.2 below, which generalizes Proposition 1.2.5.

Proposition 1.3.2 *The following properties hold for two morphisms* $f : A \longrightarrow B$ *and* $g : B \longrightarrow C$*:*

(i) f *and* g *are monomorphisms* $\Longrightarrow$ $g \circ f$ *is a monomorphism* $\Longrightarrow$ f *is a monomorphism.*

(ii) f *and* g *are epimorphisms* $\Longrightarrow$ $g \circ f$ *is an epimorphism* $\Longrightarrow$ g *is an epimorphism.*

PROOF: (i) If f and g are monomorphisms, then

$$(g \circ f) \circ u = (g \circ f) \circ v \Longrightarrow g \circ (f \circ u) = g \circ (f \circ v) \Longrightarrow f \circ u = f \circ v \Longrightarrow u = v\ .$$

If $g \circ f$ is a monomorphism, then $f \circ u = f \circ v \Longrightarrow g \circ f \circ u = g \circ f \circ v \Longrightarrow u = v$.

(ii) By the duality principle. □

As an illustration of the comment above which regards the duality principle as a common scheme for certain rigorous proofs, let us infer (ii) from (i) explicitly.

Assume $f : A \longrightarrow B$ and $g : B \longrightarrow C$ are epimorphisms in **C**. Then in the dual category $\mathbf{C}^0$ the morphisms $g : C \longrightarrow B$ and $f : B \longrightarrow A$ are monomorphisms, hence their composite $f \circ g : C \longrightarrow A$ is a monomorphism of $\mathbf{C}^0$ by (i), which means that $g \circ f : A \longrightarrow C$ is a monomorphism in **C**. A similar direct proof holds for the second implication from (ii).

It could be objected that the above inference of (ii) from (i) is not simpler than a direct proof of (ii), imitating the proof of (i). However category theory comprises many theorems with rather complicated proofs; in such cases the direct proof of the dual theorem is equally complicated, whereas the reasoning by duality (even if it would be explicitly stated as before !) is a mere translation of the formulation of the given theorem.

Remark 1.3.1 Duality is an *involutive* concept, that is, $\mathbf{C}^{00} = \mathbf{C}$, $\mathbf{c}^{00} = \mathbf{c}$ and $\mathbf{t}^{00} = \mathbf{t}$, for every category $\mathbf{C}$, every concept $\mathbf{c}$ and every theorem $\mathbf{t}$.

The concepts of monomorphism and epimorphism, introduced so far as an illustration of the duality principle, deserve attention for their own sakes. For these concepts have been suggested by the characterizations (v) of injections and surjections in Theorem 1.2.1 and Corollary 1.2.2, respectively. As a matter of fact, it is easily seen that the following more general properties hold:

Remark 1.3.2 In any category in which the morphisms are functions and composition is the composition of functions, it follows that identities coincide with identity mappings 1_A, every injective morphism is a monomorphism and every surjective mapping is an epimorphism; therefore every bijection is a bimorphism.

A natural question is to decide in which categories the converse statements are true, or equivalently, monomorphisms (epimorphisms) coincide with injective (surjective) mappings. Such problems may be more difficult or even still unsolved for certain categories. We sketch below a few answers.

Lemma 1.3.1 *If the function $f : A \longrightarrow B$ is not injective, there exist two bijective mappings $u, v : A \longrightarrow A$ such that $f \circ u = f \circ v$ but $u \neq v$.*

PROOF: Let $a, a' \in A$ fulfill $f(a) = f(a')$ but $a \neq a'$. Define $u(x) = x$ for all $x \in A$, $v(x) = x$ for all $x \in A \setminus \{a, a'\}$, $v(a) = a'$, $v(a') = a$. □

Lemma 1.3.2 *If the function $f : A \longrightarrow B$ is not surjective, there exist a set Y and two injective functions $u, v : B \longrightarrow Y$ such that $u \circ f = v \circ f$ but $u \neq v$.*

PROOF: Take $b \in B \setminus f(A)$ and $Y = B \cup \{\infty\}$, where $\infty \notin B$. Define $u(y) = y$ for every $y \in B$, $v(y) = y$ for every $y \in B \setminus \{b\}$, $v(b) = \infty$. □

Examples 1.3.18 and 1.3.19 below follow from Remark 1.3.2 and Lemmas 1.3.1 and 1.3.2, while Example 1.3.20 is based on Remark 1.3.2 alone.

Example 1.3.18 In the categories given in Examples 1.3.2 and 1.3.6, monomorphisms (epimorphisms) coincide with injective (surjective) mappings, therefore bimorphisms coincide with bijections.

Example 1.3.19 In Example 1.3.3 all morphisms are monomorphisms, while epimorphisms, bimorphisms, surjective functions and bijections coincide.

Example 1.3.20 In Example 1.3.4 all morphisms are epimorphisms, while monomorphisms, bimorphisms, injective functions and bijections coincide.

Example 1.3.21 In Examples 1.3.5 and 1.3.14–1.3.16, all morphisms are bimorphisms. For Example 1.3.5 this follows from Remark 1.3.2, while in Examples 1.3.14–1.3.16 this is a consequence of the fact that every non-empty Hom is a singleton.

Example 1.3.22 In each of the categories **Mon, Rng** and **Rng1** there exist epimorphisms that are not surjections. To prove this for **Mon**, take the inclusion $i : \mathbb{N} \longrightarrow \mathbb{Z}$ of natural number into integers. This is not a surjection, but if $u \circ i = v \circ i$ for some homomorhisms of monoids $u, v : \mathbb{Z} \longrightarrow M$ this means $u(n) = v(n)$ for every $n \in \mathbb{N}$, hence $u(-n) = -u(n) = -v(n) = v(-n)$, therefore $u = v$. A similar counterexample can be given for **Rng** anf **Rng1**, namely the inclusion $\mathbb{Z} \longrightarrow \mathbb{Q}$.

As a matter of fact, the detection of monomorphisms and epimorphisms in a category like those in Examples 1.3.7–1.3.13 is a problem pertinent to the corresponding algebraic or topological theory. Thus e.g. one can prove, using group-theoretical methods, that the monomorphisms (epimorphisms) of **Grp** coincide with the injective (surjective) homomorphims of groups.

We have already noted that monomorphisms, epimorphisms and bimorphisms generalize injections, surjections and bijections, respectively. Yet bimorphisms lack an important property of bijections, so that a stronger concept was also necessary.

Definition 1.3.4 A morphism $f : A \longrightarrow B$ is called an *isomorphism* from A to B, while A and B are said to be *isomorphic*, provided there exists a morphism $f^{-1} : B \longrightarrow A$ such that

$$f^{-1} \circ f = 1_A \text{ and } f \circ f^{-1} = 1_B \, . \tag{1.3.10}$$

Lemma 1.3.3 *In any category, if $f : A \longrightarrow B$ is an isomorphism, then each of the equations $g \circ f = 1_A$ and $f \circ g = 1_B$ has the unique solution $g = f^{-1}$.*

Proof: Suppose e.g. that $f \circ g = 1_B$. Then $g = 1_A \circ g = f^{-1} \circ f \circ g = f^{-1} \circ 1_B = f^{-1}$. □

Proposition 1.3.3 *In any category:*

(i) *Every identity morphism 1_A is an isomorphism.*

(ii) *If f is an isomorphism, then f^{-1} is an isomorphism uniquely determined by f, and $(f^{-1})^{-1} = f$.*

(iii) *The composite $g \circ f$ of two isomorphisms f, g is an isomorphism, and $(g \circ f)^{-1} = f^{-1} \circ g^{-1}$.*

(iv) *Every isomorphism is a bimorphism.*

Comment Properties (i)-(iii) show that the relation of being isomorphic is reflexive, symmetric and transitive. Thus we can say that isomorphism is an equivalence relation on the class of objects of the category. (This is a slight abuse of language, as the class **C** may be a proper class, not a set). The morphism

f^{-1} is naturally called the *inverse* of f.

PROOF: (i): Immediate, with $1_A^{-1} = 1_A$.

(ii): The inverse of an isomorphism is unique, by Lemma 1.3.3. Equation (1.3.10) can be read to the effect that f^{-1} is an isomorphism with f in the rôle of f^{-1}; so, $(f^{-1})^{-1} = f$ by the uniqueness of the inverse.

(iii): If $f : A \longrightarrow B$ and $g : B \longrightarrow C$ are isomorphisms, then

$$(f^{-1} \circ g^{-1}) \circ (g \circ f) = f^{-1} \circ (g^{-1} \circ g \circ f) = f^{-1} \circ f = 1_A ,$$

and similarly $(g \circ f) \circ (f^{-1} \circ g^{-1}) = 1_C$, proving that $g \circ f$ is an isomorphism.

(iv): We have

$$f \circ u = f \circ v \Longrightarrow u = 1_A \circ u = f^{-1} \circ f \circ u = f^{-1} \circ f \circ v = 1_A \circ v = v ,$$

therefore f is a monomorphism. By duality, f is also an epimorphism. □

The equivalence (i) $\Longleftrightarrow$ (vii) in Theorem 1.2.3 implies also the following:

Remark 1.3.3 In any category in which the morphisms are functions and composition is the composition of functions, every isomorphism is a bijection. The following partial converse holds: if both a bijection f and its inverse f^{-1} are morphisms, then f is an isomorphism.

The next result is based on Remark 1.3.3, alone or together with one of Examples 1.3.18–1.3.20.

Example 1.3.23 In Examples 1.3.2, 1.3.3, 1.3.4 and 1.3.6, isomorphisms coincide with bijections, hence with bimorphisms.

Example 1.3.24 In Examples 1.3.5 and 1.3.16 all morphisms are isomorphisms, while in Example 1.3.15 a stronger property holds: identities are the only morphisms. This follows by Remark 1.3.3 for Example 1.3.5, and by easy direct observations for the other two categories. (Remark 1.3.3 does not apply to Example 1.3.16!)

Example 1.3.25 In the categories **Mon, Rng, Rng1** and in Example 1.3.14 there exist bimorphisms that are not isomorphisms. Indeed, the inclusion i in Example 1.3.22 is in fact a bimorphism because it is injective, but it is not an isomorphism because it is not surjective. For the category in Example 1.3.14 it was noted in Example 1.3.21 that all morphisms are bimorphisms, but if $a\rho b$ and $\neg(b\rho a)$, then (a, b) is not an isomorphism.

Remark 1.3.4 In any category in which the morphisms are functions and composition is the composition of functions, every bimorphism which is not a bijection cannot be an isomorphism.

An even stronger result is provided in

Example 1.3.26 In the category **Top** there exist bijective morphisms that are not isomorphisms, or, in topological terms, there exist bijective continuous functions that are not homeomorphisms. It suffices to take as object A a set X equipped with the discrete topology (i.e., all the subsets of X are open) and as object B the same set X equipped with any other topology; let f be the identity mapping of X. Then $f : A \longrightarrow B$ is a continuous mapping, hence a bijective morphism $f : A \longrightarrow B$. If f were an isomorphism, then the morphism $f^{-1} : B \longrightarrow A$ would be the inverse mapping of f by Theorem 1.2.3, (i) $\Longleftrightarrow$ (vii), therefore f^{-1} would coincide with f as a mapping. But f is not a continuous mapping from B to A.

Fortunately enough, this unpleasant situation is not general, as shown in the next example.

Example 1.3.27 In the categories **Mon, Grp, Ab, Rng, Rng1** and **Vec**, isomorphisms coincide with bijective morphisms. For instance, let $f : G \longrightarrow H$ be a bijective homomorphism of monoids. For every $x', y' \in H$, set $x = f^{-1}(x')$ and $y = f^{-1}(y')$; then

$$f^{-1}(x')f^{-1}(y') = xy = f^{-1}(f(xy)) = f^{-1}(f(x)f(y)) = f^{-1}(x'y') ,$$

showing that f^{-1} is a homomorphism of monoids. Similar proofs hod for the other categories.

As a matter of fact, it has been noted in universal algebra that the coincidence between isomorphisms and bijective morphisms is valid in any algebraic structure consisting of totally defined operations. Note that in all these cases there is no distinction between the set-theoretical and the category-theoretical meanings of f^{-1}.

Summarizing, the concepts of category theory introduced so far, starting with the very notion of a category, originate in basic notions related to sets and functions. Further developments sketched in this section will confirm that **Set** is the origin of category theory and its standard model. From this point of view, the categories in Examples 1.3.1 and 1.3.3–1.3.6 are variations of the category **Set**, the algebraic categories in Examples 1.3.7–1.3.12 and the category **Top** in Example 1.3.13 illustrate the natural occurrence of categories in algebra and topology, while the categories in Examples 1.3.14–1.3.16 are a kind of "teratological beings", yet perfectly consistent from the logical point of view.

Subsets and quotient sets are also modelled in the theory of categories. Recall that if A is a subset of B, then the inclusion function $i : A \longrightarrow B$, defined by $i(x) = x$, is injective. If $\equiv$ is an equivalence relation on A, then with the quotient set $Q = A/\equiv$ is associated the canonical surjection $p : A \longrightarrow Q$ defined by $p(x) = \hat{x}$. In the case of algebraic categories like those mentioned above, these mappings are morphisms.

Definition 1.3.5 A *subobject* (*quotient object*) of an object A is an object P for which there exists a monomorphism $u : P \longrightarrow A$ (an epimorphism $v : A \longrightarrow P$).

Remark 1.3.5 The notions of subobject and quotient object are dual to each other.

If $u : P \longrightarrow A$ is a monomorphism and $f : P" \longrightarrow P$ an isomorphism, then $u \circ f : P' \longrightarrow A$ is a monomorphism by Propositions 1.3.3(iv) and 1.3.2(i). Therefore, taking also into account the duality principle, we obtain:

Remark 1.3.6 Every object isomorphic to a subobject (quotient object) of A is also a subobject (a quotient object).

Clearly Proposition 1.3.2 implies

Remark 1.3.7 If A_1 is a subobject (quotient object) of A, and A_2 is a subobject (quotient object) of A_1, then A_2 is a subobject (quotient object) of A.

Example 1.3.28 In the category **Set** and also in the category given in Example 1.3.6, the subobjects (quotient object) of any object A are simply the subsets (quotient sets) of A; cf. Example 1.3.18.

Example 1.3.29 In Examples 1.3.5 and 1.3.14–1.3.16, A is a subobject of B and B is a quotient object of A for every morphism $f : A \longrightarrow B$.

As to the familiar categories familiar categories of sets and mappings which are studied in various fields of mathematics, as e.g. those in Examples 1.3.7–1.3.13, it is the rôle of each corresponding mathematical theory to decide whether the subobjects (quotient objects) coincide with the usual constructions of taking subalgebras, topological subspaces etc., (quotient algebras, quotient topological spaces, etc.). The problem reduces to the determination of the monomorphisms and epimorphisms of the corresponding category.

Exercise 1.3.2 Solve the above problem for the category **Cor** of correspondences.

Now we pass to the higher level of interconnections between categories.

Definition 1.3.6 A *subcategory* of a category **C** is a category **S** satisfying the following conditions:

(i) $\mid \mathbf{S} \mid \subseteq \mid \mathbf{C} \mid$, that is, every object of **S** is an object of **C**;

(ii) for every $A, B \in \mid \mathbf{S} \mid$, $\mathrm{Hom}_{\mathbf{S}}(A,B) \subseteq \mathrm{Hom}_{\mathbf{C}}(A,B)$;

(iii) for every morphisms $f : A \longrightarrow B$ and $g : B \longrightarrow C$ in **S**, the composite $g \circ f$ is the same in **S** as in **C**;

(iv) for every $A \in \mid \mathbf{S} \mid$, the identity 1_A is the same in **S** and in **C**.

The subcategory **S** is said to be a *full subcategory* if $\mathrm{Hom}_{\mathbf{S}}(A,B) = \mathrm{Hom}_{\mathbf{C}}(A,B)$ for every $A, B \in \mid \mathbf{S} \mid$.

Example 1.3.30 The category **Set** is a subcategory of **C**. The categories in Examples 1.3.3–1.3.6 are subcategories of **Set**; the subcategory in Example 1.3.6 is full. The category **Ab** is a full subcategory of **Grp**, which is a full subcategory of **Mon**. The category of pointed sets in Example 1.3.16 is not a subcategory of **Set**, because the morphisms are different.

Example 1.3.31 In terms of universal algebra, a ring is defined as an algebra $(R, +, \cdot, 0)$ satisfying certain axioms. A ring with unit may be defined as an algebra $(R, +, \cdot, 0, 1)$, where $(R, +, \cdot, 0)$ is a ring and 1 satisfies $x \cdot 1 = 1 \cdot x = x$ for all $x \in R$. Using this definition, **Rng1** is not a subcategory of **Rng**, because the objects are different. Alternatively, a ring with unit may be defined as a ring R for which there exists an element $1 \in R$ satisfying the above identities. According to this definition, **Rng1** is a subcategory of **Rng**. The null morphism $0 : R \longrightarrow R'$ between two rings is defined by $0(x) = 0$ for all $x \in R$. In particular the null morphism is defined for $R, R' \in\mid \mathbf{Rng1} \mid$, but it is not a morphism of **Rng1**, thus showing that **Rng1** is not a full subcategory of **Rng**.

As the objects of a category are linked by morphisms, categories are linked by functors.

Definition 1.3.7 a A *covariant functor* $F : \mathbf{C} \longrightarrow \mathbf{K}$, or simply a *functor* from the category $\mathbf{C}$ from the category $\mathbf{K}$ consists of an *object function* which assigns to each object A of $\mathbf{C}$ an object FA of $\mathbf{K}$, and, for every $A, B \in\mid \mathbf{C} \mid$, a *morphism function* which assigns to each morphism $u \in \mathbf{C}(A, B)$ a morphism $Fu \in \mathbf{K}(FA, FB)$, in such a way that the following conditions are fulfilled:

(1.3.11) $$F(v \circ u) = Fv \circ Fu \ ,$$

(1.3.12) $$F(1_A) = 1_{FA} \ ,$$

for every $A, B, C \in\mid \mathbf{C} \mid$ and every $u \in \mathbf{C}(A, B)$, $v \in \mathbf{C}(B, C)$.

The same letter F denotes the object function $F :\mid \mathbf{C} \mid \longrightarrow \mid \mathbf{K} \mid$, all the morphism functions $F : \mathbf{C}(A, B) \longrightarrow \mathbf{K}(FA, FB)$ and the functor $F : \mathbf{C} \longrightarrow \mathbf{K}$ itself. Properly speaking, the object function is not a function in the sense of Definition 1.2.8, because its domain and codomain may be proper classes instead of sets. Yet the meaning of such a generalized concept is clear; besides, in each concrete situation the association $A \mapsto FA$ is determined by a perfectly rigorous law.

Definition 1.3.8 A *contravariant functor* $F : \mathbf{C} \longrightarrow \mathbf{K}$ is a covariant functor $F : \mathbf{C} \longrightarrow \mathbf{K}^0$.

In other words, a contravariant functor $F : \mathbf{C} \longrightarrow \mathbf{K}$ associates with each object A of $\mathbf{C}$ an object FA of $\mathbf{K}$ and with each morphism $u \in \mathbf{C}(A, B)$ a morphism $Fu \in \mathbf{K}(FB, FA)$, in such a way that identities

(1.3.11⁰) $$F(v \circ u) = Fu \circ Fv$$

and (1.3.12) hold.

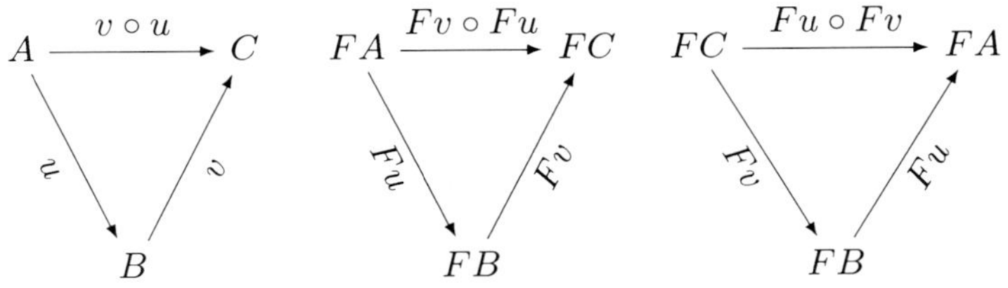

covariant contravariant

Fig.1.3

Example 1.3.32 Given a category $\mathbf{C}$, the *identity functor* $I_{\mathbf{C}} : \mathbf{C} \longrightarrow \mathbf{C}$ is the functor which leaves invariant every object A and every morphism f of $\mathbf{C}$.

Example 1.3.33 Given a subcategory $\mathbf{S}$ of a category $\mathbf{C}$, the *inclusion functor* $I_{\mathbf{SC}}$ is the functor $I_{\mathbf{SC}} : \mathbf{S} \longrightarrow \mathbf{C}$ which leaves invariant every object A and every morphism f of $\mathbf{S}$.

Example 1.3.34 The *duality functor* of a category $\mathbf{C}$ is the functor $D : \mathbf{C} \longrightarrow \mathbf{C}^0$ defined by $DA = A$ and $Df = f$ for every $A, B \in \mid \mathbf{C} \mid$ and every $f \in \mathbf{C}(A, B)$ (cf. Definitions 1.3.2 and 1.3.8).

Example 1.3.35 Let $\mathbf{C}$ be a category of sets endowed with a certain structure, the morphisms being the functions which "preserve" the structure, as for instance the categories in Examples 1.3.7–1.3.13. So an object A is a pair consisting of an *underlying set* X and the structure defined on it. The *underlying-set functor* $U : \mathbf{C} \longrightarrow \mathbf{Set}$, also called the *forgetful functor*, is the functor defined by $UA = X$ on objects and defined on morphisms as follows. Assume $A, B \in \mid \mathbf{C} \mid$ have underlying sets X and Y, respectively, and $f \in \mathbf{C}(A, B)$. Then f is a function $f : X \longrightarrow Y$ which preserves the structure and $Uf = f \in \mathbf{Set}(X, Y)$.

Example 1.3.36 More generally, let $\mathbf{C}$ be a category as in the previous Example and $\mathbf{S}$ a full subcategory of $\mathbf{C}$, consisting of all the objects that have a certain richer structure. The forgetful functor $U : \mathbf{S} \longrightarrow \mathbf{C}$ sends each object with underlying set X into the set X endowed with the poorer structure which characterizes $\mathbf{C}$, and each morphism $u \in \mathbf{S}(A, B)$ into $u \in \mathbf{C}(UA, UB)$. Thus, e.g., $U : \mathbf{Ab} \longrightarrow \mathbf{Grp}$ sends each Abelian group G into G regarded simply as a group, and $Uu = u$. The forgetful functor $U : \mathbf{Rng} \longrightarrow \mathbf{Ab}$ sends each ring into its additive group and again $Uu = u$, etc.

See also Examples 1.3.41–1.3.44 below. Functors arise naturally in algebra and topology. Further examples will be given in the subsequent chapters.

Exercise 1.3.3 Describe the concept of a functor in terms of arrows only; cf. Exercise 1.3.1.

As a functor is a pair of functions, the composition of functors is naturally defined in terms of the composition of these functions.

Definition 1.3.9 Given two functors $F : \mathbf{C} \longrightarrow \mathbf{C}'$ and $G : \mathbf{C}' \longrightarrow \mathbf{C}''$, the *composite*

$$G \circ F : \mathbf{C} \longrightarrow \mathbf{C}'' \tag{1.3.13}$$

is the functor defined by $(G \circ F)A = G(F(A))$ and $(G \circ F)u = G(F(u))$ for every $A, B \in \mid \mathbf{C} \mid$ and every $u \in \mathbf{C}(A, B)$.

In the above definition the two functors may be both covariant, or both contravariant, or one of them covariant and the other contravariant.

Proposition 1.3.4 *Let $F : \mathbf{C} \longrightarrow \mathbf{C}'$, $G : \mathbf{C}' \longrightarrow \mathbf{C}''$ and $H : \mathbf{C}'' \longrightarrow \mathbf{C}'''$. Then the following hold:*

(i) *The composite $G \circ F$ is a covariant functor if F and G are both covariant or both contravariant, and it is a contravariant functor otherwise.*

(ii) $H \circ (F \circ G)) = (H \circ F) \circ G$.

(iii) $F \circ I_{\mathbf{C}} = I_{\mathbf{C}'} \circ F = F$.

PROOF: Routine verification left to the reader. □

The next two concepts might be regarded as injectivity and surjectivity of functors.

Definition 1.3.10 A covariant functor $F : \mathbf{C} \longrightarrow \mathbf{C}'$ is said to be *faithful* (*full*) if for every $A, B \in \mid \mathbf{C} \mid$, the morphism function $F : \mathbf{C}(A, B) \longrightarrow \mathbf{C}'(FA, FB)$ is injective (surjective). A faithful (full) contravariant functor F is defined similarly, using the morphism function $F : \mathbf{C}(A, B) \longrightarrow \mathbf{C}'(FB, FA)$.

In other words, F is: (i) faithful, if for every $u, v : A \longrightarrow B$ in $\mathbf{C}$, the equality $Fu = Fv$ implies $u = v$; (ii) full, if every $u' : FA \longrightarrow FB$ in$\mathbf{C}$' can be written as $u' = Fu$ for some $u; A \longrightarrow B$ in $\mathbf{C}$. For a contravariant functor similar interpretations are obtained by reversing arrows.

Example 1.3.37 The identity functors, the inclusion functors, the duality functors and the forgetful functors are faithful (cf. Examples 1.3.32–1.3.36).

Example 1.3.38 The identity functors and the duality functors are full. An inclusion functor $I_{\mathbf{SC}}$ is full if and only if $\mathbf{S}$ is a full subcategory of $\mathbf{C}$ (cf. Examples 1.3.32–1.3.34).

Proposition 1.3.5 *Let $F : \mathbf{C} \longrightarrow \mathbf{C}'$ and $G : \mathbf{C}' \longrightarrow \mathbf{C}''$ be functors. Then the following hold:*

(i) *F and G are faithful $\Longrightarrow$ $G \circ F$ is faithful $\Longrightarrow$ F is faithful.*

(ii) *F and G are full $\Longrightarrow$ $G \circ F$ is full.*

PROOF: If F and G are covariant, apply Proposition 1.2.35 to the morphism functions $F : \mathbf{C}(A, B) \longrightarrow \mathbf{C}'(FA, FB)$ and $G : \mathbf{C}'(FA, FB) \longrightarrow \mathbf{C}''(GFA, GFB)$. The other three cases concerning the variances (i.e., covariant or contravariant) of F and G are treated in the same way. □

Warning The symmetry of conditions (i) and (ii) in the original Proposition 1.2.5 is broken in Proposition 1.3.5, where the implication $G \circ F$ is faithful $\Longrightarrow F$ is faithful, from (i), has no analogue in (ii). This is due to the fact that an arbitrrary pair A', B' of objects of $\mathbf{C}'$ need not be of the form $A' = FA, B' = FB$, with A, B objects of $\mathbf{C}$.

Exercise 1.3.5 Let F be a functor and f a morphism in the domain of F. (i) If f is an isomorphism then Ff is an isomorphism. (ii) If F is faithful and full and f is a monomorphism (an epimorphism), then Ff is a monomorphism (an epimorphism).

The following definition is a natural extension of Definition 1.3.4.

Definition 1.3.11 An *isomorphism* (a *dual isomorphism*) of categories is a covariant (contravariant) functor $F : \mathbf{C} \longrightarrow \mathbf{C}'$ with the property that there exists a functor $F^{-1} : \mathbf{C}' \longrightarrow \mathbf{C}$ such that

(1.3.14) $$F^{-1} \circ F = I_{\mathbf{C}} \text{ and } F \circ F^{-1} = I_{\mathbf{C}'} .$$

The categories $\mathbf{C}$ and $\mathbf{C}$' are then said to be *isomorphic* (*dually isomorphic*), while F^{-1} is known as the *inverse* of F.

Example 1.3.39 The identities $I_{\mathbf{C}}$ are isomorphisms with $I_{\mathbf{C}}^{-}1 = I_{\mathbf{C}}$, while the dualities $D : \mathbf{C} \longrightarrow \mathbf{C}^0$ are dual isomorphisms, the inverse of D being the duality functor of $\mathbf{C}^0$.

Further examples will be studied in the subsequent chapters.

Lemma 1.3.4 *If $F : \mathbf{C} \longrightarrow \mathbf{C}'$ is an isomorphism (a dual isomorphism) and a functor $G : \mathbf{C}' \longrightarrow \mathbf{C}$ satisfies $G \circ F = I_{\mathbf{C}}$ or $F \circ G = I_{\mathbf{C}'}$, then $G = F^{-1}$.*

PROOF: Suppose e.g that $F \circ G = I_{\mathbf{C}}$. Then $G = I_{\mathbf{C}} \circ F = F^{-1} \circ F \circ G = F^{-1} \circ I_{\mathbf{C}'} = F^{-1}$. □

Proposition 1.3.6 *If F is an isomorphism (a dual isomorphism), then F^{-1} is also an isomorphism (a dual isomorphism) and $(F^{-1})^{-1} = F$.*

PROOF: Since $I_{\mathbf{C}}$ and $I_{\mathbf{C}'}$ are covariant functors, it follows from (1.3.14) that F and F^{-1} have the same variance (both covariant or both contravariant). The symmetry of conditions (1.3.14) immediately implies that F^{-1} is an isomorphism (a dual isomorphism) with inverse F, and since the inverse is unique by Lemma 1.3.4, this means that $(F^{-1})^{-1} = F$. □

Proposition 1.3.7 *Let $F : \mathbf{C} \longrightarrow \mathbf{C}'$ and $G : \mathbf{C}' \longrightarrow \mathbf{C}''$ be two isomorphisms (dual isomorphisms, one of them an isomorphism and the other a dual isomorphism). Then $G \circ F$ is an isomorphism (an isomorphism, a dual isomorphism) and*

(1.3.15) $$(G \circ F)^{-1} = F^{-1} \circ G^{-1} .$$

PROOF: In view of Proposition 1.3.6, it suffices to notice that $(F^{-1} \circ G^{-1}) \circ (G \circ F) = I_{\mathbf{C}}$ and $(G \circ F) \circ (F^{-1} \circ G^{-1}) = I_{\mathbf{C}''}$. □

Let us relate isomorphism to the previous properties.

Remark 1.3.8 An isomorphism or a dual isomorphism of categories is both a faithful and a full functor, because $Fu = Fv \Longrightarrow u = F^{-1}Fu = F^{-1}Fv = v$ and $v' = FF^{-1}v'$. The converse does not hold, because faithfulness and fullness refer only to morphisms, whereas isomorphism concerns also objects. This is illustrated in the next example.

Example 1.3.40 The inclusion functor $I_{\mathbf{SC}} : \mathbf{S} \longrightarrow \mathbf{C}$ of a full subcategory $\mathbf{S}$ of a category $\mathbf{C}$ is both faithful and full, but not an isomorphism, because the object function is not surjective unless $\mathbf{S} = \mathbf{C}$.

The next Proposition is a variant of Definition 1.3.11.

Proposition 1.3.8 *A functor is an isomorphism if and only if both its object function and all of its morphism functions are bijections.*

PROOF: If $F : \mathbf{C} \longrightarrow \mathbf{C}'$ is an isomorphism, then it follows from (1.3.14) that the object function is a bijection, while Remark 1.3.8 shows that for every objects A, B of $\mathbf{C}$, F establishes a bijection between $\mathbf{C}(A, B)$ and $\mathbf{C}'(FA, FB)$.

Conversely, assume the object function and the morphism functions of F are bijections. Let F^{-1} denote the inverse functions of all these bijections. Then relations (1.3.14) hold and it remains to prove that the functions F^{-1} form a functor.

Take two objects A', B' of $\mathbf{C}'$; they are of the form $A' = FA, B' = FB$. If F is a covariant functor, then F^{-1} maps $\mathbf{C}'(A', B') = \mathbf{C}'(FA, FB)$ to $\mathbf{C}(A, B) = \mathbf{C}(F^{-1}A', F^{-1}B')$. Furthermore, take $u', v' \in \mathbf{C}'(B', C')$, where $C' = FC$. Setting $F^{-1}u' = u, F^{-1}v' = v$, we have $F(v \circ u) = Fv \circ Fu$, therefore

$$F^{-1}v' \circ F^{-1}u' = v \circ u = F^{-1}(Fv \circ Fu) = F^{-1}(v' \circ u') .$$

Similarly we obtain $F^{-1}1_{A'} = 1_{F^{-1}A'}$. If the functor F is covariant we get $F^{-1}(v' \circ u') = F^{-1}u' \circ F^{-1}v'$. □

We have just seen how functors interconnect categories. Now we introduce natural transformations, which link functors.

Definition 1.3.12 Let $F, G : \mathbf{C} \longrightarrow \mathbf{C}'$ be two covariant functors. A *natural transformation*, also called a *functorial morphism*,

$$\varphi : F \longrightarrow G .$$

is a function φ which assigns to each object A of $\mathbf{C}$,[2] a morphism $\varphi_A \in \mathbf{C}'(FA, GA)$ in such a way that

[2]Like the object function of a functor, this is not properly speaking a function, because the domain may be a proper class.

$$\varphi_B \circ Ff = Gf \circ \varphi_A \tag{1.3.16}$$

for every $A, B \in \mid \mathbf{C} \mid$ and every $f \in \mathbf{C}(A, B)$. If, moreover, φ_A is an isomorphism for every object A, we say that φ is a *natural isomorphism* or a *functorial isomorphism*, while F and G are said to be *isomorphic*.

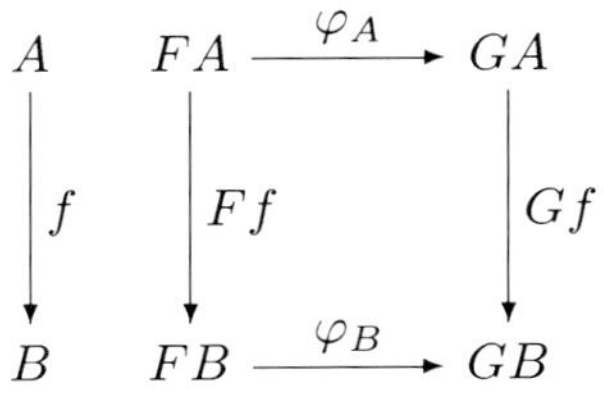

Fig. 1.4

Examples 1.3.41–1.3.44 below are taken from Mac Lane [1971].

Example 1.3.41 Let **CRng1** be the full subcategory of **Rng1** (cf. Example 1.3.11) having as objects all commutative rings with unit, and **Grp** the category of (cf. Example 1.3.8). Take a natural number $n \geq 2$.

For every $A \in \mid \mathbf{CRng1} \mid$ let $S_n A$ be the group of all non-singular matrices of order n with entries in A and for every morphism $f : A \longrightarrow B$ in **CRng1**, let $S_n f$ be the mapping which transforms each matrix $\| m_{ij} \| \in S_n A$ into the matrix $\| f(m_{ij}) \|$. Since $\det \| f(m_{ij}) \| = f(\det \| m_{ij} \|)$ and f transform every unit of A into a unit of B, it follows that the matrix $\| f(m_{ij}) \|$ is non-singular, whence one sees easily that $S_n f : S_n A \longrightarrow S_n B$. Therefore $S_n : \mathbf{CRng1} \longrightarrow \mathbf{Grp}$, which obviously satisfies (1.3.11) and (1.3.12), is a functor.

Let further $U : \mathbf{CRng1} \longrightarrow \mathbf{Grp}$ be the functor such that for every $A, B \in \mid \mathbf{CRng1} \mid$ and every morphism $f : A \longrightarrow B$, UA is the group of units of A and $Uf = f|_{UA}$.

Now a functorial morphism $\det : S_n \longrightarrow U$ can be defined as follows: for every commutative ring with unit A, $\det_A$ is the group homomorphism which associates each matrix $\| m_{ij} \| \in S_n A$ with its determinant $\det \| m_{ij} \|$.

Example 1.3.42 For every group A, the *commutator* $\mathrm{Com}A = \{xyx^{-1}y^{-1} \mid x, y \in A\}$ is a normal subgroup of A. A functor $\mathrm{Com} : \mathbf{Grp} \longrightarrow \mathbf{Grp}$ can be defined which associates each group A with $\mathrm{Com}A$ and each homomorphism of groups $f : A \longrightarrow B$ with the homomorphism $\mathrm{Com}f : \mathrm{Com}A \longrightarrow \mathrm{Com}B$ defined by $\mathrm{Com}f = f|_{\mathrm{Com}A}$.

For every group A, let $\varphi_A : A \longrightarrow \mathrm{Com}A$ be the constant homomorphism $\varphi_A(x) = e$ (the unit of A) for every $x \in A$. Clearly $\mathrm{Com}f \circ \varphi_A = \varphi_B \circ f$ for every homomorphism $f : A \longrightarrow B$, so that $\varphi : I_{\mathbf{Grp}} \longrightarrow \mathrm{Com}$ is a natural transformation.

Example 1.3.43 Let $P : \mathbf{Grp} \longrightarrow \mathbf{Grp}$ be the functor defined by $PA = A/\mathrm{Com}A$ on the objects of **Grp**, while a morphism $f : A \longrightarrow B$ is sent to $Pf : \mathrm{Com}A \longrightarrow \mathrm{Com}B$ defined by $Pf(p_A(x)) = p_B(f(x))$ for every $x \in A$.

For every group A, let $\psi_A : A \longrightarrow A/\mathrm{Com}A$ be the canonical surjection. Then it is immediately seen that the ψ_As are the components of a functorial morphism $\psi : I_{\mathbf{Grp}} \longrightarrow P$.

Example 1.3.44 Let **FVec** be the full subcategory of **Vec** (cf. Example 3.1.13) having as objects all vector spaces of finite dimension. Let* : $\mathbf{FVec} \longrightarrow \mathbf{FVec}$ be the contravariant functor which assigns to each $A \in\mid \mathbf{FVec} \mid$ the space A^* of all linear forms over A and to each linear mapping $f : A \longrightarrow A'$ the linear mapping $f^* : A'^* \longrightarrow A^*$ defined by $f^*(\alpha') = \alpha' \circ f$ for every $\alpha' \in A'^*$.

A natural isomorphism $\varphi : I_{\mathbf{FVec}} \longrightarrow^{**}$ is then defined by taking, for each $A \in\mid \mathbf{FVec} \mid$, the isomorphism $\varphi_A : A \longrightarrow A^{**}$ defined as follows: for every $x \in A$, the image $\varphi_A(x) \in A^{**}$ is the linear form $\varphi_A(x) : A^* \longrightarrow F$ (the field of scalars) given by $\varphi_A(x)(\alpha) = \alpha(x)$ for every $\alpha \in A^*$. The routine verifications are left to the reader.

The next natural step is the introduction of the composition of functorial morphisms.

Definition 1.3.13 Consider three covariant functors $F, G, H : \mathbf{C} \longrightarrow \mathbf{C}'$. The *composite* of two natural transformations $\varphi : F \longrightarrow G$ and $\psi : G \longrightarrow H$ is the natural transformation $\psi \circ \varphi : F \longrightarrow H$ defined by $(\psi \circ \varphi)_A = \psi_A \circ \varphi_A$ for every $A \in\mid \mathbf{C} \mid$. The *identity* $1_F : F \longrightarrow F$ is defined by $(1_F)_A = 1_{FA}$.

Proposition 1.3.9 *The composite of two natural transformations (isomorphisms) is a natural transformation (isomorphism) and*

(1.3.17) $$\chi \circ (\psi \circ \varphi) = (\chi \circ \psi) \circ \varphi \,,$$

provided $\chi \circ (\psi \circ \varphi)$ *exists, or equivalently, provided* $(\chi \circ \psi) \circ \varphi$ *exists. Besides, if* $\varphi : F \longrightarrow G$, *then*

(1.3.18) $$\varphi \circ 1_F = \varphi \text{ and } 1_G \circ \varphi = \varphi \,.$$

PROOF: Routine. □

Proposition 1.3.10 (i) *A functorial morphism* $\varphi : F \longrightarrow G$ *is a functorial isomorphism if and only if there is a functorial morphism* $\varphi^{-1} : G \longrightarrow F$ *such that*

(1.3.19) $$\varphi^{-1} \circ \varphi = 1_F \text{ and } \varphi \circ \varphi^{-1} = 1_G \,.$$

(ii) *When this is the case,* φ^{-1} *is a functorial isomorphism uniquely determined by* φ, *and* $(\varphi^{-1})^{-1} = \varphi$.

PROOF: Similar to that of Proposition 1.3.6. □

Proposition 1.3.11 *Let* $\varphi : F \longrightarrow G$ *and* $\psi : G \longrightarrow H$ *be functorial isomorphisms. Then* $\psi \circ \varphi : F \longrightarrow H$ *is a functorial isomorphism and*

(1.3.20) $$(\psi \circ \varphi)^{-1} = \varphi^{-1} \circ \psi^{-1} \,.$$

PROOF: Similar to that of Proposition 1.3.7. □

Natural transformations enable the introduction of the concept of equivalent categories, which is weaker than isomorphic categories. Yet two equivalent categories are close enough: certain results obtained for one of the categories can be automatically transported to the other category. So, the equivalence of categories can be used as a strong proving device.

Definition 1.3.14 Two categories $\mathbf{C}$ and $\mathbf{C}'$ are said to be *equivalent* (*dually equivalent*) if there exist two covariant (contravariant) functors $F : \mathbf{C} \longrightarrow \mathbf{C}'$ and $G : \mathbf{C}' \longrightarrow \mathbf{C}$ and two natural isomorphisms $\varphi : I_{\mathbf{C}} \longrightarrow G \circ F$ and $\psi : I_{\mathbf{C}'} \longrightarrow F \circ G$. When this is the case, we also say that F and G establish an *equivalence* (a *dual equivalence*) between $\mathbf{C}$ and $\mathbf{C}'$.

Thus, for instance, the natural isomorphism φ in Example 1.3.14 can be regarded as establishing a dual equivalence between **FVec** and itself, with * in the double rôle of F and G, and φ in the double rôle of φ and ψ. This is, of course, a rather peripheral example, because the interest lies in the equivalence of different categories. There is plenty of such equivalences in mathematics. In Ch.5§5 we will establish dual equivalences between certain categories of lattices and certain categories of topological spaces.

Remark 1.3.9 In other words, Definition 1.3.14 means that for every $A \in \mid \mathbf{C} \mid$ there is an isomorphism $\varphi_A : A \longrightarrow GFA$ and for every $A' \in \mid \mathbf{C}' \mid$ there is an isomorphism $\psi_{A'} : A' \longrightarrow FGA'$, with the properties that $\varphi_B \circ f = GFf \circ \varphi_A$ for every morphism $f : A \longrightarrow B$ in $\mathbf{C}$ and $\psi_{B'} \circ f' = FGf' \circ \psi_{A'}$ for every morphism $f' : A' \longrightarrow B'$ in $\mathbf{C}'$.

We have noted in Proposition 1.3.8 that an isomorphism of categories is characterized by the fact that its object functions and all of its morphism functions are bijections. It is natural to expect that a functor which establishes an equivalence (a dual equivalence) of categories has similar but weaker properties. Indeed:

Proposition 1.3.12 (A) *If the functor F establishes an equivalence between two categories $\mathbf{C}$ and $\mathbf{C}'$, then the following hold:*

(ia) *If the objects A and B satisfy $FA = FB$, then they are isomorphic.*

(ib) *Every object A' of $\mathbf{C}'$ is isomorphic to FA for some object A of $\mathbf{C}$.*

(iia) *If the morphisms f and g of $\mathbf{C}$ fulfill $Ff = Fg$, then $g = j \circ f \circ i$ for some isomorphisms i and j of $\mathbf{C}$.*

(iib) *For every morphism f' of $\mathbf{C}'$ there exist a morphism f of $\mathbf{C}$ and two isomorphisms i' and j' of $\mathbf{C}'$ such that $f' = j' \circ Ff \circ i'$.*

(B) *Similar statement for dual equivalences.*

PROOF: (A) Let $G : \mathbf{C}' \longrightarrow \mathbf{C}$ be a functor such that there exist natural isomorphisms $\varphi : I_{\mathbf{C}} \longrightarrow G \circ F$ and $\psi : I_{\mathbf{C}'} \longrightarrow F \circ G$. So the identities in Remark 1.3.9 hold.

(ia): We have the isomorphisms $A \overset{\varphi_A}{\longrightarrow} GFA = GFB \overset{\varphi_B^{-1}}{\longrightarrow} B$.

(ib): The isomorphism $\psi_{A'}$ maps A' to FGA', where $GA' \in \mid \mathbf{C} \mid$.

(iia): Let $f : A \longrightarrow B$ and $g : C \longrightarrow D$ fulfill $Ff = Fg$. Then $FA = FC$ and $FB = FD$, consequently $GFf = Gfg$, $GFA = GFC$ and $GFB = GFD$. Now from $\varphi_B \circ f = GFf \circ \varphi_A$ and $\varphi_D \circ g = GFg \circ \varphi_C$ we infer

$$g = \varphi_D^{-1} \circ GFf \circ \varphi_C = \varphi_D^{-1} \circ \varphi_B \circ f \circ \varphi_A^{-1} \circ \varphi_C \, , \tag{1.3.21}$$

where $\varphi_D^{-1} \circ \varphi_B$ and $\varphi_A^{-1} \circ \varphi_C$ are isomorphisms of $\mathbf{C}$.

(iib): If $f' : A' \longrightarrow B'$ is a morphism in $\mathbf{C}'$, then from $\psi_{B'} \circ f' = FG\,f' \circ \psi_{A'}$ we infer $f' = \psi_{B'}^{-1} \circ FG\,f' \circ \psi_{A'}$, where $\psi_{B'}^{-1}$ and $\psi_{A'}$ are isomorphisms of $\mathbf{C}'$ and $G\,f'$ is a morphism of $\mathbf{C}$.

(B): Similar proof. □

Corollary 1.3.1 *For every $A, B \in \mid \mathbf{C} \mid$, F establishes an injection from $\mathbf{C}(A, B)$ into $\mathbf{C}'(FA, FB)$.*

PROOF: Let $f, g : A \longrightarrow B$ satisfy $Ff = Fg$. Then by applying (1.3.21) with $A = C$ and $B = D$ we get $f = g$. □

We conclude this section with a little supplement, for which we adopt a rather informal style.

Recall the forgetful functor $U : \mathbf{C} \longrightarrow \mathbf{Set}$ in Example 1.3 35 and the familiar constructions of free monoids, free groups, free rings, etc., generated by a given set of free generators. In all these cases we have a category $\mathbf{C}$ of sets equipped with a structure, and for every set A there exists an object $UA \in \mid \mathbf{C} \mid$ whose underlying set includes A and with the property that for every $X \in \mid \mathbf{C} \mid$ and every function $f : A \longrightarrow UX$ there is a unique morphism $\overline{f} : UA \longrightarrow X$ in$\mathbf{C}$ such that $\overline{f}|_A = f$. If we introduce the inclusion mapping $\eta_A : A \longrightarrow UA$, the latter condition can be stated more properly as $U\overline{f} \circ \eta_A = f$.

Furthermore, for every function $u : A \longrightarrow B$, take $f = \eta_B \circ u$ and $X = FB$. Then $f : A \longrightarrow UX$ and therefore $\overline{f} \in \mathbf{C}(UA, UB)$. Setting $Fu = \overline{f}$, we get a commutative square: $UFu \circ \eta_A = \eta_B \circ u$, and Fu is unique with this property. In particular $UF1_A = 1_A$ and since $U1_{FA} = 1_A$, it follows that $F1_A = 1_{FA}$. Also, if $v : B \longrightarrow C$, we have the commutative square $UFv \circ \eta_B = \eta_C \circ v$. By concatenating these commutative squares we obtain a "bigger" commutative square $\eta_C \circ v \circ u = UFv \circ UFu \circ \eta_A$, and since U is a functor, this reads $\eta_X \circ (v \circ u) = UF(v \circ u) \circ \eta_A$. Thus $F : \mathbf{Set} \longrightarrow \mathbf{C}$ is a functor which we may call a *free functor*, while $\eta : I_{\mathbf{Set}} \longrightarrow UF$ is a natural transformation.

The following general situation subsumes the above example. Two functors $F : \mathbf{A} \longrightarrow \mathbf{C}$ and $G : \mathbf{C} \longrightarrow \mathbf{A}$ are given which satisfy the following property. There is a natural transformation $\eta : I_{\mathbf{C}} \longrightarrow GF$ such that for every object A of $\mathbf{A}$, every object X of $\mathbf{C}$ and every morphism $f : A \longrightarrow GX$ in$\mathbf{A}$ there is a unique morphism $\overline{f} : FA \longrightarrow X$ such that $G\overline{f} \circ \eta_A = f$.

It can be proved that that the above property is equivalent to the following one. There is a natural transformation $\varepsilon : FG \longrightarrow I_{\mathbf{A}}$ such that for every object

X of **C**, every object A of **A** and every morphism $g : FA \longrightarrow X$ in **C** there is a unique morphism $\overline{g} : A \longrightarrow GX$ such that $\varepsilon_X \circ F\overline{g} = g$.

When the above situation holds, one says that F and G form a pair pf *adjoint functors* ; to be specific, F is a *left adjoint* of G and G is a *right adjoint* of F.

Fig. 1.5

Another example in which a forgetful functor has a left adjoint is the following. Take again **A**=**Set**, and **C**=**Top**, the category of topological spaces and continuous functions. Define a functor T : **Set** $\longrightarrow$ **Top** as follows: for every set X, TX is the set X endowed with the discrete topology (all of the sets are open), and for every function $f : X \longrightarrow Y$, $Tf : TX \longrightarrow TY$ is defined by $(Tf)(x) = f(x)$ for all $x \in X$. Let U : **Top** $\longrightarrow$ **Set** be the forgetful functor. The adjointness is immediately checked: we have $UT = I_{\mathbf{Set}}$, so that we can use the identical functorial morphism $\eta : I_{\mathbf{Set}} \longrightarrow I_{\mathbf{Set}}$ defined by $\eta_X = 1_X$ for every set X. If $X, Y \in \mid$ **Top** $\mid$ and f is a function $f : X \longrightarrow UY$, then the function $\overline{f} : TX \longrightarrow Y$ defined by $\overline{f}(x) = f(x)$ satisfies $\overline{f} \circ 1_X = f$.

We have referred in §1 to the direct product of sets. In everyday mathematics we denote the direct product of a family of sets $(X_i)_{i\in I}$ by $\prod_{i\in I} X_i$; its elements are the families of the form $(x_j)_{j\in I}$ with $x_j \in X_j$ for all $j \in I$. Denote this product by P for short. The functions $p_i : P \longrightarrow X_i$ defined by $p_i((x_j)_{j\in I}) = x_i$ $(\forall\, i \in I)$ are called the *canonical projections* $(\forall\, i \in I)$ and it is easily seen that they satisfy the following *universality property*: for any set X and any family of functions $q_i : X \longrightarrow A_i$ $(i \in I)$ there is a unique function $u : X \longrightarrow A$ such that $p_i \circ u = q_i$ $(\forall\, i \in I)$. In many familiar categories, like those mentioned in §1, direct products do exist, with the same universality property.

In category theory the above concept has been generalized as follows. The *direct product* of a family of objects $(A_i)_{i\in I}$ is a pair $(A, (u_i)_{i\in I})$, where A is an object and $p_i : A \longrightarrow A_i$ are morphisms such that for any pair $(X, (q_i)_{i\in I})$, where q_i are morphisms $q_i : X \longrightarrow A_i$, there is a unique morphism $u : X \longrightarrow A$ such that $p_i \circ u = q_i$ for all $i \in I$. The dual concept is called *direct sum* or *direct coproduct* and is defined as follows: a pair $((p_i)_{i\in I}), A)$, where A is an object and $p_i : A_i \longrightarrow A$ are morphisms such that for any pair $((q_i)_{i\in I}, X)$, where X is an object and $q_i : A_i \longrightarrow X$ are morphisms, there is a unique morphism $u : A \longrightarrow X$ such that $u \circ p_i = q_i$ for all $i \in I$. The existence of direct products and/or of direct sums is one of the problems much studied in category theory.

Exercise 1.3.4 Prove that in any category the direct product (direct sum), whenever it exists, is unique up to an isomorphism. In other words, two direct products (direct sums) A and A' of the same family of objects $(A_i)_{i\in I}$ are isomorphic.

Example 1.3.45 In the category **Set** the direct sum is realized as follows. The *disjoint union* $\sum_{i\in I} A_i$ of a family $(A_i)_{i\in I}$ of sets is the set $\bigcup\{A_i^* \mid i \in I\}$, where $A_i^* = \{i\} \times A_i = \{(i,x) \mid x \in A_i\}$ $(i \in I)$. Note that the mapping $x \mapsto (i,x)$ $(x \in A_i)$ establishes a bijection between A_i and A_i^*, while $A_i^* \cap A_j^* = \varnothing$ for $i,j \in I$ and $i \neq j$. The *canonical injections* $\ell_i : A_i \longrightarrow \sum_{j\in I} A_j$ are defined by extending the bijection $A_i \longrightarrow A_i^*$, that is, $\ell(x) = (i,x)$ $(\forall x \in A_i)$. The reader can easily check that the canonical injections satisfy the required universality property which makes the disjoint union of sets a direct sum in the category **Set**.

Chapter 2

Ordered Sets

As explained in the Preface, the concept of order is a tool used in mathematics both in its more general form and under certain supplementary conditions. This chapter first deals with partially ordered sets at the general level, then with totally ordered sets and well-ordered sets, thus beginning the specialization line which leads to ordinals (to be studied in the next chapter).

The first section is devoted to arbitrary posets (partially ordered sets), described in terms of each of the relations $\leq, <, \geq, >$; the relationships between them is explicitly pointed out. The next section introduces the various distinguished elements which a poset may posses : least/greatest element, minimal/maximal elements, etc., and constructs their complete existential theory, i.e., all logical implications between them. The important subclass of chains, or totally ordered sets, is studied in §3. The next section follows Szele [1950] in presenting several conditions equivalent to the axiom of choice and currently used in mathematics. The last section studies the basic properties of well-ordered sets, including several properties which are usually presented as properties of ordinal numbers, but which don't actually need the concept of ordinal itself, being in fact properties of well-ordered sets.

Much more about partially ordered sets can be found in the monographs by Erné [1982] and Gaspard, Leclerc and Montjardet [2007].

Keywords: Poset, chain, duality, distinguished elements, complete existential theory, isotone mapping, axiom of choice, Kuratowski-Zorn lemma, well-ordered set, segment.

2.1 Partially Ordered Sets

This section is essentially a study of the relationships between the order relations $\leq, <, \geq, >$, independently of their various concrete meanings which appear in mathematics.

Sergiu Rudeanu

Definition 2.1.1 If a binary relation $\leq$ on a set P satisfies the properties: for every $x, y, z \in P$,

(2.1.1) $$x \leq x \qquad \text{(reflexivity)} ,$$

(2.1.2) $$x \leq y \;\&\; y \leq z \Longrightarrow x \leq z \qquad \text{(transitivity)} ,$$

(2.1.3) $$x \leq y \;\&\; y \leq x \Longrightarrow x = y \qquad \text{(antisymmetry)} ,$$

then $\leq$ is called a *partial order*, and $(P, \leq)$ is a *partially ordered set* or simply a *poset*. If $\leq$ is reflexive and transitive, we say that $\leq$ is a *quasi-order* and $(P, \leq)$ is a *quasi-ordered set*. $(P, \leq)$ is often identified with its *underlying set* P.

Definition 2.1.2 Let $(P, \leq)$ be a poset. For every $x, y \in P$, we write

(2.1.4) $$x \geq y \Longleftrightarrow y \leq x ,$$

(2.1.5) $$x < y \Longleftrightarrow x \leq y \;\&\; x \neq y ,$$

(2.1.6) $$x > y \Longleftrightarrow y < x; ,$$

(2.1.7) $$x \# y \Longleftrightarrow x \not\leq y \;\&\; y \not\leq x ,$$

where $x \not\rho\, y$ means the negation of $x \,\rho\, y$, for any binary relation ρ. The conditions $x \leq y$, $x \geq y$, $x < y$, $x > y$, and $x \# y$ are read "x less than or equal to y, x greater than or equal to y, x less than y, x greater than y", and "x incomparable to y", respectively. The binary relation $\geq$ and $>$ are said to be the *duals* of the relations $\leq$ and $<$, respectively.

Remark 2.1.0 If $(P, \leq)$ is a poset (quasi-ordered set), then $(P, \geq)$ is also a poset (quasi-ordered set), called the *dual* of $(P, \leq)$.

Example 2.1.1 The empty set $\varnothing$ is a poset under the empty relation $\perp$.

Example 2.1.2 Every set X is a poset under the identity relation $=$.

Example 2.1.3 Every equivalence relation is a quasi-order.

Example 2.1.4 The divisibility relation $\mid$ is a quasi-order on the set $\mathbb{Z}$ of integers, but not a partial order, because $n \mid -n$ and $-n \mid n$ although $n \neq -n$ for $n \neq 0$.

Example 2.1.5 The set $\mathbb{N} = \{0, 1, 2, \ldots\}$ of natural numbers endowed with the divisibility relation $\mid$ is a poset.

Example 2.1.6 The set of continuous real functions $f : \mathbb{R} \longrightarrow \mathbb{R}$ becomes a poset if we define, for all continuous real functions f, g,

$$f \leq g \Longleftrightarrow f(x) \leq g(x) \; (\forall\, x \in \mathbb{R}) .$$

Example 2.1.7 Let X be an arbitrary set and $P \neq \varnothing$ a poset. The set P^X of all functions $f : X \longrightarrow P$ becomes a poset under the *pointwise order*

$$f \leq g \iff f(x) \leq g(x) \quad (\forall x \in X) .$$

To provide further examples, let us introduce a suggestive graphical representation of finite (and some infinite) posets. It is based on

Definition 2.1.3 Let $(P, \leq)$ be a poset and $x, y \in P$. We say that *x is covered by y* or *y covers x* and we write $x \prec y$ or $y \succ x$, and we write $x \prec y$ or $y \succ x$, if $x < y$ and there is no element $z \in P$ such that $x < z < y$. We also say that x is a *predecessor* of y, while y is a *successor* of x.

The *Hasse diagram* of a finite poset P is obtained by representing the elements of P as points of a plane, in such a way that $x \prec y$ if and only if the points (representing) x and y are connected by a (straight) line which contains no other point of P, x being situated under y.

Warning The above explanation is not a rigorous definition. since the term "under" remains undefined. However, from an intuitive point of view, the use of Hasse diagrams turns out to be a very efficient tool, both didactical and heuristic (especially in providing counter-examples). The situation is quite analogous to the use of figures in geometry. Note also that certain peculiarities of the drawing are irrelevant (cf. Example 2.1.8).

Example 2.1.8 Figure 2.1 represents the Hasse diagrams of seven posets. If the K-shaped diagram is changed so as to suggest the letter X, the poset remains the same. Also, a drawing suggesting the Cyrillic letter I would represent the same poset as the N-shaped diagram.

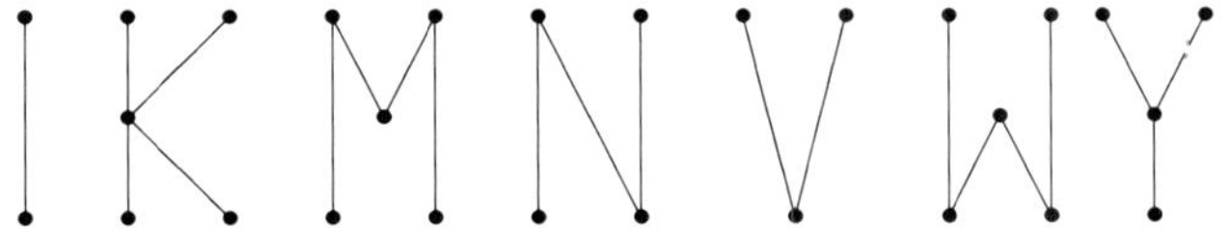

Fig. 2.1

Example 2.1.9 The Hasse diagram of the set of all subsets of a set $\{a, b, c\}$, ordered by set inclusion, is depicted in Fig. 2.2. It includes the three similar diagrams for the subsets of the sets $\{a, b\}, \{a, c\}$ and $\{b, c\}$.

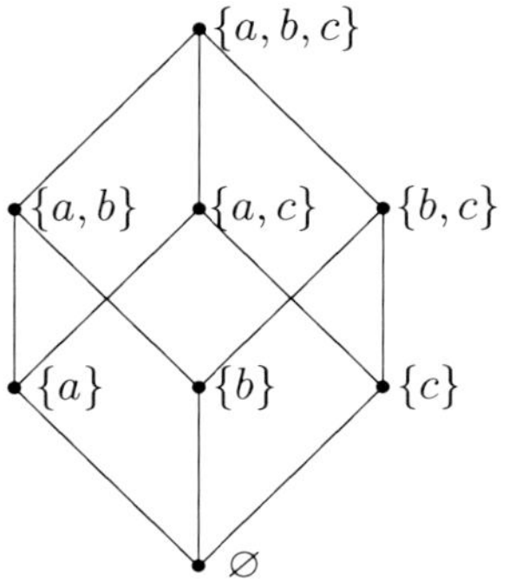

Fig. 2.2

Exercise 2.1.1 Let $(Q, \leq)$ be a quasi-ordered set and let x, y denote arbitrary elements of Q. Set

(2.1.8) $$x \equiv y \Longleftrightarrow x \leq y \;\&\; y \leq x \; .$$

Show that (2.1.8) is an equivalence on Q and the quotient set $Q/\equiv$ is partially ordered by the relation

(2.1.9) $$\hat{x} \leq \hat{y} \Longleftrightarrow x \leq y \; .$$

Exercise 2.1.2 What is obtained by applying the previous construction to an equivalence relation (cf. Example 2.1.2) ? to the quasi-ordered set $(\mathbb{Z}, |)$ in Example 2.1.3 ?

Exercise 2.1.3 Prove the independence of axioms (2.1.1)–(2.1.3). (Hint: Use the empty relation, a convenient three-element set, a quasi-ordered set).

Exercise 2.1.4 Show that axioms (2.1.1)–(2.1.3) can be replaced by (2.1.2) and

(2.1.10) $$x \leq t \;\&\; y \leq x \Longleftrightarrow x = y \; .$$

We begin our study by showing that relations $\leq$ and $<$ behave exactly as in the well-known case of the order relation of $\mathbb{R}$.

Definition 2.1.4 A relation of *strict partial order* on a set P is a binary relation $<$ satisfying the following properties: for every $x, y, z \in P$,

(2.1.11) $$x \not< x \; ,$$

(2.1.12) $$x < y \;\&\; y < z \Longrightarrow x < z \; ,$$

(2.1.13) $$x < y \Longrightarrow y \not< x \; .$$

Proposition 2.1.1 *For every partially ordered set $(P, \leq)$, the relation $<$ defined by (1.5) is a strict partial order.*

PROOF: For every $x \in P$, the property $x \neq x$ is false, hence so is $x \leq x \,\&\, x \neq x$, i.e., $x < x$ is false. Now we prove (2.1.13). Indeed, if $x < y \,\&\, y < x$ for some $x, y \in P$, this would imply $x \leq y \,\&\, y \leq x$, hence $x = y$, therefore $x < x$, a contradiction. To prove (2.1.12) we assume $x < y \,\&\, y < z$. Then $x \leq y \,\&\, y \leq z$, therefore $x \leq z$ and it remains to prove that $x \neq z$. Indeed, $x = z$ would imply $x < y \,\&\, y < x$, which contradicts (2.1.13). □

Proposition 2.1.2 *Let $<$ be a strict partial order on a set P. Then the relation $\leq$ defined by*

$$x \leq y \Longleftrightarrow x < y \text{ or } x = y \tag{2.1.14}$$

for every $x, y \in P$, is a partial order on P.

PROOF: The relation is reflexive, because $x = x$ implies $x < x$ or $x = x$, that is, $x \leq x$. To prove antisymmetry, assume $x \leq y \,\&\, y \leq x$; then $x \neq y$ would imply $x < y \,\&\, y < x$, a contradiction. Finally assume $x \leq y \,\&\, y \leq z$. If $x = y$ or $y = z$, then the hypothesis reduces to $x \leq z$; therefore we can assume $x \neq y$ and $y \neq z$, which transforms the hypothesis into $x < y \,\&\, y < z$, thus implying $x < z$. This implies $x \leq z$ because $p \Longrightarrow p$ or q. □

Proposition 2.1.3 *Let P be a set. Definitions* (2.1.5) *and* (2.1.14) *establish a bijection between the set of all partial orders and the set of all strict partial orders that can be defined on P.*

PROOF: We use the technique suggested in Corollary 1.2.3 and prove the following two facts. A) If $\leq$ is a partial order, $<$ is associated with $\leq$ by (2.1.5) and $\leq'$ is associated with $<$ by (2.1.14), then $\leq' = \leq$. B) If $<$ is a strict partial order, $\leq$ is associated with $<$ by (2.1.14) and $<'$ is associated with $\leq$ by (2.1.5), then $<' = <$.

A) We have

$$x \leq' y \Longleftrightarrow x < y \text{ or } x = y \Longleftrightarrow (x \leq y \,\&\, x \neq y) \text{ or } x = y\,.$$

If $x \leq' y$, then $x \leq y$ or $x = y$, hence $x \leq y$ (because $x = y$ implies $x \leq y$). Conversely, if $x \leq y$, then either $x = y$, which implies $x \leq' y$, or $x \neq y$, which implies $x \leq y \,\&\, x \neq y$, hence again $x \leq' y$.

B) We have

$$x <' y \Longleftrightarrow x \leq y \text{ and } x \neq y \Longleftrightarrow (x < y \text{ or } x = y) \,\&\, x \neq y\,.$$

If $x < y$, then $x < y$ or $x = y$, and since $x \neq y$, it follows that $x <' y$. Conversely, $x <' y$ implies $x < y$ or $x = y$; but $\leq'$ is a strict partial order, hence $x \neq y$, therefore $x < y$. □

It follows from Proposition 2.1.3 that in a poset we can work simultaneously with relations $\leq$ and $<$, using properties (2.1.1)–(2.1.3) (2.1.5) and (2.1.14). These and properties (2.1.15)–(2.1.21) below are very useful and will be tacitly used in the subsequent sections and chapters.

Exercise 2.1.5 Prove that

(2.1.15) $$x < y \Longrightarrow x \neq y\ ,$$

(2.1.16) $$x < y \Longrightarrow x \leq y\ ,$$

(2.1.17) $$x \leq y\ \&\ y < z \Longrightarrow x < z\ ,$$

(2.1.18) $$x < y\ \&\ y \leq z \Longrightarrow x < z\ ,$$

(2.1.19) $$x \# y \Longleftrightarrow x \neq y\ \&\ x \not< y\ \&\ y \not< x\ ,$$

(2.1.20) $$x \not\leq y \Longleftrightarrow y < x \text{ or } y \# x\ ,$$

(2.1.21) $$x \not< y \Longleftrightarrow y \leq x \text{ or } x \# y\ .$$

Warning In particular it is important to note properties (2.1.19)–(2.1.21), which are the major difference between arbitrary posets and totally ordered sets (which in this eBook will be introduced in §2.3). Under the influence of the totally ordered set $(\mathbb{R}, \leq)$, beginners often believe that in a poset $x \not\leq y \Longleftrightarrow y < x$ and $x \not< y \Longleftrightarrow y \leq x$, which is wrong.

Exercise 2.1.6 Prove that: (i) (2.1.11) & (2.1.12) $\Longrightarrow$ (2.1.13), while the axiom system $\{(2.1.11), (2.1.12)\}$ is independent; (ii) (2.1.13) $\Longrightarrow$ (2.1.11), while the axiom system $\{(2.1.12), (2.1.13)\}$ is independent.

Definition 2.1.5 Let $(P, \leq)$ and $(Q, \leq)$ be two posets (there is no risk of confusion caused by the use of the same symbol $\leq$ in both posets). An *order homomorphism*, or an *increasing mapping*, or an *isotone mapping* from $(P, \leq)$ to $(Q, \leq)$ is a function $f : P \longrightarrow Q$ such that for every $x, y \in P$,

(2.1.22) $$x \leq y \Longrightarrow f(x) \leq f(y)\ ,$$

As $x = y \Longrightarrow f(x) = f(y) \Longrightarrow f(x) \leq f(y)$, we infer:

Remark 2.1.1 A mapping $f : P \longrightarrow Q$ is isotone if and only if

(2.1.23) $$x < y \Longrightarrow f(x) \leq f(y)\ .$$

Definition 2.1.6 Let **Pos** be the *category of posets*, having all posets as objects and all isotone mappings as morphisms.

The proof of the fact that **Pos** is actually a category reduces to the following simple remarks: 1) the composite gf of two isotone mappings is also an isotone mapping, because

$$x \leq y \Longrightarrow f(x) \leq f(y) \Longrightarrow g(f(x)) \leq g(f(y))\ ,$$

and 2) the identity mapping 1_P of a poset P is isotone, because $x \leq y$ can be written $1_P(x) \leq 1_P(y)$.

The category **Pos** is versatile in producing new objects from given ones. Here are a few examples.

Example 2.1.10 The *direct sum* of a family of posets $(P_i)_{i \in I}$ is constructed in the case when the index set I is also a poset. In that case the set-theoretical disjoint union $\sum_{i \in I}$ (cf. Example 1.3.45) is equipped with the strict partial order $<$ defined as follows: $(i, x) < (j, y)$ if either $i < j$ or $i = j$ and $x < y$ in P_i. The poset $\sum_{i \in I} P_i$ becomes the sum of the family $(P_i)_{i \in I}$ in the category **Pos**; the easy verification is left to the reader. If $I = \{1, 2\}$, the sum is also written $P_1 + P_2$.

Example 2.1.11 There are two natural ways of endowing the Cartesian product $P_1 \times P_2$ of two posets P_1, P_2 with a partial order. The first on uses the so-called *componentwise* definition: $(x_1, x_2) \leq (y_1, y_2) \Longleftrightarrow x_1 \leq y_1 \;\&\; x_2 \leq y_2$. The poset obtained in this way is the *direct product* of P_1 and P_2 in the category **Pos**. The second construction is the *lexicographic product*, corresponding to the *lexicographic order*, defined by $(x_1, x_2) < (y_1, y_2)$ if either $x_1 < y_1$ or $x_1 = y_1 \;\&\; x_2 < y_2$. Both constructions can easily be extended to a finite number o posets. The easy verifications are left to the reader.

Now we determine the monomorphisms, epimorphisms and isomorphisms in the category **Pos**.

Proposition 2.1.4 *In the category* **Pos**:

(i) *Monomorphisms coincide with injective morphisms.*

(ii) *Epimorphisms coincide with surjective morphisms.*

PROOF: (i) (also providing a new proof for Lemma 1.3.1) In view of Remark 1.3.2, it suffices to show that a non-injective morphism is not a monomorphism. Let $f : (P, \leq) \longrightarrow (Q, \leq)$ be an isotone mapping such that $f(a) = f(a')$ for some $a, a' \in P$ with $a \neq a'$. Take the set $\{0, 1\}$ endowed with $=$ as partial order. Define $u, v : \{0, 1\} \longrightarrow P$ by $u(0) = u(1) = a$, $v(0) = a, v(1) = a'$. Then u and v are isotone mappings, $fu = fv$, but $u \neq v$.

(ii) In view of Remark 1.3.2, it suffices to show that a non-surjective morphism is not an epimorphism. Let $f : (P, \leq) \longrightarrow (Q, \leq)$ be an isotone mapping and $q \in Q \setminus f(P)$. The sets $Q^- = \{y \in Q \mid y < q\}, Q^- = \{y \in Q \mid q < y\}$ and $Q^\# = \{y \in Q \mid y \# q\}$ are pairwise disjoint and their union is $Q \setminus \{q\}$. Take the set $\{0, \frac{1}{2}, 1\}$ endowed with the usual ordering $0 < \frac{1}{2} < 1$ and define $u, v : Q \longrightarrow \{ 0, \frac{1}{2}, 1\}$ by $u(y) = v(y) = 0$ for $y \in Q^-$, $u(y) = v(y) = \frac{1}{2}$ for $y \in Q^\#$, $u(y) = v(y) = 1$ for $y \in Q^+$, and $u(q) = \frac{1}{2}, v(q) = 1$. Then $uf = vf$ but $u \neq v$, therefore it remains to show that u and v are isotone.

Take $x, y \in Q$ with $x < y$; in view of Remark 2.1.1, we have to prove that $u(x) \leq v(x)$ and $v(x) \leq v(y)$.

If $x \in Q^+$, we have $y \in Q^+$, hence the inequalities are satisfied in the stronger form $u(x) = u(y)$ and $v(x) = v(y)$. If $x \in Q^-$, then $u(x) = 0 \leq u(y)$ and $v(x) = 0 \leq v(y)$. The cases $y \in Q^-$ and $y \in Q^+$ are treated similarly.

So the remaining cases are $x, y \in Q^\# \cup \{q\}$. But $x = q$ would imply $y \in Q^+$, while $y = q$ would imply $x \in Q^-$; both cases are contradictory. Therefore $x, y \in Q^\#$ and the inequalities are again satisfied as equations. □

Proposition 2.1.5 *If* $(P, \leq), (Q, \leq)$ *are posets, then the following conditions are equivalent for a function* $f : P \longrightarrow Q$*:*

(i) f *is an isomorphism in the category* **Pos***;*

(ii) f *is a bijection and for every* $x, y \in P$*,*

$$x \leq y \Longleftrightarrow f(x) \leq f(y); \tag{2.1.24}$$

(iii) f *is a surjection and* (2.1.24) *holds for every* $x, y \in P$*;*

(iv) f *is a bijection and for every* $x, y \in P$*,*

$$x < y \Longleftrightarrow f(x) < f(y). \tag{2.1.25}$$

PROOF: (i)$\Longleftrightarrow$(ii): By Remark 1.3.3.

(ii)$\Longrightarrow$(iv): If $x < y$ then $f(x) \leq f(y)$ by Remark 2.1.1. But $x \neq y$, hence $f(x) \neq f(y)$, therefore $f(x) < f(y)$. One proves similarly that $f(x) < f(y) \Longrightarrow x < y$.

(iv)$\Longrightarrow$(iii): We use the injectivity of f. If $x = y$, both sides of (2.1.24) are true. If $x \neq y$, then $f(x) \neq f(y)$ and

$$x \leq y \Longleftrightarrow x < y \Longleftrightarrow f(x) < f(y) \Longleftrightarrow f(x) \leq f(y)\ .$$

(iii)$\Longrightarrow$(ii): It remains to prove that f is injective. Indeed, it follows from (2.1.24) that $f(x) = f(y)$ implies $x \leq y$ and $y \leq x$, therefore $x = y$. □

Corollary 2.1.1 *In the category* **Pos** *bimorphisms coincide with bijective morphisms, but there exist bimorphisms which are not isomorphisms.*

PROOF: The first statement is immediate from Proposition 2.1.4. Furthermore, let $f : \{o, a, b\} \longrightarrow \{0, \frac{1}{2}, 1\}$, where the order is given by $o < a$, $o < b$ and $a \# b$ (the V-poset in Example 2.1.8) and $0 < \frac{1}{2} < 1$, be defined by $f(o) = 0$, $f(a) = \frac{1}{2}$, $f(b) = 1$. Then f is a bimorphism by the first statement, but it is not an isomorphism because (2.1.25) is not fulfilled. □

Definition 2.1.7 The isomorphisms of the category **Pos** are called *order isomorphism*.

Having in view Proposition 2.1.4, the example in the proof of Corollary 2.1.1 shows that a bijective bimorphism between two posets need not be an isomorphism, even if the posets are finite. The next Proposition provides sufficient conditions in order that a monomorphism between two finite posets be an isomorphism. Note that in view of Proposition 2.1.4 and of the fact that every injection between two finite sets is a bijection, every monomorphism between two finite posets is a bijective bimorphism.

Proposition 2.1.6 (Abian [1970]) A) *If* P *is a finite poset, then every monomorphism* $f : P \longrightarrow P$ *is an order isomorphism.*

B) *If* P *and* Q *are finite posets, and the maps* $f : P \longrightarrow Q$ *and* $g : Q \longrightarrow P$ *are monomorphisms, then* f *and* g *are order isomorphisms.*

PROOF: A) The monomorphism f is a bijection, so that using Proposition 2.1.5 it suffices to prove that $f(x) < f(y) \Longrightarrow f(x) < f(y)$. Let $S = \{(u,v) \in P^2 \mid u < v\}$ and define an injective mapping $f^* : S \longrightarrow S$ by setting $f^*(u,v) = (f(u), f(v))$. If $f(x) < f(y)$, then $(f(x), f(y)) \in S$ and since f^* is also surjective, it follows that $(f(x), f(y)) = f^*(u,v)$ for some $(u,v) \in S$. Therefore $f(x) = f(u)$ and $f(y) = f(v)$, hence $x = u$ and $y = v$, implying that $(x,y) = (u,v) \in S$.

B) The injections f, g and gf are isotone and bijections, hence gf is an isomorphism by A). Besides, $f^{-1} = (gf)^{-1}g$, where $(gf)^{-1}$ and g are morphisms, therefore f^{-1} is a morphism. □

Exercise 2.1.7 Prove that every monomorphism fulfills (2.1.25). Is the converse true ?

Exercise 2.1.8 Prove that the quotient of the quasi-ordered set $(\mathbb{Z}, |)$ by the equivalence relation $m \equiv n \Longleftrightarrow m \mid n \ \& \ n \mid m$ is isomorphic to the poset $(\mathbb{N}, |)$ (cf. Examples 2.1.3, 2.1.4 and Exercises 2.1.1, 2.1.2).

Exercise 2.1.9 Provide formula-descriptions of the posets in Fig. 2.1 (like the description of the V-poset given in the proof of Corollary 2.1.1). Find an X-shaped Hasse diagram for the K-poset.

Exercise 2.1.10 Prove that if each P_i is isomorphic to Q_i $(i \in I)$, then $\sum_{i\in I} P_i$ is isomorphic to $\sum_{i\in I} Q_i$ (cf. Example 2.1.10). A similar result holds for $P_1 \times P_2$ (cf. Example 2.1.11).

Remark 2.1.2 In view of Definition 1.3.5 and Proposition 2.1.4, a *subobject* of a poset P is a poset S for which there is an injective isotone mapping $f : S \longrightarrow P$. Note that S need not be isomorphic to the subset $f(S)$ of P equipped with the restriction of the order of P to the elements of $fS)$ (see e.g. Corollary 2.1.1). Thus in the case of the category **Pos**, the concept of a subobject is weaker than the natural idea of a subset of P endowed with the *induced* order. A similar comment holds for the *quotient objects*, with the same counter-example.

We conclude this section with a brief study of the relations $\geq, >$ dual to $\leq, <$ (cf. Definition 2.1.2).

Proposition 2.1.7 *The dual of a (strict) partial order is also a (strict) partial order.*

PROOF: Obvious. □

It follows that with each concept of the theory of posets we can associate a *dual concept* obtained by replacing in the original concept $\leq$ by $\geq$ and $<$ by $>$, and conversely. Besides, if in a statement concerning posets we replace each concept by its dual, we obtain a new concept called the *dual* of the original statement. The dual of a theorem concerning posets is also a theorem, known as the *dual theorem*. This easy remark, known as the *duality principle* for posets, follows from Proposition 2.1.7. Indeed, the dual theorem is nothing but the original theorem applied to the dual poset and then expressed in terms of the original poset. See Ch.1§3 for similar comments concerning the duality principle in categories.

Proposition 2.1.7 entitles us to introduce

Definition 2.1.8 The *order duality* is the functor $OD : \mathbf{Pos} \longrightarrow \mathbf{Pos}$ whose object function is defined by $OD(P, \leq) = (P, \geq)$ and whose morphism function is given by $ODf = f$.

Remark 2.1.3 The covariant functor OD is an *involutive automorphism*, that is, an isomorphism from **Pos** to itself, which coincides with its inverse; cf. Proposition 1.3.6. This is the categorical ground for the duality principle explained above. The order duality OD is not to be confused with the (contravariant) duality functor $D : \mathbf{Pos} \longrightarrow \mathbf{Pos}^0$; cf. Example 1.3.34.

Definition 2.1.9 Let $(P, \leq), (Q, \leq)$ be two posets. A *decreasing mapping*, or an *antitone mapping*, sometimes also called a *dual homomorphism*, from $(P, \leq)$ to $(Q, \leq)$, is a function $f : P \longrightarrow Q$ such that for every $x, y \in P$,

$$x \leq y \Longrightarrow f(y) \leq f(x) \ . \tag{2.1.26}$$

Let **APos** be the category of all posets as objects and all antitone mappings as morphisms. The isomorphisms of the category **APos** are said to be *anti-isomorphisms*, sometimes also called *dual isomorphisms*.

The previous results about the category **Pos** are easily extended, mutatis mutandis, to the category **APos**. Thus, in particular the antitone mappings are also characterized by the condition

$$x < y \Longrightarrow f(y) \leq f(x) \ , \tag{2.1.27}$$

while the anti-isomorphisms are those bijections, or equivalently, those surjective mappings which satisfy

$$x \leq y \Longleftrightarrow f(y) \leq f(x) \ ; \tag{2.1.28}$$

also, a bijective mapping is an anti-isomorphism if and only if

$$x < y \Longleftrightarrow f(y) < f(x) \ . \tag{2.1.29}$$

All these results follow from the previous ones and

Remark 2.1.4 Given two posets $(P, \leq), (Q, \leq)$, a mapping $f : P \longrightarrow Q$ is antitone if and only if it is an isotone mapping $f : (P, \leq) \longrightarrow (Q, \geq)$.

Warning The duality of posets (order duality), the dual category $\mathbf{Pos}^0$ of **Pos**, and the category **APos** are not to be confused with each other. So, the concept of isotone mapping is *self-dual* (coincides with its dual): a mapping $f : (P, \geq) \longrightarrow (Q, \geq)$ is a morphism of **Pos** if and only if it satisfies Definition 2.1.5, that is, $x \geq y \Longrightarrow f(x) \geq f(y)$ $(\forall x, y \in P)$, which coincides with (2.1.22). Furthermore, the morphisms of $\mathbf{Pos}^0$ are not even mappings, because $\mathbf{Pos}^0(P, Q) = \mathbf{Pos}(Q, P)$, and a **Pos**–morphism from Q to P is not a mapping from P to Q. Therefore the concept of an antitone mapping is neither the dual of the concept of isotone mapping, nor the concept of a morphism in $\mathbf{Pos}^0$. In particular, the anti-isomorphisms are neither the duals of the isomorphisms of **Pos**, nor the isomorphisms of $\mathbf{Pos}^0$.

For this reason we avoid the terms "dual homomorphism" and "dual isomorphism", and adopt the terminology antitone mapping–anti-isomorphism (both terminologies are encountered in the literature).

Exercise 2.1.11 Which mappings are both isotone and antitone?

Exercise 2.1.12 Prove that the posets M and W in Fig. 2.1 are anti-isomorphic, and they are the single pair of posets with this property in Fig. 2.1.

2.2 Distinguished Elements in a Poset

In this section we define the various distinguished elements which a (subset of a) poset may posses and establish their complete existential theory, that is, all logical implications between these concepts.

In the sequel we indicate dual statements by notations like $2.2.n'$ and $2.2.n''$, while $2.2.n'$ & $2.2.n''$ is simply denoted by $2.2.n$.

Definition 2.2.1 Let x, y be two elements of a poset P. If $x \leq y$ (if $x < y$) we say that x is a *(strict) lower bound* of y, while y is a *(strict) upper bound* of x.

More generally, let A be a subset of P. Recall first that according to Definition 1.2.3, for every $x \in P$,

$$x \leq A \ (x < A) \text{ means } x \leq a \ (x < a) \text{ for all } a \in A \ ,$$

$$A \leq x \ (A < x) \text{ means } a \leq x \ (a < x) \text{ for all } a \in A \ .$$

Exercise 2.2.1 Prove that $x \leq \varnothing$ and $\varnothing \leq x$. Do $x \leq A$ and $A \leq y$ imply $x \leq y$?

Definition 2.2.2′ Let A be a subset of a poset P. We say that an element $x \in P$ is:

(i) a *(strict) lower bound* of A, if

(2.2.1′) $$x \leq A$$

(if

(2.2.2″) $$x < A,$$

respectively);

(ii) the *least element* of A or the *first element* of A, if

(2.2.3′) $$x \in A \ \& \ x \leq A \ ;$$

(iii) a *minimal element* of A, if

(2.2.4′.1) $$x \in A$$

and

(2.2.4′.2) for every $a \in A$, if $a \leq x$ then $a = x$;

(iv) the *greatest lower bound* of A, denoted g.l.b. A, or $\inf A$, or $\bigwedge\{a \mid a \in A\}$, or $\bigwedge A$, if

(2.2.5′.1) $x \in A$

and

(2.2.5′.“2) if $z \leq A$ then $z \leq x$;

(v) the *predecessor* of A, if

(2.2.6′.1) $x < A$

and

(2.2.6′.2) if $z < A$ then $z \leq x$.

Definition 2.2.2″. The concepts dual to those in Definition 2.2.2′. are called (i) *(strict) upper bound* of A, (ii) *greatest element* of A, or *last element* of A, (iii) *maximal element* of A, (iv) *least upper bound* of A, denoted l.u.b. A, or $\sup A$, or $\bigvee\{a \mid a \in a\}$, or $\bigvee A$, and (v) *successor* of A.

In other words, the greatest lower bound (the least upper bound) of a set A is exactly what its name indicates: the greatest element (the least element) of the set of all lower bound (upper bounds) of the set A. Also, the predecessor (successor) of A is the greatest (least) element of the set of all strict lower bounds (strict upper bounds) of A.

Warning A subset A of P, and in particular the set P itself, may or may not have some of the distinguished elements in Definition 2.2.2. Note that when one says that a set A *has* some of the elements mentioned in Definition 2.2.2, this means in fact the existence of those elements, which, however, may or may not belong to A. The strict lower bounds, the predecessor, the strict upper bounds and the successor of A, if any, do not belong to A. The least element of A and the greatest element of A, whenever it exists, belongs to A.

If a singleton $\{y\}$ has a predecessor (successor) in the sense of Definition 2.2.2, this is unique by Theorem 2.2.1 below, and it is also the unique predecessor (successor) of y in the sense of Definition 2.1.3. The converse does not hold.

Example 2.2.1 The set $\mathbb{R}$ of real numbers, endowed with the usual ordering, as well as the subsets $\mathbb{Q}$ of rational numbers and $\mathbb{Z}$ of integers, has no element possessing any of the properties in Definition 2.2.2.

Example 2.2.2 The subset $A = \{1, 2, 3\}$ of the set $\mathbb{Z}$ of integers has the following distinguished elements: the predecessor 0, the least element 1 which is also a minimal element and g.l.b. A, the greatest element 3 which is also a maximal element and l.u.b. A, and the successor 4.

Example 2.2.3 A poset P, regarded as its own subset $A = P$, has no strict lower bounds and no strict upper bounds, hence no predecessor and no successor.

Example 2.2.4′ Consider again the ordered set $\mathbb{R}$. The closed interval $[0, 1]$ has strict strict lower bounds and least element 0, but no predecessor and no minimal element different from the least element. The open interval $(0, 1)$ has strict lower bunds and the predecessor 0, but no minimal elements and hence no least element.

Example 2.2.5′ Consider the poset $(2^X, \subseteq)$, where the set X is not empty. Let $Y \in 2^X$ fulfill $\varnothing \neq Y \neq X$ and let $\mathcal{A} = \{Z \in 2^X \mid Y \subset Z\}$. Then $\mathcal{A}$ has the predecessor Y and the minimal elements $Y \cup \{z\}$, where $z \in X \setminus Y$, but no least element if the set $X \setminus Y$ has at least two elements.

Example 2.2.6′ Modify the above example by taking the poset$(2^X \setminus \{Y\}, \subseteq)$ and the same subset $\mathcal{A}$, which has the same minimal elements as in Example 2.2.5′. The strict lower bounds of $\mathcal{A}$ are the sets strictly included in Y, but there is no predecessor if the set Y has at least two elements.

Example 2.2.7″ The I-shaped poset in Fig. 2.1 has greatest element, which is the unique maximal element. Each of the other posets in Fig. 2.1 has two maximal elements, but no greatest element.

Examples 2.2.1″-2.2.6″ and **2.2.7′**. Duals of the above examples, left to the reader.

Exercise 2.2.2 Study the existence of all the distinguished elements in Definition 2.2.2 for all of the posets in Fig. 2.1.

The reader is urged to provide other examples by himself/herself. However the possibility of constructing such examples is somewhat limited by some links which exist between the above concepts and which we are going to establish below. These properties will show in particular that certain coincidences arising in the above examples (for instance in Example 2.2.2) are not accidental.

Remark 2.2.1′ A lower (upper) bound of A either is strict or it is the least (greatest) element of A.

Remark 2.2.2′ An element $x \in P$ is a minimal element of a set $A \subseteq P$ if and only if

(2.2.7′) $$x \in A \ \& \text{ there is no } a \in A \text{ with } a < x \, .$$

This follows from (2.1.14).

Remark 2.2.3′ Two distinct minimal elements of a set A are incomparable (because the existence of two minimal elements x' and x'' satisfying $x' < x''$ would contradict Remark 2.2.1′).

Theorem 2.2.1′ *Let A be a non-empty subset of a poset P. Then each of the following distinguished elements associated with A either does not exist or is unique: the least element of A,* g.l.b.A, *the predecessor of A. Moreover, the least element of A, if any, is the unique minimal element of A.*

PROOF: Let x' and x'' be least elements of A. Then $x' \leq A$ and since $x'' \in A$ it follows that $x' \leq x''$; interchanging the rôles of x' and x'', we obtain $x'' \leq x'$, therefore $x' = x''$. Moreover, if y is a minimal element of A, then $y \in A$, hence $x' \leq y$, therefore $x' = y$ by the definition of minimality.

By duality, the greatest element is unique, therefore g.l.b.A, the greatest element of the set of lower bounds of A, is unique, and the predecessor of A, which is the greatest element of the set of strict lower bounds of A, is also unique. □

The converse of the last statement in Theorem 2.2.1′ is false: if a set A has a unique minimal element, this need not be the least element of A, as shown in the following

Example 2.2.8′ Let $A = P = \{a, 0, -1, -2, \ldots\}$, where the subset $\{0, -1, -2, \ldots\}$ is equipped with the usual ordering of integers, while a is an element incomparable with any negative integer, but $a < 0$ (see Fig. 2.3). Then A is the unique minimal element, but there is no least element.

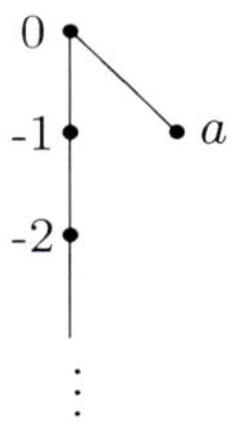

Fig. 2.3

However:

Exercise 2.2.3′ If x is the unique minimal element of a finite poset A, then x is the least element of A. (Hint. Use Proposition 2.2.1′ below).

Proposition 2.2.1′ *Every finite non-empty subset A of a poset P has minimal elements : for every $a \in A$ there is a minimal element $x \in A$ such that $x \leq a$.*

Comment This property is important and its proof is typical: many proofs concerning finite ordered sets are based on the fact that a decreasing (or an increasing) sequence stops after a finite number of steps.

PROOF: If a is a minimal element of A, the conclusion is verified for $x = a$ itself. Otherwise Remark 2.2.2′ implies the existence of an element $a_1 \in A$ satisfying $a_1 < a$. If a_1 is a minimal element, the desired conclusion holds for $x = a_1$; otherwise the above reasoning shows there exists an element $a_2 < a_1$ etc.

We thus obtain a sequence

$$(2.2.8') \qquad a > a_1 > a_2 > \dots$$

of elements of A. The relation $>$ is transitive and $x > y$ implies $x \neq y$, therefore the elements of (2.2.8′) are pairwise distinct. Since on the other hand A is finite, it follows that the construction of the sequence (2.2.8′) stops after a finite number of steps, $a > a_1 > a_2 \cdots > a_m$. This implies that a_m is a minimal element, for otherwise the construction could have been continued with $a_m > a_{m+1}$ for some $a_{m+1} \in A$. Thus the conclusion of the Proposition holds for $x = a_m$. □

The above properties refer to the various distinguished elements that a *given set* $A \subseteq P$ may possess. Now from the point of view of a *given element* $x \in P$ we are interested in establishing the relationships between the various properties introduced in Definition 2.2.2 and which that element may possess. What we are actually going to do is an exhaustive study of all logical implications between the concepts in Definition 2.2.2′, that is, the *complete existential theory* of these concepts, according to the terminology of E.H. Moore [1910]. For a brief presentation of the literature discussing complete existential theories see Padmanabhan and Rudeanu [2008].

We first deal with the concepts of strict lower bound, predecessor, least element and minimal element. Clearly the intersection of the spheres of the concepts of "strict lower bound" and "minimal element" is empty, because a minimal element of the set A belongs to A, while a strict lower bound does not ($x < A$ and $x \in A$ would imply $x < x$, a contradiction). Then the predecessor of A is a strict lower bound by definition, while the least element of A is a minimal element by Theorem 2.2.1′. It is easily seen that the converses of these two statements are false. For instance, in the poset depicted in Fig. 2.4, the set $A = \{a, b, 1\}$ has the minimal elements a and b, but no least element. Also, besides the predecessor c, the set A has also the strict lower bound 0.

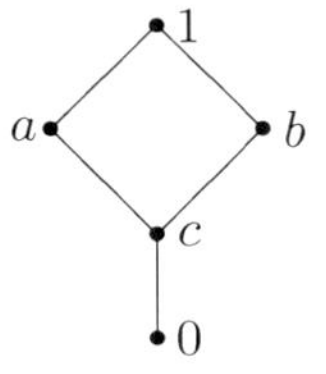

Fig. 2.4

The above preliminary results are visualised in Fig. 2.5, where each rectangle represents th extent of a consistent concept, or, in other terms, the non-empty "sphere" of that concept. We have thus established:

Lemma 2.2.1′ *Figure* 2.5 *represents the complete existential theory of the four concepts involved in it.*

<table>
<tr><td>1^0</td><td>strict l.b., not predecessor</td><td rowspan="2">strict l.b. = 1^0 or 2^0</td></tr>
<tr><td>2^0</td><td>predecessor</td></tr>
<tr><td>3^0</td><td>least element</td><td rowspan="2">minimal element = 3^0 or 4^0</td></tr>
<tr><td>4^0</td><td>minimal, not least element</td></tr>
</table>

Fig. 2.5

To complete the discussion with the concepts of lower bound and greatest lower bound, we need a few preliminary results which have also an intrinsic interest.

Warning The proofs of the three lemmas below work with elements $a \in A$, which means that the set A is tacitly assumed to be non-empty. This amounts to the necessity to have separate proofs for the case $A = \varnothing$. It is common, but dangerous practice, to pay no attention to this case; for sometimes the result is not valid for the empty set.

Exercise 2.2.3′ The reader is urged to provide the missing proofs of the next three lemmas for the case $A = \varnothing$. This is easy, provided one notes that sentences of the form "for every $x \in \varnothing$, $p(x)$" mean in fact "if $x \in \varnothing$ then $p(x)$", hence they are true because the premise is false. In particular the empty set $\varnothing$ has g.l.b. if and only if the set P has least element 0, in which case g.l.b.$\varnothing = 0$.

Lemma 2.2.2′ $x < A \Longleftrightarrow x \leq A \;\&\; x \notin A$.

PROOF: Let $x < A$. This means that for every $a \in A$ we have $x < a$, hence $x \leq a$; therefore $x \leq A$. On the other hand, $x \in A$ together with $x < A$ would imply $x < x$, a contradiction. Conversely, let $x \leq A$ and $x \notin A$. Then for every $a \in A$ we have $x \leq a$ and also $x \neq a$ (otherwise $x \in A$); therefore $x < A$. □

Lemma 2.2.3′ *Let* $x =$ g.l.b.A
. *Then if* $x \in A$ *it follows that*

(2.2.9′) x is the least element of A ,

whereas if $x \notin A$ *it follows that*

(2.2.10′) A has no least element and x is the predecessor of A .

Conversely, if the element x *satisfies* (2.2.9′) *or* (2.2.10′), *then* $x =$ g.l.b.A.

PROOF: Let $x =$ g.l.b.A ; then $x \leq A$. If $x \in A$, this implies (2.2.9′) by definition (2.2.3′). Now assume $x \notin A$. Then $x < A$ by Lemma 2.2.2′ and if $z < A$ then $z \leq A$, hence $z \leq x$ by the property of the g.l.b.; thus x is the predecessor of A by definition (2.2.6′). Besides, the existence of a first element b of A would imply $b \in A$, hence $x < b$ by the property of the predecessor, and also $b \leq A$, hence $b \leq x$ by the property of the g.l.b.; but $b \leq x$ contradicts $x < b$.

Conversely, let x fulfill (2.2.9′) or (2.2.10′). If x is the least element of A, then $x \leq A$ and since $x \in A$, it follows that $z \leq A$ implies $z \leq x$; thus $x =$ g.l.b.A.

Now assume (2.2.10′). Then $x < A$, hence $x \leq A$; besides, for every $z \leq A$ it follows that $z \notin A$ (otherwise z would be the least element of A), hence $z < A$ by Lemma 2.2.2′, therefore $z \leq x$ by the property of the predecessor. Thus $x = \text{g.l.b.}A$. □

Lemma 2.2.4′ *The following properties are equivalent:*

(i) *x is the least element of the set A ;*

(ii) *x is both a lower bound and a minimal element of the set A ;*

(iii) *x is both the g.l.b. and a minimal element of A .*

PROOF: (i)$\Longrightarrow$(iii): By Lemma 2.2.3′ and Theorem 2.2.1′.

(iii)$\Longrightarrow$(ii): Trivial.

(ii)$\Longrightarrow$(i): $x \leq A$ by the first half of the hypothesis and $x \in A$ by the minimality of x. □

Theorem 2.2.2′ *The complete existential theory of the six concepts in Definition* 2.2.2′ *is indicated in* Fig. 2.6.

1^0	strict l.b., not predecessor	strict l.b. = 1^0 or 2^0 or 3^0
2^0	predecessor, not g.l.b.	predecessor = 2^0 or 3^0
3^0	predecessor and g.l.b.	l.b. = 1^0 or 2^0 or 3^0 or 4^0
4^0	least element	g.l.b. = 3^0 or 4^0
5^0	minimal, not least element	minimal element = 4^0 or 5^0

Fig. 2.6

PROOF: The spheres 3^0 and 4^0 are not empty: see e.g. Fig. 2.4, where 0 is the predecessor but not the g.l.b. of the set $\{c, a, b, 1\}$, while c is the predecessor and the g.l.b. of the set $\{a, b, 1\}$.

Trivially, predecessor = 2^0 or 3^0. It was already noted in Fig. 2.5 that minimal = 4^0 or 5^0. Lemma 2.2.3′ says that g.l.b. is least element or (both g.l.b. and) predecessor, hence g.l.b. = 3^0 or 4^0.

It was noted in Fig. 2.5 that strict l.b. = 1^0 or predecessor, hence stict l.b. = 1^0 or 2^0 or 3^0.

Since l.b. = strict l.b. or least element, it follows that l.b. = 1^0 or 2^0 or 3^0 or 4^0. □

The complete existential theory of a set of properties enables one to quickly obtain various relationships between those properties. For instance, we can re-prove Lemma 2.2.4′ as follows:

l.b. & minimal = (1^0 or 2^0 or 4^0) & (4^0 or 5^0) = 4^0 ,
g.l.b. & minimal = (3^0 or 4^0) & (4^0 or 5^0) = 4^0 .

The statements dual to the results in this section are left to the reader. As we know, dual theorems can be proved either directly or by applying the duality principle.

2.3 Totally Ordered Sets

The general properties of posets studied in the previous two sections are in particular valid for the ordering of real numbers, which, however, has an important supplementary property: every two elements are comparable. The posets having this property are said to be totally ordered sets or chains, and in this section we examine the specialization of the previous properties of posets to the case of totally ordered sets.

Definition 2.3.1 A poset $(T, \leq)$ which satisfies, for every $x, y \in T$,

$$x < y \text{ or } x = y \text{ or } y < x \ . \tag{2.3.1}$$

is said to be *totally ordered* or a *toset.* A non-empty toset is called a *chain.*

Example 2.3.1 The set $\mathbb{R}$ of real numbers, as well as any subset of it, is a chain under the usual ordering of reals.

Example 2.3.2 The empty set is linearly ordered (cf. Example 2.1.1).

Example 2.3.3 The set I in Fig. 2.1 from Example 2.1.8 is a toset.

Example 2.3.4 The direct sun $\sum_{i \in I} T_i$ of a family $\{T_i \mid i \in I\}$ of totally ordered sets T_i is also totally ordered (cf. Examples 1.3.45 and 2.1.10).

Example 2.3.5 The lexicographic product of two totally ordered sets T_1, T_2 is also a toset, whereas the direct product of T_1 and T_2 is not totally ordered, unless one of the factors is a singleton (cf. Example 2.1.11).

See also Section 2.5 and Chapter 3.

Remark 2.3.1 The property of being a toset is *hereditary*, i.e., every subset S of a toset T is a toset under the restriction to S of the order of T.

Proposition 2.3.1 *The following conditions are equivalent for a partially ordered set T:*

(i) *T is a totally ordered set ;*

(ii) $x \leq y$ *or* $y \leq x$ $(\forall x, y \in T)$;

(iii) $x \not\leq y \Longrightarrow y < x$ $(\forall x, y \in T)$;

(iv) $x \not< y \Longrightarrow y \leq x$ $(\forall x, y \in T)$;

(v) $x \not\leq y \Longleftrightarrow y < x$ $(\forall x, y \in T)$;

(vi) $x \not< y \Longleftrightarrow y \leq x$ $(\forall x, y \in T)$;

(vii) *for every subset $A \subseteq T$, if A has a minimal element, then this is the least element of A ;*

(viii) *for every subset $A \subseteq T$, if A has a minimal element, then this is the* g.l.b *of A ;*

(ix) *for every subset* $A \subseteq T$, *if* A *has a maximal element, then this is the greatest element of* A ;

(x) *for every subset* $A \subseteq T$, *if* A *has a maximal element, then this is the* l.u.b. *of* A .

PROOF: (i)$\Longrightarrow$(ii): Immediate from (2.3.1) and (2.1.14).

(ii)$\Longrightarrow$(iii): If $x \not\leq y$ then $y \leq x$. But $y \neq x$ (otherwise $x \leq y$, therefore $y < x$ by (2.1.5).

(iii)$\Longrightarrow$(v): The implication $y < x \Longrightarrow x \not\leq y$ converse to (iii) holds in any poset. Indeed, if $y < x$ then $x \neq y$ by (2.1.15), therefore $x \leq y$ would imply $x < y$ by (2.1.5), thus contradicting (2.1.13). So $x \not\leq y$.

(v)$\Longrightarrow$(iv): Trivial.

(iv)$\Longrightarrow$(i): Given $x, y \in T$, either $x < y$ or $x \not< y$. In the latter case $y \leq x$ by (iv), therefore $y < x$ or $y = x$ by (2.1.14).

(ii)$\Longrightarrow$(vii): Let m be a minimal element of A. Then $m \in A$ and for every $a \in A$ we have $m \leq a$ or $a \leq m$ by (ii). In the latter case $m = a$ by the minimality of m. So $m \leq a$ in both cases, therefore m is the least element of A.

(vii)$\Longrightarrow$(viii): By Lemma 2.2.3$'$.

(viii)$\Longrightarrow$(i): Let us show that property (viii) fails in a poset P which is not totally ordered. Indeed, in such a set we have $a\#b$ for some $a, b \in P$ and taking $A = \{a, b\}$, it follows that a and b are minimal elements of A which, however, has no g.l.b.

So properties (i)–(viii) are equivalent, therefore (i)$\Longleftrightarrow$(ix)$\Longleftrightarrow$(x) follows by duality. □

Exercise 2.3.1 Show that (2.1.2), (2.1.3) and property (ii) in Proposition 2.3.1 form an independent set of axioms for totally ordered sets. (Hint: Take a "circular order" on a three-element set, and the set $\mathbb{Z}$ endowed with the relation $x\rho y \Longleftrightarrow | x| \leq |y|$).

The above Proposition 1.3.1 is, in fact, a theorem on partially ordered sets and this entitled us to use the duality principle in the proof. Now it is natural to ask whether or not the duality principle is valid for totally ordered sets.

Generally speaking, if **K** is a subclass of **Pos**, since the dual of a theorem concerning posets is valid by the duality principle, in particular it is valid when applied to posets in **K**; however this dual theorem may not hold in **K**, because the dual of a poset in **K** need not belong to **K** (for instance, if **K** is the subclass of all psets with least element). The above reasoning, however, implies that in the particular case when the dual of a poset in **K** also belongs to **K**, then the duality principle holds for **K**. Moreover, the membership in **K** being itself a property, we have thus established:

Remark 2.3.2 Given a subclass **K** of **Pos**, the duality principle holds in **K** if and only if the dual of every poset from **K** belongs also to **K**.

In particular Definition 2.3.1 immediately implies that the dual of a chain is a chain. Therefore:

Remark 2.3.3 The duality principle is also valid for the class of totally ordered sets.

Definition 2.3.2 A mapping $f : P \longrightarrow Q$ between two posets $(P, \leq), (Q, \leq)$ is *strictly increasing* (*strictly decreasing*), if for every $x, y \in P$,

$$x < y \Longrightarrow f(x) < f(y) \tag{2.3.2}$$

(if

$$x < y \Longrightarrow f(y) < f(x) \tag{2.3.3}$$

respectively).

Remark 2.3.4 Clearly every strictly increasing (strictly decreasing) mapping is increasing (decreasing). (If f is strictly increasing and $x \leq y$, then either $x < y$ and $f(x) < f(y)$, or $x = y$ and $f(x) = f(y)$; in both cases $f(x) \leq f(y)$).

We have seen in Proposition 2.1.5 that the isomorphisms (anti-isomorphisms) of the category **Pos** are some of the strictly increasing (strictly decreasing) bijections. In the case when the domain P is totally ordered, this result can be refined as follows:

Lemma 2.3.1 *Let $(T, \leq)$ be a totally ordered set, $(Q, \leq)$ a partially ordered set and $f : T \longrightarrow Q$. Then:*

(i) *If f is strictly increasing or strictly decreasing, then f is injective.*

(ii) *If, moreover, f is surjective, then Q is totally ordered and f is an isomorphism (anti-isomorphism).*

PROOF: Let f be strictly increasing (if it is strictly decreasing, the proof will be similar).

(i) If $x \neq y$, then $x < y$ or $y < x$, thus implying $f(x) < f(y)$ or $f(y) < f(x)$, respectively; in both cases $f(x) \neq f(y)$.

(ii) Let $x', y' \in Q$. Then $f^{-1}(x'), f^{-1}(y') \in T$, hence they are comparable, say $f^{-1}(x') < f^{-1}(y')$, which implies $x' < y'$ by (2.3.2). So Q is totally ordered and it remains to prove that f^{-1} is increasing.

Let $x', y' \in Q$ satisfy $x' \leq y'$. This actually implies $f^{-1}(x') \leq f^{-1}(y')$, for otherwise $f^{-1}(y') < f^{-1}(x')$ by condition (iii) in Proposition 2.3.1, hence (2.3.2) would imply $y' < x'$, which contradicts $x' \leq y'$. □

Definition 2.3.3 Let **Tos** denote the category of totally ordered sets and increasing mappings.

It is clear that **Tos** is actually a category, namely a full subcategory of **Pos**.

Proposition 2.3.2 *Let $(T, \leq)$ and $(V, \leq)$ be totally ordered sets. The following conditions are equivalent for a mapping $f : T \longrightarrow V$:*

(i) *f is a monomorphism in* **Tos** *;*

(ii) *f is an injective morphism in* **Tos** *;*

(iii) *f is a strictly increasing mapping.*

PROOF: (i)$\Longrightarrow$(iii): Let f be a monomorphism in **Tos** and suppose by reductio ad absurdum that f is not strictly increasing. Then there exist $a, a' \in T$ such that $a < a'$ and $f(a) \not< f(a')$; but $f(a) \leq f(a')$, therefore $f(a) = f(a')$. Let the set $\{a, a'\}$ be equipped with the total order $a < a'$. Define $u, v : \{a, a'\} \longrightarrow T$ by $u(a) = u(a') = a$ and $v(a) = a, v(a') = a'$. These functions are increasing, hence they are morphisms of **Tos**, and $f \circ u = f \circ v$, therefore $u = v$ because f is a monomorphism of **Tos**. So $a = u(a') = v(a') = a'$, a contradiction.

(iii)$\Longrightarrow$(ii): By Lemma 2.3.1.

(ii)$\Longrightarrow$(i): By Remark 1.3.2. □

Proposition 2.3.3 *In the category* **Tos** *epimorphisms coincide with surjective morphisms.*

PROOF: By Remark 1.3.2 and the counter-example in the proof of (ii) from Proposition 2.1.4 with P and Q totally ordered. □

The fact that **Tos** is a full subcategory of **Pos** immediately implies that every isomorphism of **Tos** is also an isomorphism in **Pos**. It is natural to look for intrinsic characterizations of these isomorphisms.

Proposition 2.3.4 *Let* $(T, \leq)$ *and* $(V, \leq)$ *be totally ordered sets. The following conditions are equivalent for a mapping* $f : T \longrightarrow V$*:*

(i) f *is an isomorphism in* **Tos** *;*

(ii) f *is a bimorphism in* **Tos** *;*

(iii) f *is a bijective morphism in* **Tos** *;*

(iv) f *is a surjective strictly increasing mapping ;*

(v) f *is a surjective mapping fulfilling*

$$x < y \Longleftrightarrow f(x) < f(y) \ . \tag{2.1.25}$$

PROOF: (i)$\Longrightarrow$(ii): By Proposition 1.3.3 (iv).

(ii)$\Longleftrightarrow$(iii): By Propositions 2.3.2 and 2.3.3.

(iii)$\Longrightarrow$(iv): If $x < y$ then $f(x) \leq f(y)$ because f is a morphism and $f(x) \neq f(y)$ because f is injective; therefore $f(x) < f(y)$.

(iv)$\Longrightarrow$(v): f is an isomorphism by Lemma 2.3.1, therefore (2.1.15) holds by Proposition 2.1.5.

(v)$\Longrightarrow$(i): f is strictly increasing, therefore it is injective by Proposition 2.3.2. Now f is an isomorphism by Proposition 2.1.5. □

We conclude this section with two properties of finite tosets.

Proposition 2.3.5 *Every finite non-empty chain has least and greatest elements.*

PROOF: Immediate from Propositions 2.2.1 and the equivalences (i)$\Longleftrightarrow$(vii) $\Longleftrightarrow$(ix) in Proposition 2.3.1. □

Proposition 2.3.6 *For every natural number* $n > 0$*, any chain with* n *elements is isomorphic to the set* $\{0, 1, \ldots, n-1\}$ *endowed with the usual ordering of* $\mathbb{N}$.

PROOF: Let C be a chain with n elements. Then C has a first element c_0 by Proposition 2.3.5; set $f(c_0) = 0$. If the set $C \setminus \{c_0\}$ is empty, the conclusion is already established. Otherwise this set is not empty and totally ordered by Remark 2.3.1, hence it has a first element c_1; set $f(c_1) = 1$, etc. It is easy to see that this construction establishes an isomorphism $f : C \longrightarrow \{0, 1, \ldots, n-1\}$; the details are left to the reader. □

2.4 Some Properties Equivalent to the Axiom of Choice

In this section we state a few important properties and prove their equivalence with the axiom of choice. Each of these properties turns out to be an efficient tool for proving several theorems from various fields of mathematics. The theorem to be proved in this section shows that admitting the axiom of choice is the same as admitting each of these theorem-proving tools. Our presentation essentially follows that of Szele [1950]. For further reading the reader is referred to Rubin and Rubin [1963].

We first state and explain these properties, which we denote by (a), (a′), (b), (b*), ..., i) . Then we prove their equivalence.

(a) *Axiom of choice.*

(a′) *General principle of choice.*

Already explained in Section 1.1, ZF9.

Properties (b), (b*), (c) and (c*) below are variants of a result discovered by Kuratowski in 1922 and independently by Zorn in 1935 and which, therefore, is to be called the *Kuratowski-Zorn lemma* (versus the name *Zorn lemma*, much used in the literature).

(b) *If $P \neq \varnothing$ is a poset such that every chain of P has an upper bound, then P has a maximal element.*

(b*) *If $P \neq \varnothing$ is a poset such that every chain of P has an upper bound, then for every $x \in P$ there is a maximal element m of P such that $x \leq m$.*

(c) *If $P \neq \varnothing$ is a poset such that every chain of P has a* l.u.b., *then P has a maximal element.*

(c*) *If $P \neq \varnothing$ is a poset such that every chain of P has a* l.u.b., *then for every $x \in P$ there is a maximal element m of P such that $x \leq m$.*

Definition 2.4.1 Let S be a set, $\mathcal{P} \subseteq 2^S$ and p a property defined for the sets in $\mathcal{P}$. Then p is said to be *consistent* if there exist sets from $\mathcal{P}$ having property p, in which case these sets are called *p-sets*. In particular the sets in $\mathcal{P}$ are said to be *$\mathcal{P}$-sets*, and a maximal element of the poset $(\mathcal{P}, \subseteq)$ will be called a *maximal $\mathcal{P}$-set*. A *maximal chain* of a poset P is a maximal element of the set of all chains of P

The next four properties are apparent particularizations of (b), (b*), (c) and (c*), respectively.

(d) *If S is a set, $\varnothing \neq \mathcal{P} \subseteq 2^S$ and every chain of $(\mathcal{P}, \subseteq)$ has an upper bound, then there is a maximal $\mathcal{P}$-set.*

(d*) *If S is a set, $\varnothing \neq \mathcal{P} \subseteq 2^S$ and every chain of $(\mathcal{P}, \subseteq)$ has an upper bound, then for every set $X \in \mathcal{P}$ there is a maximal $\mathcal{P}$-set M such that $X \subseteq M$.*

(e) *If S is a set, $\varnothing \neq \mathcal{P} \subseteq 2^S$ and the union of the sets of any chain of $(\mathcal{P}, \subseteq)$ is again a $\mathcal{P}$-set, then there is a maximal $\mathcal{P}$-set.*

(e*) *If S is a set, $\varnothing \neq \mathcal{P} \subseteq 2^S$ and the union of the sets of any chain of $(\mathcal{P}, \subseteq)$ is again a $\mathcal{P}$-set, then for every set $X \in \mathcal{P}$ there is a maximal $\mathcal{P}$-set M such that $X \subseteq M$.*

Definition 2.4.2 A property p defined for the subsets of a certain set S is said to be *of finite character* if for every $X \subseteq S$,

$$X \text{ has property } p \iff \text{all finite non-empty subsets of } X \text{ have property } p.$$

Thus, for instance – and this is precisely the case which interests us – the property of being a chain, which is defined for the subsets X of a poset, is of finite character (the above implication $\Longrightarrow$ is trivial, while $\Longleftarrow$ follows using the two-element subsets of X).

(f) *If S is a set, $\varnothing \neq \mathcal{P} \subseteq 2^S$, the union of the sets of any chain of $(\mathcal{P}, \subseteq)$ is again a $\mathcal{P}$-set and p is a consistent property of finite character defined for all $\mathcal{P}$-sets, then there is a maximal p-set.*

(f*) *If S is a set, $\varnothing \neq \mathcal{P} \subseteq 2^S$, the union of the sets of any chain of $(\mathcal{P}, \subseteq)$ is again a $\mathcal{P}$-set and p is a consistent property of finite character defined for all $\mathcal{P}$-sets, then for every p-set X there is a maximal p-set M such that $X \subseteq M$.*

The next two properties, known as the *Tuckey lemma*, are apparent particularizations of (f) and (f*), respectively.

(g) *If S is a set and p a consistent property of finite character defined for the subsets of S, then there is a maximal p-set.*

(g*) *If S is a set and p a consistent property of finite character defined for the subsets of S, then for every p-set X there is a maximal p-set M such that $X \subseteq M$.*

Now we specialize to the property of being a chain:

(h) *Every non-empty poset includes a maximal chain.*

(h*) *If $P \neq \varnothing$ is a poset, then for every chain L of P there is a maximal chain $\overline{L}$ such that $L \subseteq \overline{L}$.*

The next and last property in this section is perhaps the most widely used equivalent of the axiom of choice. We must first introduce the concept of a well-ordered set, which will be studied in some detail in the next section and, in fact, in the subsequent chapter.

Definition 2.4.3 A *well-ordered set* is a totally ordered set W such that every non-empty subset of W has a first element.

Property (i) below is the famous *Zermelo theorem*:

(i) *Every set can be well-ordered.*

Warning More exactly, this means that for every set S there exists a subset $\leq$ of $S \times S$ such that $(S, \leq)$ is a well-ordered set. However the Zermelo theorem does not offer a procedure for actually constructing such a well-ordering whatever the set S may be (although in some particular cases, e.g. when the set S is finite, such a well-order can be easily constructed).

Theorem 2.4.1 *The above properties* (a), (a′), (b), (b*), ... , (h), (h*), (i) *are equivalent.*

PROOF: (a)$\Longleftrightarrow$(a′): Already proved in Section 1.1.

(b)$\Longrightarrow$(c): The implications (b) and (c) have the same conclusion, but the premise of (c) is stronger than that of (b).

(c)$\Longrightarrow$(c*): Assume that every chain of the poset P has a l.u.b. and take $x \in P$. The set $[x) = \{y \in P \mid x \leq y\}$, endowed with the order induced by that of P, fulfills the hypotheses of (c). For $x \in [x) \neq \varnothing$ and if L is a chain in $[x)$, then L is also a chain in P, hence $s =$ l.u.b. of L exists and clearly $s \in [x)$, so that s is the l.u.b. of L in $[x)$ as well. It follows from (c) that $[x)$ has a maximal element m. Then $m \in [x)$, that is $x \leq m$, and it remains to prove that m is a maximal element of P. But $m \leq m' \in P$ implies $m' \in [x)$, hence $m = m'$ by the maximality of m in $[x)$.

(c*)$\Longrightarrow$(e*): For (e*) is a specialization of (c*).

(e*)$\Longrightarrow$(e): Trivial.

(e)$\Longrightarrow$(f*): Suppose $\mathcal{P}$, p and X satisfy the hypotheses of (f*). We first prove that the set $\mathcal{Q}$ of those p-sets which include X verifies the hypotheses of (e). Let $\mathcal{L}$ be a chain of $\mathcal{P}$ and $U = \bigcup\{L \mid L \in \mathcal{L}\}$; we have to prove that $U \in \mathcal{Q}$. But $U \in \mathcal{P}$ and to prove that U is a p-set it suffices to show that every finite subset $F = \{x_1, \ldots, x_n\}$ of U is a p-set. The inclusion $F \subseteq U$ means that each x_i belongs to some $L_i \in \mathcal{L}$ $(i = 1, \ldots, n)$. Since $\mathcal{L}$ is a chain, so is the set $\{L_1, \ldots, L_n\}$ by Remark 2.3.1, hence there is an $i_0 \in \{1, \ldots, n\}$ such that $L_i \subseteq L_{i_0}$ $(i = 1, \ldots, n)$ e.g. by Proposition 2.3.6. Therefore $F \subseteq L_{i_0}$ and as F is finite and L_{i_0} is a p-set, it follows that F is a p-set, too.

In view of (e), the poset $(\mathcal{Q}, \subseteq)$ has a maximal element M. But $M \in \mathcal{Q}$ means that $X \subseteq M$ and M is a p-set. Moreover, M is a maximal p-set by the same argument as in the last part of the proof of (c)$\Longrightarrow$(c*).

(f*)$\Longrightarrow$(f): Trivial.

(f)$\Longrightarrow$(g*): Using the notation and hypotheses from (g*), let $\mathcal{P} = \{Y \in 2^S \mid X \subseteq Y\}$. Clearly $\mathcal{P}$ fulfills the hypotheses from (f), hence there is a set M maximal among all p-sets of $\mathcal{P}$. It follows immediately as above that M is a maximal p-set.

(g*)$\Longrightarrow$(h*): Because being a chain is a property of finite character.

(h*)$\Longrightarrow$(h): Trivial.

(h)$\Longrightarrow$(i): The proof that every non-empty set S can be well-ordered will consist of three steps. Let $\mathcal{P}$ be the set of all pairs of the form $(X, \leq)$, where $X \subseteq S$ and $\leq$ is a well-order on X; clearly $\mathcal{P} \neq \varnothing$. We first prove that the following relation is a partial order on $\mathcal{P}$: $(X_1, \leq_1)\ i\ (X_2, \leq_2)$ means: $\alpha) X_1 \subseteq X_2$, $\beta) \leq_1$ is the restriction of $\leq_2$ to the elements of $X_1 \times X_1$, and $\gamma)\ a \leq_2 b \in X_1 \Longrightarrow a \in X_1$.

The reflexivity of i is immediate. Furthermore, if $(X_1, \leq_1)\ i\ (X_2, \leq_2)$ and $(X_2, \leq_2)\ i\ (X_1, \leq_1)$, then $X_1 \subseteq X_2$, $\leq_1 \subseteq \leq_2$ and $X_2 \subseteq X_1$, $\leq_2 \subseteq \leq_1$, therefore $X_1 = X_2$ and $\leq_1 = \leq_2$. Finally, if $(X_1, \leq_1)\ i\ (X_2, \leq_2)$ and $(X_2, \leq_2)\ i\ (X_3, \leq_3)$, it follows immediately that $X_1 \subseteq X_3$, $\leq_1$ is the restriction of $\leq_3$ to $X_1 \times X_1$, and if $a \leq_3 b \in X_1$ then $b \in X_2$, hence $a \in X_2$ by γ, therefore $a \leq_2 b \in X_1$ by β, finally $a \in X_1$ again by γ.

Second step. In view of (h), there is a maximal chain $\mathcal{L}$ in $\mathcal{P}$. Define $\mathcal{X} = \{X \in 2^S \mid (\exists \leq_L \subseteq X \times X)\ (X, \leq_L) \in \mathcal{L}\}$ and $M \bigcup_{X \in \mathcal{X}} X$. Then $(\mathcal{X}, \subseteq)$ is a chain by α and a well-known argument (like in the proof of (e)$\Longrightarrow$(f*)) shows that every finite subset $F \subseteq M$ fulfills $F \subseteq X$ for some $(X, \leq_X) \in \mathcal{L}$; in the sequel we will tacitly apply this property. We now define $x \leq y$ in M to mean $x \leq_L y$, where $x, y \in X$ and $(X, \leq_X) \in \mathcal{L}$, and proceed to show that $(M \leq)$ is a well-ordered set.

The first thing to prove is the consistency of the definition: if $x, y \in X'$ where $(X', \leq_{X'}) \in \mathcal{L}$, then either $(X, \leq_X)\ i\ (X', \leq_{X'})$ or $(X', \leq_{X'})\ i\ (X, \leq_X)$ and in both cases $x \leq_X y$ holds if and only if $x \leq_{X'} y$. Then we have to prove that $\leq$ is a partial order. Clearly $x \leq x$. If $x \leq y$ and $y \leq x$, take $(X, \leq_X) \in \mathcal{L}$ such that $x, y \in X$; then $x \leq_X y$ and $y \leq_X x$, therefore $x = y$. Also, if $x \leq y$ and $y \leq x$, take $(X, \leq_X) \in \mathcal{L}$ such that $x, y, z \in X$; then $x \leq_x y$ and $y \leq_X z$, hence $x \leq_X z$, therefore $x \leq z$. Note also that the order $\leq$ is total, because for every $x, y \in M$ we have $x, y \in X$ where $(X, \leq_X) \in \mathcal{L}$, and the well-order $\leq_X$ is total.

Finally let $\varnothing \neq A \subseteq M$; we must prove the existence of the least element of A. Take $a_0 \in A$; then $a_0 \in X_0$, where $(X_0, \leq_0) \in \mathcal{L}$, and set $A_0 = \{a \in A \mid a \leq a_0\}$. If we succeed in proving that $A_0 \subseteq X_0$, then, as $(X_0, \leq_0)$ is a well-ordered set, it will follow that A_0 has a first element a_1, which will be the first element of A, because if $a \in A$ then either $a \in A_0$, in which case $a_1 \leq_0 a$ hence $a_1 \leq a$, or $a \not\leq a_0$, in which case, since $\leq$ is a total order, it follows that $a_0 < a$, therefore $a_1 < a$ by (2.1.17).

So take an arbitrary $a \in A_0$ and prove that $a \in X_0$. But $a \in X$ for some $(X, \leq_X) \in \mathcal{L}$ and either $(X, \leq_X)\ i\ (X_0, \leq_0)$ or $(X_0, \leq_0)\ i\ (X, \leq_X)$. In the former case $a \in X_0$ follows from $X \subseteq X_0$, while in the latter case we have $a, a_0 \in X$ and $a \leq a_0$, hence $a \leq_X a_0 \in X_0$, therefore $a \in X_0$ by γ.

The third step in proving that S can be well-ordered is to show that $S = M$. Otherwise take an element $s \in S \setminus M$, put $\overline{M} = M \cup \{s\}$ and extend the order of M to an order of $\overline{M}$ setting $x \leq s$ for every $x \in \overline{M}$. If we succeed in proving that $(\overline{M}, \leq)$ is a well-ordered set and $(X, \leq_X)\ i\ (\overline{M}, \leq)$ for every $(X, \leq_X) \in \mathcal{L}$, then $\mathcal{L} \cup \{(\overline{M}, \leq)\}$ is a chain in $\mathcal{P}$ distinct from $\mathcal{L}$ (because $\overline{M} \notin \mathcal{X}$ as $s \notin X$ for any $(X, \leq_X) \in \mathcal{L}$), thus contradicting the maximality of $\mathcal{L}$.

But $\overline{M}$ is obviously a toset and if $\varnothing \neq A \subseteq \overline{M}$ then either $A \cap M \neq \varnothing$, in which case $A \cap M$ has a first element which is also the first element of A because $A \cap M < s$, or $A \cap M = \varnothing$, in which case $A = \{s\}$ has trivially a least element. So $(\overline{M}, \leq)$ is well-ordered and it remains to prove that $(X, \leq_X)\ i\ (\overline{M}, \leq)$ for every

$(X, \leq_X) \in \mathcal{L}$. Since $(M, \leq)\ i\ (\overline{M}, \leq)$, it suffices to show that $(X, \leq_X)\ i\ (M, \leq)$. But $X \subseteq M$ and $\leq_X$ is the restriction of $\leq$ to $X \times X$ by the very definition of $\leq$, therefore it remains to prove that $a \leq b \in X$ implies $a \in X$. Indeed, take $(X', \leq_{X'}) \in \mathcal{L}$ such that $a, b \in X'$ and $a \leq_{X'} b$. Then either $(X, \leq_X)\ i\ (X', \leq_{X'})$ or $(X', \leq_{X'})\ i\ (X, \leq_X)$. In the former case $a \leq_{X'} b \in X$, hence $a \in X$ by γ, while in the latter case $X' \subseteq X$, implying again $a \in X$.

(i)$\Longrightarrow$(a): Let $\{S_i \mid i \in I\}$, where $I \neq \varnothing$, be a family of (disjoint) non-empty sets. Then $\bigcup_{i \in I} S_i$ can be endowed with a well-order. For each $i \in I$ let s_i be the first element of S_i with respect to this order. The mapping $\varepsilon : I \longrightarrow \bigcup_{i\in I} S_i$ defined by $\varepsilon(i) = s_i$ is a choice function.

(a)$\Longrightarrow$(h): Let $(P, \leq)$ be a poset with $P \neq \varnothing$. To obtain a maximal chain of P, we take a choice function ε on $2^P \setminus \{\varnothing\}$, i.e., $\varepsilon(x) \in X$ for every non-empty set $X \subseteq P$. Let $\mathcal{L}$ be the set consisting of all chains plus the empty set. For each $L \in \mathcal{L}$, let $\widehat{L}$ be the set of those $x \in P$ for which $L \cup \{x\}$ is a chain; in particular $\widehat{\varnothing} = P$. Define $f : \mathcal{L} \longrightarrow \mathcal{L}$ by

$$f(L) = \begin{cases} L \cup \{\varepsilon(\widehat{L} \setminus L)\}, & \text{if } \widehat{L} \setminus L \neq \varnothing, \\ L, & \text{if } \widehat{L} \setminus L = \varnothing. \end{cases} \tag{2.4.1}$$

Clearly a chain L is maximal if and only if $f(L) = L$, therefore it suffices to find a chain satisfying the latter condition.

Let us use, during this proof, the term *collection* for any subset $\mathcal{I} \subseteq \mathcal{L}$ satisfying the following conditions: I) $\varnothing \in \mathcal{L}$; II) if $L \in \mathcal{I}$, then $f(L) \in \mathcal{I}$, and III) if $\varnothing \neq \mathcal{H} \subseteq \mathcal{I}$ and $\mathcal{H}$ is totally ordered under $\subseteq$, then $\bigcup\{L \mid L \in \mathcal{H}\} \in \mathcal{I}$. Clearly such collections exist, for instance $\mathcal{L}$ itself, and the intersection of any non-empty set of collections is again a collection. This easily implies that the intersection $\mathcal{I}_0$ of all the collections is the smallest collection (included in any other collection). So $\mathcal{I}_0 \neq \varnothing$ and if we succeed in proving that $(\mathcal{I}_0, \subseteq)$ is a toset, then it will follow that $\overline{L} = \bigcup\{L \mid L \in \mathcal{I}_0\} \in \mathcal{I}_0$ by III, therefore $f(\overline{L}) \subseteq \overline{L}$ by the definition of $\overline{L}$, hence $f(\overline{L}) = \overline{L}$ (because $L \subseteq f(L)$ for every $L \in \mathcal{L}$), and this will show that $\overline{L}$ is a maximal chain.

Thus it remains to prove that $\mathcal{I}_0$ is totally ordered under set inclusion. Let us also introduce, for the convenience of this proof, the term *comparable chain* for any chain $C \in \mathcal{I}_0$ such that $C \subseteq L$ or $L \subseteq C$ for every $L \in \mathcal{I}_0$. It suffices to prove that all chains in $\mathcal{I}_0$ are comparable.

A preliminary remark is that, for every comparable chain C, the relations $L \in \mathcal{I}_0$ and $L \subset C$ imply $f(L) \subseteq C$; for otherwise $f(L) \neq C$ and $C \subseteq f(L)$, hence $C \subset f(L)$, which, together with $L \subset C$ would imply that $f(L) \setminus L$ contains more than one element, a contradiction. Now let C be an arbitrary but fixed comparable chain; we shall prove that the set

$$\mathcal{K} = \{L \in \mathcal{I}_0 \mid L \subseteq C \text{ or } f(C) \subseteq L\} \tag{2.4.2}$$

is a collection.

Indeed, $\varnothing \in \mathcal{I}_0$ by I and $\varnothing \subset C$ hence $\varnothing \in \mathcal{K}_0$. Furthermore, if $L \in \mathcal{K}$ then $L \in \mathcal{I}_0$, hence $f(L) \in \mathcal{I}_0$ by II and we have to prove that $f(L) \subseteq C$ or $f(C) \subseteq f(L)$. But $f(C) \subseteq L$ or $L \subseteq C$; in the former case it follows that

$f(C) \subseteq f(L)$, while in the latter case either $L = C$ thus implying trivially $f(C) \subseteq f(L)$, or $L \subset C$, in which case $f(L) \subseteq C$ by the preliminary remark. Finally let $\mathcal{H} \subseteq \mathcal{K} \setminus \{\varnothing\}$ be totally ordered under set inclusion and let us prove that $H = \bigcup\{L \mid L \in \mathcal{H}\}$ belongs to $\mathcal{K}$. But $\mathcal{H} \subseteq \mathcal{K} \subseteq \mathcal{I}_0$, hence $H \in \mathcal{I}_0$ by III and we have to show that $H \subseteq C$ or $f(C) \subseteq H$. If $f(C) \subseteq L$ for some $L \in \mathcal{H}$, then $f(C) \subseteq H$; otherwise $L \subseteq C$ for all $L \in \mathcal{H}$, thus implying $H \subseteq C$. Therefore $\mathcal{K}$ is a collection.

Moreover, since $\mathcal{K} \subseteq \mathcal{I}_0$ and $\mathcal{I}_0$ is the smallest collection , it follows that in fact $\mathcal{K} = \mathcal{I}_0$, hence for every comparable chain C we have

$$L \in \mathcal{I}_0 \Longrightarrow L \subseteq C \text{ or } f(C) \subseteq L\ . \tag{2.4.3}$$

Furthermore, let $\mathcal{C}$ be the set consisting of all comparable chains plus the empty set. As $\mathcal{I}_0$ is the smallest collection, if we succeed in proving that $\mathcal{C}$ is a collection, it will follow that $\mathcal{I}_0 \subseteq \mathcal{C}$, i.e., all the chains of $\mathcal{I}_0$ are comparable, as desired.

So it remains to check properties I-III for $\mathcal{C}$. But $\varnothing \in \mathcal{C}$ by the definition of $\mathcal{C}$. Furthermore, given $C \in \mathcal{C}$, we have $C \in \mathcal{I}_0$ by definition and if $L \in \mathcal{I}_0$, then (2.4.3) implies $L \subseteq f(C)$ or $f(C) \subseteq L$. We have thus proved that $f(C)$ is a comparable chain, i.e., $f(C) \in \mathcal{C}$. Finally suppose $\mathcal{H} \subseteq \mathcal{C} \setminus \{\varnothing\}$ and let $\overline{H} = \bigcup\{L \mid L \in \mathcal{H}\}$. Then, as $\mathcal{H} \subseteq \mathcal{C} \subseteq \mathcal{I}_0$, it follows that $\overline{H} \in \mathcal{I}_0$ by III. Moreover, since the elements of $\mathcal{H}$ are comparable chains, it follows that for every $L \in \mathcal{I}_0$ either $L \subseteq H$ for some $H \in \mathcal{H}$, in which case $L \subseteq \overline{H}$, or $H \subset L$ for all $H \in \mathcal{H}$, in which case $\overline{H} \subseteq L$. So $\overline{H} \in \mathcal{C}$, concluding the proof of (h).

(h)$\Longrightarrow$(b*): Suppose every chain of P has an upper bound and take $x \in P$. Let $[x) = \{y \in P \mid x \leq y\}$ be endowed with the order induced by that of P. As $x \in [x) \neq \varnothing$, it follows that $[x)$ has a maximal chain L. Then $x \leq L$ because $L \subseteq [x)$, and L is also a chain of P, hence $L \leq m$ for some $m \in P$. It follows that $x \leq m$ and it remains to prove that m is a maximal element of P. Otherwise $m < m'$ for some $m' \in P$, hence $L < m'$ and $x < m'$, therefore $L \subset L \cup \{m'\}$ and $m' \in [x)$, so that $L \cup \{m'\}$ is a chain of $[x)$ and this contradicts the maximality of L in $[x)$.

(b*)$\Longrightarrow$(b): Trivial.

We have thus proved that properties (a), (a′), (b), (b*), (c), (c*), (e), (e*), (f), (f*), (g*), (h), (h*), (i) are equivalent. It remains to include (d), (d*) and (g) into this chain of equivalences.

(b*)$\Longrightarrow$(d*): Apply (b*) to the poset $(\mathcal{P}, \subseteq)$.

(d*)$\Longrightarrow$(d): Trivial.

(d)$\Longrightarrow$(e): The union is an upper bound.

(g*)$\Longrightarrow$(g): Trivial.

(g)$\Longrightarrow$(h): The property of being a chain is a consistent property of finite character defined for the subsets of a poset P. □

2.5 Well-Ordered Sets

As explained in the introduction to this chapter, the aim of this section is to establish the basic properties of well-ordered sets, including several properties which are usually presented as properties of ordinal numbers, in such a way as to simplify the task of the next chapter, which will be devoted to ordinal and cardinal numbers.

We recall (cf. Definition 2.4.3) that a *well-ordered set* or *woset* is a totally ordered set W such that every non-empty subset of W has a least element.

Example 2.5.1 The empty set $\varnothing$ is well-ordered.

This follows from the non-existence of non-empty subsets of $\varnothing$.

Example 2.5.2 More generally, every finite totally ordered set is well-ordered.

Indeed, every non-empty subset of a finite chain is a non-empty finite chain as well, hence it has a first element e.g. by Proposition 2.3.6 or by Proposition 2.5.1 below.

Example 2.5.3 The set $\mathbb{N}$ of natural numbers, endowed with the usual ordering, is well-ordered.

For $\varnothing = A \subseteq \mathbb{N}$ implies the existence of a number $n \in A$ and the set $\{p \in A \mid p \leq n\}$ has a first element m, by the same reasoning as in Example 2.5.2, and it is immediately seen that m is the smallest number of A, too (either $p \leq n$ in which case $m \leq p$, or $n < p$, implying $m < p$ because $m \leq n$).

Example 2.5.4 More generally, every down-bounded set of integers is well-ordered. In other words, if $\varnothing = I \subseteq \mathbb{Z}$ and $z \leq I$ for some $z \in \mathbb{Z}$, then I is well-ordered under the usual ordering of integers.

The proof is as for the set $\mathbb{N}$. As a matter of fact, one can pass from Example 2.5.4 to Example 2.5.3 on account of the following

Remark 2.5.1 The property of being a well-ordered set is hereditary, i.e., every subset of a well-ordered set is also well-ordered. The dual of a woset is not always a woset (see for instance Example 2.5.3). Therefore the duality principle does not hold for wosets.

Example 2.5.5 Recall that the axiom of choice enables us to prove the existence of a well-ordering for any set (the so-called *Zermelo theorem*; cf. Theorem 2.4.1, property (i)).

Examples 2.5.6 and 2.5.7 below furnish procedures for constructing new wosets from given ones.

Example 2.5.6 The sum $\sum_{i \in I} W_i$ of a family $\{W_i\}_{i \in I}$ of well-ordered sets W_i under a well-ordered set I of indices, is also a woset (cf. Examples 1.3.45, 2.1.10 and 2.3.4).

For let $\varnothing = A \subseteq \sum_{i \in I} W_i$. The set of those $i \in I$ for which there exists $x \in W_i$ such that $(i, x) \in A$ is not empty, hence it has a first element i_0. The set of those $x \in W_{i_0}$ for which $(i_0, x) \in A$ is not empty, hence it has a first element x_0. Then (i_0, x_0) is the first element element of A.

Example 2.5.7 The lexicographic product of two wosets W' and W'' is also well-ordered (cf. Examples 2.1.11 and 2.3.5).

Indeed, suppose $\varnothing \neq A \subseteq W' \times W''$. Let A' be the set of those $x' \in W'$ for which there exists $x'' \in W''$ such that $(x', x'') \in A$. Then $A' \neq \varnothing$, hence A' has a first element x'_0. The set $A'' = \{x'' \in W'' \mid (x'_0, x'') \in A\}$ is not empty, therefore it has a first element x''_0. It follows that (x'_0, x''_0) is the first element of A.

Let us give some "negative" examples, too.

Example 2.5.8 None of the sets $\mathbb{Z}$ (integers), $\mathbb{Q}$ (rationals), $\mathbb{R}$ (reals) is well-ordered.

This follows e.g. from the following

Remark 2.5.2 Every non-empty well-ordered set has first element.

However the following example shows that the reason for $\mathbb{Z}, \mathbb{Q}, \mathbb{R}$ not being well-ordered is deeper than the non-existence of the least element.

Example 2.5.9 Let $-\infty$ be a least element added to the reals. None of the sets $\mathbb{Z} \cup \{-\infty\}, \mathbb{Q} \cup \{-\infty\}, \mathbb{R} \cup \{-\infty\}$ is well-ordered.

This follows from the following more general property:

Proposition 2.5.1 *A totally ordered set is well-ordered if and only if it contains no infinite decreasing sequence*

$$x_1 > x_2 > \cdots > x_n > \dots \quad . \tag{2.5.1}$$

PROOF: If a totally ordered set T contains a sequence of the form (2.5.1), then the set $\{x_1, x_2, \dots, x_n, \dots\}$ has no first element, therefore T is not well-ordered. Conversely, suppose T is a totally ordered set which contains no sequence of the form (2.5.1) and let $\varnothing \neq A \subseteq T$; we have to prove the existence of the first element of A. Take $a_1 \in A$; if a_1 is the first element of A, the property is established. Otherwise $a_1 \not\leq a_2$ for some $a_2 \in A$, hence $a_2 < a_1$; if a_2 is the first element of A, the property is established. Otherwise $a_2 \not\leq a_3$ for some $a_3 \in A$, etc. We thus construct a decreasing sequence $a_1 > a_2 > a_3 > \dots$, which must stop after a finite number of steps, say $a_1 > a_2 > \cdots > a_n$. It then follows that a_n is the first element of A, otherwise the sequence could be continued. □

It is also interesting to note that the condition of being totally ordered can be dispensed with in Definition 2.4.3:

Proposition 2.5.2 *A partially ordered set P is well-ordered if and only if every non-empty subset of P has least element.*

PROOF: Necessity: trivial. Sufficiency: if the condition is fulfilled, then for every two elements $x, y \in P$, the set $\{x, y\}$ has a first element, which is x or y, thus implying $x \leq y$ or $y \leq x$, respectively. □

As a matter of fact, Proposition 2.5.1 can be strengthened even more, as shown in Exercise 2.5.1 below.

Exercise 2.5.1 (Erné [1982], Satz 92). Let ρ be a binary relation on a set P, such that for any non-empty subset A of P there is a unique element $a \in A$ such that aRx for all $x \in A$. Then (P,ρ) is a well-ordered set.

Hint. Prove in turn that ρ is total, reflexive, antisymmetric and transitive.

Exercise 2.5.2 (Pierce [1968]) Let W be a set and $\leq, <$ two binary relations on W, linked by (2.1.14). Then $(W, \leq)$ is a well-ordered set if and only if $<$ fulfills the following conditions: (i) $x < y \Longrightarrow y \not< x$; (ii) for every $\varnothing \neq A \subseteq W$ there exists $a \in A$ such that $a < x$ for any $x \in A - \{a\}$. Infer a two-axiom definition for chains.

Hint. To prove transitivity apply (ii) to $\{x, y, z\}$.

We now proceed to establish various properties of well-ordered sets. An elementary technique to be used in subsequent proofs runs as follows: (i) to prove the existence of an element a satisfying a certain property, construct a as the first element of a certain non-empty subset; (ii) to prove that all the elements satisfy a certain property, assume that the set of elements which do not fulfill that property is not empty and obtain a contradiction from the existence of the first element of that set.

Proposition 2.5.3 *An element of a non-empty well-ordered set either has a successor or is the greatest element.*

Comment The converse of this proposition is false, as shown e.g. by the set $\mathbb{Z}$ of integers.

Note also that $x \prec y$ means that $x < y$ while $x < z < y$ is impossible, which in a toset is equivalent to $x < y \,\&\, (x < z \Longrightarrow y \leq z)$; in other words, in a chain y is the successor of x in the sense of Definition 2.1.3 if and only if y is the successor of $\{x\}$ in the sense of Definition 2.2.2$''$ (cf. Warning after this definition).

PROOF: Let W be a well-ordered set and $a \in W$. If a is not the greatest element of W, then $x \not\leq a$ for some $x \in W$, hence the set $A = \{x \in W \mid a < x\}$ is not empty; let b denote the least element of A. Then $b \in A$, hence $a < b$, while $a < x$ implies $b \leq x$; □

For the sake of generality, Definition 2.5.1 of segments, Remark 2.5.3 and Propositions 2.5.5 and 2.5.6 below are given for posets; however the most significant properties of segments are obtained for wosets.

Definition 2.5.1 A *segment* of a poset P is a subset $S \subseteq P$ such that for every $x, y \in P$,

$$x < y \,\&\, y \in S \Longrightarrow x \in S \,. \tag{2.5.2}$$

The segment S is said to be *proper* if $S \neq P$, whereas P is the *improper* segment of P. For every $a \in P$, the segment *generated by* a is

(2.5.3) $$P(a) = \{z \in P \mid z < a\}\ .$$

Remark 2.5.3 a) Condition (2.5.2) is equivalent to $x \leq y \,\&\, y \in S \Longrightarrow x \in S$.

b) A segment of a segment of a poset P is a segment of P. Indeed, assume S is a segment of P and T is a segment of S. If $x \in P, y \in T$ and $x < y$, then $y \in S$, hence $x \in S$, therefore $x \in T$.

c) $P(a)$ is indeed a segment and is the largest subset $X \subset P$ such that $X < a$.

d) If T is a toset, then x is the successor of $T(x)$ for every $x \in T$. For $T(x) < y \,\&\, x \not\leq y$ would imply $y \in T(x)$, hence $y < y$, a contradiction.

Proposition 2.5.4 *The proper segments of a non-empty well-ordered set W coincide with the subsets of the form $W(x)$, where $x \in W$.*

PROOF: Every set $W(x)$ is a segment of W by Remark 2.5.3.c); moreover, it is proper because

(2.5.4) $$x \notin W(x)\ ,$$

since $x \nless x$. Conversely, let S be a proper segment of W. Then $\varnothing \neq W \setminus S \subseteq W$, hence $W \setminus S$ has first element a; we shall prove that $S = W(a)$. If $x \in S$, then $x < a$ (otherwise $a \leq x \in S$, hence $a \in S$ by Remark 2.5.3.a), a contradiction), therefore $x \in W(a)$. Conversely, if $x \in W \setminus S$, then $a \leq x$, hence $x \nless a$, therefore $x \notin W(a)$. □

Proposition 2.5.5 *If P is a poset and $x, y \in P$, then $P(x), P(y)$ are posets and*

(2.5.5) $$x \leq y \Longrightarrow (P(y))(x) = P(x)\ .$$

PROOF: For any $z \in P$, the inequality $x \leq y$ implies

$$z \in (P(y))(x) \Longleftrightarrow z \in P(y) \,\&\, z < x \Longleftrightarrow z \in P \,\&\, z < y \,\&\, z < x$$

$$\Longleftrightarrow z \in P \,\&\, z < x \Longleftrightarrow z \in P(x)\ .$$

□

Remark 2.5.4 The first element a, if any, of a totally ordered set T is characterized by the property $T(a) = \varnothing$.

Theorem 2.5.1 *Let W be a well-ordered set and $S \subseteq W$ such that for every $x \in W$, if $W(x) \subseteq S$ then $x \in S$. It follows that $S = W$.*

Comment This property may be called *the principle of transfinite induction* for well-ordered sets.

PROOF: Assume $S \neq W$. Then the set $W \setminus S$ has a first element a. If $x < a$ then $x \in S$, otherwise $x \in W \setminus S$, contradicting the fact that a is the first element of $W \setminus S$. So $W(a) \subseteq S$, therefore $a \in S$ by hypothesis, and this is a contradiction. □

Corollary 2.5.1 *Assume* $S \subseteq \mathbb{N} = \{0, 1, 2, \ldots, n, \ldots\}$ *satisfies* (i) $0 \in S$, *and* (ii) *for every* $n \in \mathbb{N}$, *if* $n \in S$ *then* $n + 1 \in S$. *It follows that* $S = \mathbb{N}$.

Comment This is the *principle of mathematical induction* for natural numbers (variants of it are given in Exercise 2.5.2 below). In Peano's axiomatic construction of $\mathbb{N}$, the above property is precisely one of the axioms. However, as mentioned in Section 1.1, the set $\mathbb{N}$ can be constructed e.g. within ZF set theory without taking mathematical induction as an axiom.
PROOF: In view of Theorem 2.5.1, it suffices to show that $\mathbb{N}(n) \subseteq S$ implies $n \in S$. If $n := 0$ this follows by (i), otherwise $n - 1 \in \mathbb{N}(n)$, hence $n - 1 \in S$, therefore $n \in S$ by (ii). □

Exercise 2.5.3 Let $S \subseteq \mathbb{Z}$ and $z_0, z_1 \in \mathbb{Z}$ such that $z_0 < z_1$. Consider the following three conditions: (i) $z_0 \in S$; (ii) if $z - 1 \in S$ then $z \in S$; and (iii) if $z - 1 \in S$ and $z \leq z_1$ then $z \in S$. Prove that:

α) if S fulfills (i) and (ii) then $S = \{z \in \mathbb{Z} \mid z_0 \leq z\}$;

β) if S fulfills (i) and (iii) then $S = \{z \in \mathbb{Z} \mid z_0 \leq z \leq z_1\}$.

Exercise 2.5.4 For every element x of a poset P, the set

$$P(x] = \{z \in P \mid z \leq x\} \tag{2.5.6}$$

is a segment. If x is an element of a woset W, then either x is the last element of W, in which case $W(x] = W$, or $W(x] = W(x^+)$, where x^+ denotes the successor of x. Proposition 2.5.5 extends to the sets of the form $P(x]$. Is every segment of W of the form $W(x]$?

Now we are going to study the behaviour of well-ordered sets with respect to strictly increasing mappings and isomorphisms. It turns out that this will enable the further construction of ordinals.

Let **Wos** be the category of well-ordered sets and increasing mappings; then **Wos** is a full subcategory of **Tos** and of **Pos** as well. We first point out several properties of isomorphisms in the category **Wos**. The first result is valid in **Tos**.

Proposition 2.5.6 *Let* T, T' *be tosets and* $f : T \longrightarrow T'$ *an isomorphism. For every* $a \in T$, *the restriction of* f *to the segment* $T(a)$ *of* T *establishes an isomorphism*

$$f|_{T(a)} : T(a) \longrightarrow T'(f(a)) \tag{2.5.7}$$

between $T(a)$ *and the segment* $T'(f(a))$ *of* T.

PROOF: Recall that isomorphisms coincide with surjective strictly increasing mappings by Proposition 2.3.4. If $x \in T(a)$ then $f(x) \in T'(f(a))$ by (2.3.2). If $x' \in T'(f(a))$, then as $x' = f(x)$ for some (uniquely determined) $x \in T$, it follows that $f(x) < f(a)$, hence $x < a$ by (2.1.25) from Proposition 2.1.5, therefore $x \in T(a)$. Thus the mapping (2.5.7) is surjective and also strictly increasing because f is so. Therefore (2.5.7) is an isomorphism. □

Corollary 2.5.2 *Let W, W' be wosets and $f : W \longrightarrow W'$ an isomorphism. The restriction of f to a (proper) segment S of W establishes an isomorphism between S and a (proper) segment S' of W'.*

PROOF: By Propositions 2.5.6 and 2.5.4. □

Lemma 2.5.1 *If W is a woset and $f : W \longrightarrow W$ a strictly increasing mapping, then $x \leq f(x)$ for every $x \in W$.*

PROOF: Otherwise the set $A = \{x \in W \mid f(x) < x\}$ would not be empty, hence it would have a first element a. Then $a \in A$, hence $f(a) < a$, which implies $f(f(a)) < f(a)$, therefore $f(a) \in A$ although $f(a) < a =$ least element of A, a contradiction. □

Proposition 2.5.7 (i) *A woset W cannot be isomorphic to a proper segment $W(x)$ of it.*

(ii) *If x,y are two distinct elements of a woset W, then $W(x)$ and $W(y)$ cannot be isomorphic.*

PROOF: (i) The existence of an isomorphism $f : W \longrightarrow W(x)$ would imply $f(x) \in W(x)$, that is $f(x) < x$, in contradiction with Lemma 2.5.1.

(ii) Suppose $x \neq y$, say $x < y$. Then the above property (i) and Proposition 2.5.5 show that $W(y)$ cannot be isomorphic to $(W(y))(x) = W(x)$. □

Proposition 2.5.8 *There is at most one isomorphism between two well-ordered sets.*

PROOF: Suppose $f, g : W \longrightarrow W'$ are two distinct isomorphisms. Then there is an element $a \in W$ with the property $f(a) \neq g(a)$. Proposition 2.5.6 implies that $W(a)$ is isomorphic both with $W'(f(a))$ and with $W'(g(a))$, hence $W'(f(a))$ and $W'(g(a))$ are isomorphic by the Comment to Proposition 1.3.3 or by properties (2.5.8)–(2.5.10) below. However this contradicts Proposition 2.5.7(ii). □

The next step is the introduction of an ordering of wosets.

Definition 2.5.2 If W and W' are well-ordered sets, write:

1) $W \leq W'$ if W is isomorphic to a segment of W';

2) $W < W'$ if W is isomorphic to a proper segment of W';

3) $W \approx W'$ if W and W' are isomorphic.

For the convenience of language, in the sequel we shall refer to $\leq, <$ and $\approx$ as relations, although properly speaking they do not fulfill Definition 1.2.10 because the well-ordered sets do not form a set, but a proper class. This follows from the fact that the proper class of all sets can be embedded into the class of all wosets, on account of the Zermelo theorem.

Proposition 1.3.3 shows that the isomorphism $\approx$ behaves like an equivalence relation, that is,

$$W \approx W \,, \tag{2.5.8}$$

(2.5.9) $$W \approx W' \Longrightarrow W' \approx W \ ,$$

(2.5.10) $$W \approx W' \,\&\, W' \approx W'' \Longrightarrow W \approx W'' \ ,$$

for every three well-ordered sets W, W', W''. We are going to show that $\leq$ and $<$ behave like a relation of order and the relation of strict order associated with it (Propositions 2.5.9–2.5.11), while $\leq$ is essentially a relation of well-order (Theorem 2.5.3).

Proposition 2.5.9 *For every three wosets* W, W', W'',

(2.5.11) $$W \not< W \ ,$$

(2.5.12) $$W < W' \,\&\, W' < W'' \Longrightarrow W < W'' \ ,$$

(2.5.13) $$W < W' \Longrightarrow W' \not< W \ .$$

PROOF: Property (2.5.11) is a reformulation of Proposition 2.5.7(i).

Let further $f : W \longrightarrow W'(a')$ and $g : W' \longrightarrow W''(a'')$ be isomorphisms. Then Proposition 2.5.6 implies the existence of an isomorphism

(2.5.14) $$h : W'(a') \longrightarrow (W''(a''))(g(a')) \ ,$$

where $h(x) = g(x)$ for every $x \in W'(a')$. But $g(a') \in W''(a'')$, that is, $g(a') < a''$, hence the isomorphism (2.5.14) becomes

$$h : W'(a') \longrightarrow W''(g(a'))$$

by Proposition 2.5.5. It follows that

$$h \circ f : W \longrightarrow W''(g(a''))$$

is an isomorphism, thus proving $W < W''$.

Finally $W < W'$ and $W' < W$ would imply $W < W$ by (2.5.12), which contradicts (2.5.11). □

Proposition 2.5.10 *For every three wosets* W, W', W'',

(2.5.15) $$W \leq W' \Longleftrightarrow W < W' \text{ or } W \approx W' \ ,$$

(2.5.16) $$W < W' \Longleftrightarrow W \leq W' \,\&\, W \not\approx W' \ .$$

PROOF: Property (2.5.15), the implication $\Longleftarrow$ from (2.5.16) and the implication $W < W' \Longrightarrow W \leq W'$ follow immediately from Definition 2.5.2, while the implication $W < W' \Longrightarrow W \not\approx W'$ follows by Proposition 2.5.7(i) applied to W'. □

Proposition 2.5.11 *For every three wosets* W, W', W''

(2.5.17) $$W \leq W \ ,$$

(2.5.18) $$W \leq W' \,\&\, W' \leq W \Longrightarrow W \approx W' \,,$$

(2.5.19) $$W \leq W' \,\&\, W' \leq W'' \Longrightarrow W \leq W' \,.$$

PROOF: Property (2.5.17) follows from (2.5.15) and (2.5.8).

Suppose $W \leq W'$ and $W' \leq W$. If $W \not\approx W'$ then $W < W'$ and $W' < W$ by (2.5.15) and (2.5.9); however this contradicts (2.5.13).

Finally assume $W \leq W'$ and $W' \leq W''$, i.e., W is isomorphic to a segment S' of W', while W' is isomorphic to a segment S'' of W'' If $S' = W'$ then $W \approx S''$ by (2.5.10), therefore $W \leq W''$. If S' is a proper segment of W', then S' is isomorphic to a proper segment T'' of S'' by Corollary 2.5.2. Then $W \approx T''$ again by (2.5.10); but T'' is a segment of W'' by Remark 2.5.3.b), therefore $W \leq W''$. □

Exercise 2.5.5 (i) $W < W' \Longleftrightarrow W \leq W' \,\&\, W' \not\approx W$. (ii) If $W \leq W' \,\&\, W' < W''$ or $W < W' \,\&\, W' \leq W''$, then $W \leq W''$. (iii) If $W \leq W' \,\&\, W' \approx W''$, or $W \approx W' \,\&\, W' \leq W''$, then $W \leq W''$; (iv) If $W < W' \,\&\, W' \approx W''$ or $W \approx W' \,\&\, W' < W''$, then $W < W''$.

We have established so far that the relation $\leq$ $(<)$ between wosets is an order (a strict order). To show that this is a well-order in the sense of Theorem 2.5.3 below, we need some preparation which has its own interest.

Lemma 2.5.2 *Let W and W' be two wosets and denote by S the set of those $x \in W$ for which $W(x] = \{z \in W \mid z \leq x\}$ is isomorphic to a segment of W'. Then S is a segment of W isomorphic to a segment S' of W'.*

PROOF: Note first that in view of Proposition 2.5.7, for each $x \in S$, $W(x]$ is isomorphic to a single segment S'_x of W'. Then Proposition 2.5.8 implies that for each $x \in S$ there is a unique isomorphism

(2.5.20) $$f_x : W(x] \longrightarrow S'_x \,.$$

To show that S is a segment, take $y < x \in S$ and prove that $y \in S$. But the successor y^+ of y exists by Proposition 2.5.3 and fulfills $y \prec y^+ \leq x$, hence

(2.5.21) $$W(y] = W(y^+) = \{z \in W(x] \mid z < y^+\} = (W(x])(y^+)$$

by Exercise 2.5.4. Taking into account Corollary, we infer that $f_x|_{W(y]}$ establishes an isomorphism between $W(y]$ and a segment T' of S'_x. But T' is also a segment of W' by Remark 2.5.3.b), therefore $y \in S$.

Note also that the uniqueness of each f_x implies that for $x, y \in S$ with $y < x$,

(2.5.22) $$f_x|_{W(y]} = f_y \,.$$

The set $S' = \bigcup\{S'_x \mid x \in S\}$ is a segment of W' because each S'_x is a segment. Setting

(2.5.23) $$f = \{(x, f_x(x)) \mid x \in S\} \,,$$

we obtain a mapping $f : S \longrightarrow S'$. It remains to prove that f is an isomorphism. In so doing the point will be that $f(x) = f_x(x)$ for every $x \in S$.

If $y' \in S'$, then $y' \in S'_x$ for some $x \in S$, therefore $y' = f_x(y)$ for some $y \in W(x]$, hence $y \leq x$, which implies $y \in S$. Since $y \in W(y]$, it follows from (2.5.22) that $f_x(y) = f_y(y) = f(y)$, therefore $y' = f(y)$. This shows that f is surjective.

If $x, y \in S$ and $y < x$, then (2.5.22) implies $f(y) = f_y(y) = f_x(y) < f_x(x) = f(x)$. So f is a surjective strictly increasing mapping, hence an isomorphism by Proposition 2.3.4. □

Theorem 2.5.2 *Every two well-ordered sets W and W' fulfill $W \leq W'$ or $W' \leq W$.*

PROOF: We use the notation from Lemma 2.5.2. If $S = W$ then $W \approx S'$, hence $W \leq W'$. If $S' = W'$ then $W' \approx S$, hence $W' \leq W$. It remains to prove that the segments S and S' cannot be both proper. Otherwise $S = W(a)$, $S' = W'(a')$ and $f : W(a) \longrightarrow W'(a')$ is an isomorphism. Setting $\overline{f}(x) = f(x)$ for all $x \in S$ and $\overline{f}(a) = a'$, we obtain a mapping $\overline{f} : W(a] \longrightarrow W'(a']$ which is obviously an isomorphism. Therefore $a \in S = W(a)$, a contradiction. □

Corollary 2.5.3 *Every two well-ordered sets W and W' fulfill either $W < W'$ or $W' < W$ or $W \approx W'$.*

PROOF: Immediate from Theorem 2.5.2 and Proposition 2.5.10. The three possibilities exclude each other by (2.5.13) and (2.5.16). □

Corollary 2.5.4 *Every two well-ordered sets W and W' fulfill either $W < W'$ or $W' \leq W$.*

PROOF: By Corollary 2.5.3 and Proposition 2.5.10. □

Recall that in view of Remark 2.5.1, every subset of a woset $(W, \leq)$ is also well-ordered by the relation $\leq$. Even more:

Proposition 2.5.12 *Every subset V of a well-ordered set W is isomorphic to a segment of W, or equivalently,*

(2.5.24) $$V \subseteq W \Longrightarrow V \leq W .$$

PROOF: If $V \not\leq W$, then $V \not< W$ and $V \not\approx W$ by Proposition 2.5.10, hence $W < V$ by Corollary 2.5.3. Therefore there is an isomorphism of the form $f : W \longrightarrow V(a)$. In view of Proposition 2.3.4, $f : W \longrightarrow W$ is a strictly increasing mapping, and since $f(a) \in V(a)$, we have $f(a) < a$, which contradicts Lemma 2.5.1. □

Remark 2.5.5 It follows from (2.5.24) that $\varnothing \leq W$ for every woset W.

Exercise 2.5.6 Does $V \subset W$ imply $V < W$?

Exercise 2.5.7 Show that $W \leq W'$ if and only if there is a strictly increasing mapping $f : W \longrightarrow W'$. (Hint. Apply Proposition 2.5.12 to $f(W)$ and W').

Theorem 2.5.3 *Every family* $\{W_i\}_{i\in I}$ *of pairwise non-isomorphic well-ordered sets is well-ordered by the relation* $\leq$ *in Definition* 2.5.2.

PROOF: It follows immediately from the hypothesis that $W_i \approx W_j \Longleftrightarrow W_i = W_j$ ($\forall\, i, j \in I$), whence Proposition 2.5.11 implies that the relation $\leq$ is a partial order on the set $\{W_i\}_{i\in I}$. Therefore, taking into account Proposition 2.5.2, it remains to show that every subset $\{W_j\}_{j\in J}$, where $\varnothing \neq J \subseteq I$, has a first element W_{j_0}.

Use the Zermelo theorem to endow I with a well-order and construct the sum $W = \sum_{i\in I} W_i$ (cf. Example 2.5.6). Each W_i is isomorphic to the subset $W_i^* = \{i\} \times W_i$ of W, while W_i^* is isomorphic to a segment S_i of W by Proposition 2.5.12; so $W_i \approx S_i$ ($\forall i \in I$). If we succeed in proving that $S_{j_0} \leq S_j$ for some $j_0 \in J$ and all $j \in J$, then $W_{j_0} \leq W_j$ ($\forall j \in J$) by Exercise 2 5.5, as desired.

The segments S_i are pairwise non-isomorphic because so are the W_is. Clearly there is no loss of generality in assuming that all $S_i < W$. Therefore each S_i satisfies $S_i \approx W(x_i)$ for some pairwise distinct elements $x_i \in W$. Let x_{j_0} be the first element of the non-empty set $\{x_j\}_{j\in J} \subseteq W$. It follows that for every $j \in J$ we have $x_{j_0} \leq x_j$, hence $W(x_{j_0}) \subseteq W(x_j)$, therefore $W(x_{j_0}) \leq W(x_j)$ by Proposition 2.5.12. □

Remark 2.5.6 Theorem 2.5.2 follows immediately from Theorem 2.5.3, but the above direct proof of Theorem 2.5.2 does not use the Zermelo theorem.

We finish this section with three further properties.

Proposition 2.5.13 *Let* W *be a woset and* $x, y \in W$. *Then:*

(2.5.25) $$x \leq y \Longleftrightarrow W(x) \leq W(y) \Longleftrightarrow W(x) \subseteq W(y)\ ,$$

(2.5.26) $$x < y \Longleftrightarrow W(x) < W(y) \Longleftrightarrow W(x) \subset W(y)\ .$$

PROOF: Clearly $x \leq y$ implies $W(x) \subseteq W(y)$, while this inclusion implies $W(x) \leq W(y)$ by Proposition 2.5.12. Furthermore, $W(x) \leq W(y)$ implies $x \leq y$, otherwise $y < x$ would imply $W(y) \subseteq W(x)$, therefore $W(y) \leq W(x)$ again by Proposition 2.5.12, hence $W(y) \approx W(x)$ by (2.5.18), which contradicts Proposition 2.5.7(ii).

It follows from (2.5.25) that $x \not\leq y \Longleftrightarrow W(x) \not\leq W(y)$, which by Corollary 2.5.4 can be written in the form $y < x \Longleftrightarrow W(y) < W(x)$.

If $x < y$ then $W(x) \subseteq W(y)$; but $x \in W(y) \setminus W(x)$, hence $W(x) \subset W(y)$.

If $W(x) \subset W(y)$ then $x \neq y$ while $x \leq y$ by (2.5.25), therefore $x < y$. □

Theorem 2.5.4 *Every well-ordered set* W *is isomorphic to the set of all proper segments of* W, *ordered by set inclusion, or equivalently, by the relation* $\leq$ *in Definition* 2.5.2.

PROOF: It follows by Proposition 2.5.13 that it is immaterial whether the set $Seg(W)$ of proper segments of W is endowed with $\leq$ or with $\subseteq$. Note that $Seg(\varnothing) = \varnothing$. If $W \neq \varnothing$, the mapping $f : W \longrightarrow Seg(W)$ defined by $f(x) = W(x)$ is surjective and strictly increasing by (2.5.26), therefore f is an isomorphism. □

Theorem 2.5.5 *Every well-ordered set has a successor, uniquely determined up to an isomorphism.*

Comment In other words, if W is a woset there is a woset W^+ such that

(2.5.27) $$W < W^+ ,$$

(2.5.28) $$W \leq W' \Longrightarrow W^+ \leq W', \ for\ every\ woset\ W' .$$

PROOF: Take an element $\infty \notin W$ and set $W^+ = W \cup \{\infty\}$. Extend the order $\leq$ of W by setting $x < \infty$ for every $x \in W$. Then W^+ is a woset and the identity mapping $1_W : W \longrightarrow W = W^+(\infty)$ is an isomorphism proving (2.5.27).

If W' is a woset satisfying $W \leq W'$, then either $W' = W < W^+$ or $W < W'$. In the latter case there is an isomorphism f of the form $f : W \longrightarrow W'(a')$, where $a' \in W'$. Then the mapping $\overline{f} : W^+ \longrightarrow W'$ defined by $\overline{f}(x) = f(x)$ for $x \in W$ and $\overline{f}(\infty) = a'$ is obviously strictly increasing. Then $\overline{f} : W^+ \longrightarrow W'(a']$ is a strictly increasing surjection, hence an isomorphism, therefore $W^+ \leq W'$. □

Chapter 3

Transfinite Algebra

The theory of ordinal numbers is a natural and very powerful generalization of the order-theoretical properties of natural numbers. In particular it furnishes transfinite induction, a method for constructing rather complicated mathematical concepts and for proving properties valid beyond the natural numbers. Ordinal numbers can also serve as a basis for introducing cardinal numbers. The latter evaluate "how many elements" a set possesses, being thus a kind of "quantitative" generalization of natural numbers, widely used in mathematics.

The first section, based on the preparation on well-ordered sets in Section 2.5, introduces ordinals and gives their most important properties, both order-theoretical and arithmetical. In §2 we indicate the relationship between ordinals and cardinals, while the last section is devoted to von Neumann's elementary construction of ordinals within the framework of set theory without classes.

The reader interested in a thorough study of ordinals is referred e.g. to Bachmann [1967] and Sierpiński [1965].

Keywords: Order type, ordinal, successor, Burali-Forti antinomy, ordinal of the first kind, limit ordinal, arithmetic of ordinals, transfinite induction, cardinal, aleph, von Neumann's ordinals.

3.1 Ordinals

In this section we first show that the passage from well-ordered sets to classes of isomorphic well-ordered sets, that is, ordinals, enables a mere translation of the properties of the former into order-theoretical properties of the latter. A few supplementary properties are also given. Then the essentials of the arithmetic of ordinals are established.

The basic idea is

Definition 3.1.1 The *order type* of a poset A is an object $\overline{A}$ in such a way that $\overline{A}$ is uniquely determined by A and

Sergiu Rudeanu

(3.1.0) $$\overline{A} = \overline{A'} \Longleftrightarrow A \approx A'$$

for every two posets A, A', where $\approx$ denotes isomorphism. In particular the order type $\overline{W}$ of a well-ordered set W is called the *ordinal* of W. The order type of the woset $\varnothing$ is called the *ordinal zero*; so $0 = \overline{\varnothing}$. If α is an order type and A a poset such that $\overline{A} = \alpha$, we say that A is a *representative* of α. Ordinals will be denoted by small Greek letters $\alpha, \beta, \gamma, \ldots, \omega$.

The above definition does not specify the nature of the "object" $\overline{A}$ associated with a poset A. As a matter of fact, there is a natural doubt as to the consistency of Definition 3.1.1 : do order types actually exist?

A possible answer is that we could take as $\overline{A}$ the class of all posets isomorphic to A. This is apparently a very simple solution which, however, raises some unexpected difficulties from the point of view of set-theoretical foundations; cf. Section 1.1. Thus e.g. not only there would be no class of all ordinals, but even such simple constructions as taking finite sets of ordinals $\{\alpha\}, \{\alpha,\beta\}, (\alpha,\beta)$ etc., would be forbidden because proper classes cannot occur as members of other classes. A variant of the above solution would be to take as order type of A a distinguished representative of the class of all posets isomorphic to A; however this would require a kind of much stronger axiom of choice, raising again foundation-theoretical difficulties.

We postpone the solution of this problem until §3, where we shall present a construction of ordinals – the only order types which interest us ! – within the framework of set theory without classes.

In the remainder of this section and in §2 we adopt an axiomatic approach, to the effect that we admit Definition 3.1.1 regardless of the nature of ordinals; we just make use of the fundamental property

(3.1.1) $$\overline{W} = \overline{W'} \Longleftrightarrow W \approx W' .$$

We first extend the ordering of wosets to ordinals.

Definition 3.1.2 Let α and β be ordinals. We set

(3.1.2) $$\alpha < \beta \Longleftrightarrow A < B ,$$

(3.1.3) $$\alpha \leq \beta \Longleftrightarrow A \leq B ,$$

where A and B are wosets such that $\alpha = \overline{A}, \beta = \overline{B}$ and the signs $<, \leq$ in the right sides of (3.1.2), (3.1.3) have the meanings in Definition 2.5.2.

We are now going to prove that relations (3.1.2) and (3.1.3) are well defined and satisfy

(3.1.4) $$\alpha \not< \alpha ,$$

(3.1.5) $$\alpha < \beta \,\&\, \beta < \gamma \Longrightarrow \alpha < \gamma ,$$

(3.1.6) $$\alpha < \beta \Longrightarrow \beta \not< \alpha ,$$

$$\text{either } \alpha < \beta \text{ or } \alpha = \beta \text{ or } \beta < \alpha \,, \tag{3.1.7}$$

$$\alpha \le \alpha \,, \tag{3.1.8}$$

$$\alpha \le \beta \,\&\, \beta \le \gamma \Longrightarrow \alpha \le \gamma \,, \tag{3.1.9}$$

$$\alpha \le \beta \,\&\, \beta \le \alpha \Longrightarrow \alpha = \beta \,, \tag{3.1.10}$$

$$\alpha \le \beta \text{ or } \beta \le \alpha \,, \tag{3.1.11}$$

$$\alpha \le \beta \Longleftrightarrow \alpha < \beta \text{ or } \alpha = \beta \,, \tag{3.1.12}$$

$$\alpha < \beta \Longleftrightarrow \alpha \le \beta \,\&\, \alpha \ne \beta \,, \tag{3.1.13}$$

for every three ordinals α, β, γ; moreover, every set of ordinals is well-ordered. Using the same abuse of language as in the previous section, we can summarize these properties in

Theorem 3.1.1 *The relations $\le$ and $<$ in Definition 3.1.1 are a total order and the corresponding strict total order. Besides, every set of ordinals is well-ordered under $\le$.*

PROOF: The first thing to be proved is that Definition 3.1.1 does not depend on the choice of the representatives of α and β. If $\alpha = \overline{A_1}$ and $\beta = \overline{B_1}$, then $\overline{A} = \overline{A_1}$ and $\overline{B} = \overline{B_1}$, hence $A_1 \approx A$ and $B_1 \approx B$, therefore $A \le B \Longrightarrow A_1 \le B_1$ and $A < B \Longrightarrow A_1 < B_1$ by Exercise 2.5.5. Now our theorem follows immediately from Propositions 2.5.9–2.5.11, Theorem 2.5.2, its Corollary 2.5.3 and Theorem 2.5.3. □

Remark 3.1.1 It follows by Remark 2.5.5 that 0 is the least ordinal, that is, $0 \le \alpha$ for every ordinal α.

Exercise 3.1.1 Prove that if $\alpha < \beta \le \gamma$ or $\alpha \le \beta < \gamma$ then $\alpha < \gamma$.

The axiomatic theory admits also the existence of the class of all ordinals.

Definition 3.1.3 Let **Ord** be the class of all ordinals. For each ordinal α, the *segment generated by* α is the set

$$\mathbf{Ord}(\alpha) = \{\beta \in \mathbf{Ord} \mid \beta < \alpha\} \,. \tag{3.1.14}$$

Recalling the informal discussion in Section 1.1, we can sketchy justify as follows the fact that **Ord** is a class and $\mathbf{Ord}(\alpha)$ is a set. **Ord** can be identified with a subclass of the class **Wos** of all wosets, because each ordinal corresponds to several wosets. The bijection established in the proof of Theorem 3.1.2 below shows in particular that $\mathbf{Ord}(\alpha)$ is a set.

Theorem 3.1.2 *Every ordinal α fulfills*

$$\alpha = \overline{\mathbf{Ord}(\alpha)} \,. \tag{3.1.15}$$

PROOF: It follows by Remark 3.1.1 and Definition 3.1.1 that $\overline{\mathbf{Ord}(0)} = \overline{\varnothing} = 0$. Now suppose $\alpha > 0$. If $\beta < \alpha = \overline{A}$ then β has a representative $A(x)$ where $x \in A$ is uniquely determined in view of Proposition 2.5.7(ii). Therefore we can map $\mathbf{Ord}(\alpha)$ onto the set $Seg(A)$ of proper segments of A by setting $f(x) = A(x)$. If $\gamma = \overline{A(y)}$ then

$$\beta < \gamma \Longleftrightarrow A(x) < A(y) \Longleftrightarrow x < y \tag{3.1.16}$$

by Definition 3.1.2 and Proposition 2.5.13, therefore f is an isomorphism by Proposition 2.3.4. Thus $\overline{\mathbf{Ord}(\alpha)} = \overline{Seg(A)}$, while $\overline{Seg(A)} = \overline{A} = \alpha$ by Theorem 2.5.4. □

The next two corollaries are very important in practice.

Corollary 3.1.1 *Each well-ordered set W can be isomorphically indexed by the set $\mathbf{Ord}(\alpha)$, where $\alpha = \overline{W}$; that is,*

$$W = \{x_\beta\}_{\beta<\alpha}\ , \tag{3.1.17}$$

where the mapping $\beta \mapsto x_\beta$ is an isomorphism.

PROOF: This is a paraphrase of $W \approx \mathbf{Ord}(\alpha)$. □

Corollary 3.1.1 is true even for the empty set, because there is no $\beta < \overline{\varnothing}$.

Corollary 3.1.2 *Every set can be indexed in the form* (3.1.17).

PROOF: From Corollary 3.1.1 via the Zermelo theorem. □

Warning The above corollaries do not claim that the isomorphism $\beta \mapsto x_\beta$ can actually be constructed. This only happens in certain particular cases; see e.g. Proposition 3.1.7 below.

Proposition 3.1.1 *Each ordinal α is the successor of* $\mathbf{Ord}(\alpha)$ (cf. (2.2.6″) and Definition 2.2.2″).

PROOF: Property $\mathbf{Ord}(\alpha) < \alpha$ is (3.1.14); in particular $\mathbf{Ord}(0) < 0$ because $\beta \in \mathbf{Ord}(0) = \varnothing \Longrightarrow \beta < \alpha$. If $\mathbf{Ord}(\alpha) < \gamma$ then $\alpha \leq \gamma$, otherwise $\gamma < \alpha$ would imply $\gamma \in \mathbf{Ord}(\alpha)$, hence $\gamma < \gamma$, a contradiction. □

More generally, we have:

Theorem 3.1.3 *Every set of ordinals has a successor.*

Comment Theorems 3.1.1 and 2.5.5 immediately imply that every set A of ordinals has a successor which is a woset, but the point in Theorem 3.1.3 is the existence of an ordinal which is the successor of A.

PROOF: The ordinal 0 is the successor of the empty set of ordinals by Remark 3.1.1. Now let A be a non-empty set of ordinals. Then A is a woset by Theorem 3.1.1. Setting $W_\beta = \mathbf{Ord}(\beta)$ for every $\beta \in A$, the direct sum $W = \sum_{\beta\in A} W_\beta = \bigcup_{\beta\in A} W^*_\beta$ is a woset by Example 2.5.6. It is clear that each $W^*_\beta = \{\beta\} \times W_\beta$ is a woset isomorphic to W_β; besides, $W^*_\beta \leq W$ by Proposition 2.5.12. Hence for all $\beta \in A$ we have $W_\beta \leq W$, and since $\overline{W_\beta} = \beta$ by Theorem 3.1.2, it follows that $\beta \leq \overline{W}$.

In Theorem 2.5.5 we have constructed the successor W^+ of W. It follows that $\beta < \overline{W^+}$ for all $\beta \in A$, therefore $\overline{A} = \{\gamma \in \mathbf{Ord} \mid A < \gamma \leq \overline{W^+}\}$ is a non-empty set of ordinals, hence it has a first element α. Consequently $A < \alpha$ and if an ordinal γ satisfies $A < \gamma$, then either $\gamma \in \overline{A}$, in which case $\alpha \leq \gamma$, or $\overline{W^+} < \gamma$, in which case from $A \leq \overline{W^+}$ we infer $\alpha < \gamma$. □

Corollary 3.1.3 *Every ordinal has a successor.*

Theorem 3.1.3 implies the so-called *Burali-Forti antinomy*, showing that there is no set of all ordinals: otherwise this set would have a successor ς, hence $\alpha < \varsigma$ for all ordinals; but ς being itself an ordinal, we would have $\varsigma < \varsigma$, a contradiction. In foundation-theoretical terms (cf. Section 1.1), this is simply a proof of the fact that **Ord** *is a proper class.*

Theorems 3.1.2 and 3.1.3 enable us to begin what might be called an actual construction of ordinals. The starting point is

Definition 3.1.4 An ordinal α is said to be *finite* (*infinite*) if its representatives are finite (infinite).

The definition is consistent because an infinite woset cannot be isomorphic to a finite one.

Let us choose a canonical representative for each finite ordinal. Recall that the set

$$\mathbb{N} = \{0, 1, 2, \ldots, n, \ldots\} \tag{3.1.18}$$

of natural numbers is well-ordered under its natural ordering (cf. Example 2.5.3 or Proposition 2.5.1). Moreover, let us prove that

$$m < n \Longleftrightarrow \{0, 1, \ldots, m-1\} < \{0, 1, \ldots, n-1\} \tag{3.1.19}$$

for every $m, n \in \mathbb{N}$, where the order relation in the right-hand side is given by Definition 2.5.2. Take an integer $p \geq m, n$ and set $W = \{0, 1, \ldots, p\}$. Then $m, n \in W$ and $\{0, 1, \ldots, m-1\} = W(m)$, $\{0, 1, \ldots, n-1\} = W(n)$, therefore (3.1.19) follows by Proposition 2.5.13. As suggested by (3.1.19), in this and the subsequent section *we identify each natural number n with the ordinal of the woset* $\{0, 1, \ldots, n-1\}$, that is,

$$\begin{gathered} 0 = \overline{\varnothing}\,, \\ n = \overline{\{0, 1, \ldots, n-1\}} \quad (n = 1, 2, 3, \ldots)\,. \end{gathered} \tag{3.1.20}$$

This identification is actually very convenient because, as shown by (3.1.19), *the natural ordering of* $\mathbb{N}$ *coincides with the order induced by Definition* 3.1.2. Moreover, it is immediately seen that *the successor of n is* $n+1$. Clearly *the set* $\mathbb{N}$ *exhausts all finite ordinals* by Proposition 2.3.6. Finally $n < \alpha$ *for every* $n \in \mathbb{N}$ *and every infinite ordinal* α, otherwise $\alpha \leq n$ for some n and some α, hence $A \leq \{0, 1, \ldots, n-1\}$ for some infinite woset A, thus implying the existence of an injection $f : A \longrightarrow \{0, 1, \ldots, n-1\}$ by Exercise 2.5.7; but this is impossible.

The successor of the set $\mathbb{N}$ is denoted by ω. It follows from the previous paragraph that *ω is the least infinite ordinal*, hence $\mathbb{N} = \mathbf{Ord}(\omega)$, therefore

$$\omega = \overline{\mathbb{N}} \tag{3.1.21}$$

by Theorem 3.1.2.

We postpone the construction of further ordinals to the end of this section. We need the already constructed ordinals $0, 1, \ldots, n, \ldots, \omega$ to illustrate

Definition 3.1.5 An ordinal α is said to be *of the first kind* if it is the successor of some other ordinal β, and *of the second kind* if it is $\neq 0$ and it is not of the first kind.

Example 3.1.1 All finite ordinals $\neq 0$ are of the first kind.

Example 3.1.2 The least ordinal of the second kind is ω. (Assume ω is the successor of some β; since the successor of n is $n+1$, it follows that $\beta \notin \mathbb{N}$, hence $\omega \leq \beta$, a contradiction).

Remark 3.1.2 Since 0 is the least ordinal, it follows that 0 is the unique ordinal which is neither of the first kind nor of the second kind.

Another description of the ordinals of second kind turns out to be very useful.

Definition 3.1.6 By a *limit ordinal* is meant the successor of a non-empty set Ω of ordinals without greatest element. This may be written in the form

$$\alpha = \lim \Omega \; ; \tag{3.1.22}$$

if Ω is of the form $\Omega = \mathbf{Ord}(\zeta)$, we write

$$\alpha = \lim_{\beta<\zeta} \beta \; , \tag{3.1.23}$$

while if $\Omega = \{f(\xi) \mid \xi < \zeta\}$, where $f : \mathbf{Ord} \longrightarrow \mathbf{Ord}$ and $\zeta \in \mathbf{Ord}$, we write

$$\alpha = \lim_{\xi<\zeta} f(\xi) \; . \tag{3.1.24}$$

Proposition 3.1.2 *If $\alpha > 0$ is an ordinal, the following conditions are equivalent:*

(i) *α is of the second kind;*

(ii) *α is a limit ordinal;*

(iii) *α fulfills a relation of the form* (3.1.24)*;*

(iv) *α fulfills $\alpha = \lim_{\beta<\alpha} \beta$;*

(v) *α cannot be the successor of a set Ξ of ordinals having greatest element.*

PROOF: (i)$\Longrightarrow$(v): If α is the successor of Ξ which has the greatest element ξ, it is immediately seen that α is the successor of ξ, which contradicts (i).

(v)$\Longrightarrow$(iv): Recall that α is the successor of $\mathbf{Ord}(\alpha)$ by Proposition 3.1.1. Now the hypothesis implies that $\mathbf{Ord}(\alpha)$ has no greatest element.

(iv)$\Longrightarrow$(iii)$\Longrightarrow$(ii): Trivial.

(ii)$\Longrightarrow$(i): Assume, by way of contradiction, that α is the successor of some ordinal β. But either $\Omega < \beta$ or $\beta \leq \gamma$ for some $\gamma \in \Omega$. Since $\beta < \alpha$ and α is the successor of Ω, the former case is contradictory. In the latter case $\beta \leq \gamma < \alpha$ and since α is the successor of β, it follows that $\beta = \gamma \in \Omega$. Since Ω has no greatest element, there is $\delta \in \Omega$ such that $\delta \not\leq \beta$, hence $\beta < \delta$, and since $\delta < \alpha$, this contradicts the fact that α is the successor of β. □

Now we are going to characterize the ordinals of the first and second kind in other terms.

Definition 3.1.7 Let $\{\xi_\beta\}_{\beta<\alpha}$ be a family of ordinals ξ_β. Then their *sum* is

$$\sum_{\beta<\alpha} \xi_\beta = \overline{\sum_{\beta<\alpha} W_\beta}\,, \tag{3.1.25}$$

where each $\xi_\beta = \overline{W_\beta}$ $(\beta < \alpha)$; cf. Corollary 3.1.1 and Example 2.5.6. In particular for $\alpha := n$ the sum (3.1.25) is written in the form $\xi_1 + \cdots + \xi_n$ and has the meaning

$$\xi_1 + \cdots + \xi_n = \overline{W_1 + \cdots + W_n};, \tag{3.1.25$'$}$$

where $\xi_i = \overline{W_i}$ $(i = 1, \ldots, n)$.

This definition is unambiguous by Exercise 2.1.10.

Proposition 3.1.3 (i) *The successor of an ordinal α is $\alpha + 1$.*

(ii) *The successor of a set A of ordinals with greatest element α is $\alpha + 1$.*

PROOF: (i) The successor W^+ of a woset W was constructed in the proof of Theorem 2.5.5 as $W^+ = W + \{\infty\}$, therefore $\overline{W^+} = \overline{W} + 1$.

(ii) Obvious from (i). □

Corollary 3.1.4 *The ordinals of the first kind are those of the form $\alpha + 1$.*

Corollary 3.1.5 $\xi < \eta \Longleftrightarrow \xi + 1 \leq \eta$ *and* $\xi < \eta + 1 \Longleftrightarrow \xi \leq \eta$.

PROOF: If $\xi < \eta$ then $\xi + 1 \leq \eta$ because $\xi + 1$ is the successor of ξ. If $\xi < \eta + 1$ then $\eta \not< \xi$, because $\eta < \xi$ would contradict the fact that $\eta + 1$ is the successor of η. Therefore $\xi \leq \eta$ by (3.1.7) and (3.1.12). The converse implications follow from $\alpha < \alpha + 1$ and Exercise 3.1.1 □

Proposition 3.1.4 *The relation* $\alpha = \lim \Omega$ *holds if and only if* $\Omega \subseteq \mathbf{Ord}(\alpha)$ *and every* $\beta \in \mathbf{Ord}(\alpha)$ *has a strict upper bound* $\xi \in \Omega$.

Comment This necessary and sufficient condition is expressed by saying that Ω is *cofinal* in $\mathbf{Ord}(\alpha)$.

PROOF: Let $\alpha = \lim \Omega$. Then $\Omega < \alpha$, that is,$\Omega \subseteq \mathbf{Ord}(\alpha)$. If some $\beta \in$

$\mathbf{Ord}(\alpha)$ has no strict upper bound in Ω, then $\xi \leq \beta$ for all $\xi \in \Omega$; but Ω has no greatest element, hence$\Omega < \beta < \alpha$, which contradicts the fact that α is the successor of Ω.

Conversely, Let Ω be cofinal in $\mathbf{Ord}(\alpha)$. Assume β is the greatest element of Ω; then$\Omega \leq \beta \in \Omega \subseteq \mathbf{Ord}(\alpha)$ and so the element $\beta \in \mathbf{Ord}(\alpha)$ has no strict upper bound in Ω, a contradiction. Furthermore,$\Omega \subseteq \mathbf{Ord}(\alpha)$ means$\Omega < \alpha$. Finally if$\Omega < \gamma$ then $\alpha \leq \gamma$, otherwise $\gamma \in \mathbf{Ord}(\alpha)$, hence $\gamma < \xi$ for some $\xi \in \Omega$, thus contradicting$\Omega < \gamma$. □

So far we have established properties related to the ordering of ordinals. We now relate these properties to the arithmetic of **Ord**, that is, properties of certain operations which can be defined for ordinals. The operation of *addition* has already been introduced in Definition 3.1.7 and a first property has been established in Proposition 3.1.3. Furthermore, we have:

Proposition 3.1.5 *The following properties hold for ordinals:*

(3.1.26) $$\alpha + \beta + \gamma = (\alpha + \beta) + \gamma = \alpha + (\beta + \gamma)\ ,$$

(3.1.27) $$\alpha + 0 = 0 + \alpha = \alpha\ ,$$

(3.1.28*) $$\alpha < \beta \Longleftrightarrow (\exists \delta\ \delta \neq 0\ \&\ \alpha + \delta = \beta\ ,$$

(3.1.28**) $$\alpha \leq \beta \Longleftrightarrow (\exists \delta)\ \alpha + \delta = \beta\ ,$$

(3.1.29*) $$\alpha < \beta \Longrightarrow (\forall \gamma)\ \gamma + \alpha < \gamma + \beta\ ,$$

(3.1.29**) $$\alpha \leq \beta \Longrightarrow (\forall \gamma)\ \gamma + \alpha \leq \gamma + \beta\ ,$$

(3.1.30*) $$\alpha < \beta \Longrightarrow (\forall \gamma)\ \alpha + \gamma \leq \beta + \gamma\ ,$$

(3.1.30**) $$\alpha \leq \beta \Longrightarrow (\forall \gamma)\ \alpha + \gamma \leq \beta + \gamma\ ,$$

(3.1.31*) $$\gamma + \alpha < \gamma + \beta \Longrightarrow \alpha < \beta\ ,$$

(3.1.31**) $$\gamma + \alpha \leq \gamma + \beta \Longrightarrow \alpha \leq \beta\ ,$$

(3.1.32*) $$\alpha + \gamma < \beta + \gamma \Longrightarrow \alpha < \beta\ ,$$

(3.1.33) $$\gamma + \alpha = \gamma + \beta \Longrightarrow \alpha = \beta\ ,$$

(3.1.34*) $$\alpha_1 \leq \beta_1\ \&\ \alpha_2 < \beta_2 \Longrightarrow \alpha_1 + \alpha_2 < \beta_1 + \beta_2\ ,$$

(3.1.34**) $$a_1 \leq \beta_1\ \&\ a_2 \leq \beta_2 \Longrightarrow \alpha_1 + a_2 \leq \beta_1 + \beta_2\ .$$

PROOF: Let A, B, C be wosets with $\overline{A} = \alpha$, $\overline{B} = \beta$, $\overline{C} = \gamma$.

To prove (3.1.26), note that $A + B + C$ consists of all pairs $(0, a)$ with $a \in A$ and all pairs $(1, b)$ with $b \in B$ and all pairs $(2, c)$ with $c \in C$, where each $(0, a) <$ each $(1, b) <$ each $(2, c)$. Further, $(A + B) + C$ consists of all pairs $(0, (0, a))$ with $a \in A$, then $(0, (1, b))$ with $b \in B$ and finally $(1, c)$ with

$c \in C$, where each $(0,(0,a))$ < each $(0,(1,b))$ < each $(1,c)$. The mapping $(0,a) \mapsto (0,(0,a))$, $(1,b) \mapsto (0,(1,b))$, $(2,c)) \mapsto (1,c)$ is obviously an isomorphism establishing $\alpha+\beta+\gamma = (\alpha+\beta)+\gamma$. The other equality (3.1.26) is proved similarly.

Property (3.1.27) is immediate.

If $\alpha + \delta = \beta$ with $\delta \neq 0$, take D such that $\overline{D} = \delta$. Then $D \neq \varnothing$, hence A is a proper segment of $A + D$, therefore $\alpha < \alpha + \delta = \beta$.

Now assume $\alpha < \beta$, that is, the existence of an isomorphism of the form $f : A \longrightarrow B(b)$. Set $D = B \setminus B(b)$; then $D \neq \varnothing$ and $B = B(b) + D$, hence setting $\delta = \overline{D}$ it follows that $\delta \neq 0$ and $\beta = \alpha + \delta$. This completes the proof of (3.1.28*). Further, define $g : C + A \longrightarrow C + B$ by $g(0,c) = (0,c)$ for every $c \in C$ and $g(1,a) = (1, f(a))$ for every $a \in A$; clearly g is an isomorphism between $C + A$ and $C + B(b) = (C+B)((1,b))$, therefore $\gamma + \alpha < \gamma + \beta$, thus proving (3.1.29*). Furthermore, $\alpha + \gamma = \overline{A} + \overline{C} = \overline{B(b)} + \overline{C} = \overline{B(b) + C}$. But $B(b) + C \subseteq B + C$, therefore $B(b) + C \leq B + C$ by Proposition 2.5.12, that is, $\alpha + \gamma \leq \beta + \gamma$, thus proving (3.1.30*).

Clearly (3.1.28**)–(3.1.30**) follow immediately from (3 1.28*)–(3.1.30*), respectively. Also, (3.1.31*), (3.1.31**) and (3.1.32*) follow immediately from (3.1.29**), (3.1.29*) and (3.1.30**), respectively. Now the non-trivial implication from (3.1.33) is a consequence of (3.1.31**).

If $\alpha_1 \leq \beta_1$ and $\alpha_2 < \beta_2$, then $\alpha_1 + \alpha_2 \leq \beta_1 + \alpha_2$ and $\beta_1 + \alpha_2 < \beta_1 + \beta_2$ by (3.1.30*) and (3.1.29*), respectively, therefore $\alpha_1 + \alpha_2 < \beta_1 + \beta_2$, thus proving (3.1.34*), while (3.1.34**) is a consequence of (3.1.30**) and (3.1.29**). □

It is important to note that for every two finite ordinals m and n, *the sum* $m+n$ *in the sense of Definition* 3.1.7 *coincides with the usual sum* $m+n$ *of natural numbers.* This follows from the obvious isomorphism between $\{0,\ldots,m-1\} + \{0,\ldots,n-1\}$ and $\{0,\ldots,m+n-1\}$.

The properties "missing" from Proposition 3.1.5 fail, as shown by the following examples.

Example 3.1.1 We have shown that $\omega + 1$ is the successor of ω, whereas $1 + \omega = \overline{\{-1,0,1,2,\ldots\}} = \omega$. Therefore the addition is not commutative.

Example 3.1.2 The relation $\alpha < \beta$ does not imply $\alpha + \gamma < \beta + \gamma$: take $\alpha := 0$, $\beta := 1$, $\gamma := \omega$. Taking further $\alpha := 1$, $\beta := 0$, $\gamma := \omega$, we see that none of the relations $\alpha + \gamma \leq \beta + \gamma$ and $\alpha + \gamma = \beta + \gamma$ entitles us to infer which of the relations $\alpha < b$, $\alpha = \beta$ or $\alpha > \beta$ holds.

Example 3.1.3 Taking $\alpha_1 := 0$, $\beta_1 := 1$ and $\alpha_2 := \beta_2 := \omega$, we see that $\alpha_1 < \beta_1$ and $a_2 \leq \beta_2$ do not imply $\alpha_1 + \alpha_2 < \beta_1 + \beta_2$.

Exercise 3.1.2 Prove that $\alpha < \beta$ and $\gamma + \alpha = \delta + \beta$ imply $\delta < \gamma$.

Exercise 3.1.3 Prove that $n + \alpha = \alpha$ for every finite ordinal n and every infinite ordinal α.

Exercise 3.1.4 Define and study a partial operation of *subtraction* of ordinals. (Hint. Use (3.1.28)).

Exercise 3.1.5 Prove that $\xi_\beta \le \eta_\beta$ $(\forall\, \beta < \alpha)$ implies

$$\sum_{\beta<\alpha} \xi_\beta \le \sum_{\beta<\alpha} \eta_\beta \,.$$

Exercise 3.1.6 Prove the following property of associativity. Let $\{\gamma_\beta\}_{\beta<\alpha}$ be a family of ordinals, $\gamma = \sum_{\beta<\alpha} \gamma_\beta$ and for each $\beta < \alpha$ let $\{\xi_{\beta\delta}\}_{\delta<\gamma_\beta}$ be a family of ordinals. Then

(3.1.26′) $$\sum_{\beta<\alpha} \sum_{\delta<\gamma_\beta} \xi_{\beta\delta} = \sum_{\delta<\gamma} \xi_{f(\delta)} \,,$$

where f is the isomorphism between $\mathbf{Ord}(\gamma)$ and $\sum_{\beta<\alpha} \mathbf{Ord}(\gamma_\beta)$.

Now we can relate Proposition 3.1.2 to arithmetic..

Proposition 3.1.6 *Let α and $\beta \neq 0$ be ordinals. Then:*
(i) *$\alpha + \beta$ is the successor of the set $\{\alpha + \gamma\}_{\gamma<\beta}$.*
(ii) *If β is a limit ordinal, so is $\alpha + \beta$.*

PROOF: (i) For every $\gamma < \beta$, (3.1.29*) implies $\alpha + \gamma < \alpha + \beta$. Now let ξ fulfill $\alpha + \gamma < \xi$ for every $\gamma < \beta$. This implies $\alpha < \xi$, hence $\xi = \alpha + \delta$ for some $\delta \neq 0$, by (3.1.28*). If $\beta \le \delta$, then (3.1.29**) implies $\alpha + \beta \le \alpha + \delta = \xi$, as desired. Now suppose $\delta < b$. Assuming $\xi < \alpha + \beta$, it follows that $\alpha + \gamma < \alpha + \delta < \alpha + \beta$, hence (3.1.31*) implies $\gamma < \delta < \beta$ for every $\gamma < \beta$; in particular $\delta < \delta$, a contradiction. Therefore $\alpha + \beta \le \xi$.

(ii) If β is a limit ordinal, then $\beta > 1$. Assuming $\alpha + \beta = \varepsilon + 1$, it follows that $\alpha < \varepsilon$ by Exercise 3.1.2, hence $\varepsilon = \alpha + \eta$ for some $\eta \neq 0$ by (3.1.28*); so $\alpha + \beta = \alpha + \eta + 1$, finally $\beta = \eta + 1$ by (3.1.33), a contradiction. □

The next result points out a case in which the indexing stated in the Corollaries of Theorem 3.1.2 can actually be constructed.

Proposition 3.1.7 *Let α and ζ be ordinals, $\alpha < \zeta$. Then the set $[\alpha,\zeta\,)$ of all ordinals ξ such that $\alpha \le \xi < \zeta$ can be isomorphically indexed in the form*

(3.1.35) $$[\alpha,\zeta\,) = \{\alpha + \gamma\}_{\gamma<\delta} \,,$$

where $\delta \neq 0$ is the unique ordinal such that $\zeta = \alpha + \delta$.

PROOF: Note first that $[\alpha,\zeta\,)$ is actually a set, because it is a subclass of the set $\mathbf{Ord}(\zeta)$. Furthermore, properties (3.1.28*) and (3.1.33) imply the existence and uniqueness of δ and also of γ satisfying $\xi = \alpha + \gamma$, for each ξ with $\alpha \le \xi < \zeta$, where $\gamma \neq 0$ for $\xi \neq \alpha$; moreover, $\alpha + \gamma < \alpha + \delta$ implies $\gamma < \delta$ by (1.31*). Therefore the mapping $\xi \mapsto \gamma$ is well defined. Moreover, this mapping is an isomorphism by Proposition 2.1.5. Indeed,

$$\xi = \alpha + \gamma \le \alpha + \gamma' = \xi' \iff \gamma \le \gamma'$$

by (1.31**) and (1.30**), and the mapping is surjective because $\gamma < \delta$ implies $\alpha \le \alpha + \gamma < \alpha + \delta = \zeta$ by (1.29*). □

We now introduce the *product of ordinals.*

Definition 3.1.8 If α and β are ordinals and $\alpha = \overline{A}, \beta = \overline{B}$, we set

(3.1.36) $$\alpha \cdot \beta = \alpha\beta = \overline{A \times B} \ ,$$

where $A \times B$ denotes the lexicographic product of the wosets A and B : cf. Example 2.5.7.

This definition is unambiguous by Exercise 2.1.10.

Proposition 3.1.8 *The following properties hold for ordinals :*

(3.1.37) $$(\alpha\beta)\gamma = \alpha(\beta\gamma) \ ,$$

(3.1.38) $$\alpha \cdot 0 = 0 \cdot \alpha = 0 \ ,$$

(3.1.39) $$\alpha \cdot 1 = 1 \cdot \alpha = \alpha \ ,$$

(3.1.40) $$(\textstyle\sum_{\beta<\alpha} \eta_\beta)\xi = \sum_{\beta<\alpha} \eta_\beta \xi \ ,$$

(3.1.41) $$\alpha\beta = \textstyle\sum_{\gamma<\alpha} \beta \ ,$$

(3.1.42*) $$\alpha < \beta \,\&\, \gamma \neq 0 \Longrightarrow \alpha\gamma < \beta\gamma,$$

(3.1.42**) $$\alpha \leq \beta \Longrightarrow (\forall\gamma)\ \alpha\gamma \leq \beta\gamma,$$

(3.1.43**) $$\alpha \leq \beta \Longrightarrow (\forall\gamma)\ \gamma\alpha \leq \gamma\beta,$$

(3.1.44*) $$\alpha\gamma < \beta\gamma \Longrightarrow \alpha < \beta \ ,$$

(3.1.45*) $$\gamma\alpha < \gamma\beta \Longrightarrow \alpha < \beta \ ,$$

(3.1.46) $$\alpha\gamma = \beta\gamma \,\&\, \gamma \neq 0 \Longrightarrow \alpha = \beta \ .$$

PROOF: Let A, B, C be wosets with $\overline{A} = \alpha$, $\overline{B} = \beta$, $\overline{C} = \gamma$.

Properties (3.1.37)-(3.1.39) reduce to the isomorphisms $(A \times B) \times C \approx A \times (B \times C)$, $A \times \{0\} \approx \{0\} \times A \approx A$ and to the equality $A \times \varnothing = \varnothing \times A = \varnothing$.

To prove (3.1.40), let W and W_β be wosets fulfilling $\overline{W} = \xi$ and $\overline{W_\beta} = \eta_\beta$; we have to show that $(\sum_{\beta<\alpha} W_\beta) \times W$ is isomorphic to $\sum_{\beta<\alpha}(W_\beta \times W)$. The mapping which sends each element $((\beta, y), x)$, where $y \in W_\beta, x \in W$, of the former set, to the element $(\beta, (y, x))$ of the latter set, is obviously a bijection. In view of Proposition 2.1.5, it remains to check condition (2.1.25). Indeed, it is easy to see that both conditions $((\beta, y), x) < ((\beta', y'), x')$ and $(\beta, (y, x)) < (\beta', (y', x'))$ are equivalent to

$$\beta < \beta' \text{ or } \beta = \beta' \,\&\, y < y' \text{ or } \beta = \beta' \,\&\, y = y' \,\&\, x < x \ .$$

Formula (3.1.41) is an abbreviation for $\alpha\beta = \sum_{\gamma<\alpha} \xi_\gamma$, where all $\xi_\gamma = \beta$; this reduces to the obvious equality

(3.1.47) $$\mathbf{Ord}(\alpha) \times B = \textstyle\sum_{\gamma<\alpha} B \ .$$

To prove (3.1.42*), assume $C \neq \varnothing$ and $A \approx B(b)$ for some $b \in B$. Then clearly

$$A \times C \approx B(b) \times C = \{(x,c) \mid x \in B \,\&\, x < b \,\&\, c \in C\} = (B \times C)((b,0)) .$$

Property (3.1.42**) follows from (3.1.42*) and (3.1.38).

To prove (3.1.43**), assume $\alpha \leq \beta$ and $\gamma \neq 0$ (as the case $\gamma = 0$ is trivial). Then there is a strictly increasing mapping $f : A \longrightarrow B$ by Exercise 2.5.7. It follows immediately that if $C \neq \varnothing$, the mapping $g : C \times A \longrightarrow C \times B$ defined by $g((c,a)) = (c, f(a))$ is strictly increasing, therefore $C \times A \leq C \times B$, again by Exercise 2.5.7.

Property (3.1.44*) holds because $\beta \leq \alpha \Longrightarrow \beta\gamma \leq \alpha\gamma$ by (3.1.42**). Similarly, (3.1.45*) follows from (3.1.43**), while (3.1.46) is a consequence of (3.1.42*). □

It is important to note that for every two finite ordinals m and n, *the product* mn *in the sense of Definition* 3.1.8 *coincides with the usual product* mn *of natural numbers.* This follows from the obvious isomorphism between $\{0,\dots,mn-1\}$ and $\{0,\dots,m-1\} \times \{0,\dots,n-1\}$, where $(0,0) < \cdots < (0,n-1) < (1,0) < \cdots < (1,n-1) < \cdots < (m-1,0) < \cdots < (m-1,n-1)$.

The properties "missing" from Proposition 3.1.8 fail, as shown by the following examples.

Example 3.1.4 The set $\mathbb{N} \times \{0,1\}$ consists of all pairs $(n,0)$ and $(n,1)$ with $n \in \mathbb{N}$, where the order is

$$(0,0) < (0,1) < (1,0) < (1,1) < \cdots < (n,0) < (n,1) < \dots ,$$

which implies easily that $\mathbb{N} \times \{0,1\} \approx \mathbb{N}$. Therefore

$$\omega \cdot 2 = \omega , \tag{3.1.48}$$

whereas (3.1.40) implies

$$2\omega = (1+1)\omega = 1 \cdot \omega + 1 \cdot \omega = \omega + \omega . \tag{3.1.49}$$

As $\omega < \omega + \omega$ by (1.28*), the above equalities show that the multiplication of ordinals is not commutative.

Example 3.1.5 Since $\omega(1+1) = w \cdot 2 = \omega$, whereas $\omega \cdot 1 + \omega \cdot 1 = \omega + \omega = 2\omega$, it follows that multiplication is not left distributive with respect to addition

Example 3.1.6 Since $1 < 2$ but $\omega \cdot 1 = \omega = \omega \cdot 2$, the relation $\alpha < \beta$ does not imply $\gamma\alpha < \gamma\beta$ even if $\gamma \neq .0$.

Exercise 3.1.7 Prove that $\alpha\gamma = \beta\delta \neq 0$ and $\alpha < \beta$ imply $\delta < \gamma$.

The well-known division algorithm for natural numbers extend to ordinal numbers.

Proposition 3.1.9 *Let* α *and* $\beta \neq 0$ *be ordinals. Then there exist two uniquely determined ordinals* χ *and* ρ *such that*

(3.1.50) $$\alpha = \chi\beta + \rho \text{ and } \rho < \beta\ .$$

PROOF: Set $\alpha + 1 = \gamma$ and remark first that

(3.1.51) $$\gamma\beta = (\alpha+1)\beta = \alpha\beta + \beta > \alpha\beta \geq \alpha$$

by (3.1.40), (3.1.28*) and (3.1.43**), respectively. Let $\alpha = \overline{A}, \beta = \overline{B}, \gamma = \overline{C}$. It follows from (3.1.51) that $A \approx (C \times B)((c^*, b^*))$ for some $c^* \in C$, $b^* \in B$. The sets $C(c^*) \times B$ and $\{c^*\} \times B(b^*)$ are disjoint are their union is $(C \times B)((c^*, b^*))$ because $c < c^*$ and $c = c^* \,\&\, b < b^*$ are the two disjoint possibilities for $(c, b) < (c^*, b^*)$. Since, moreover, $c < c^*$ implies $(c, b) < (c^*, b')$ for every $b, b' \in B$, it follows that

(3.1.52) $$(C \times B)((c^*, b^*)) \approx C(c^*) \times B + \{c^*\} \times B(b^*)\ ,$$

therefore $\alpha = \chi\beta + \rho$ with $\chi = \overline{C(c^*)}$ and $\rho = \overline{\{c^*\} \times B(b^*)} = \overline{B(b^*)} < \beta$.

To prove uniqueness, let $\alpha = \chi'\beta + \rho'$ with $\rho' < \beta$; then $\chi\beta + \rho = \chi'\beta + \rho'$. Assume $\chi \neq \chi'$, say $\chi < \chi'$. Then $\chi + 1 \leq \chi'$, hence

$$\chi\beta + \beta + \rho' = (\chi+1)\beta + \rho' \leq \chi'\beta + \rho' = \chi\beta + \rho$$

by (3.1.40), (3.1.42**) and (3.1.30**), therefore $\beta + \rho' \leq \rho$ by (3.1.31**), thus implying $\beta \leq \rho$ by (3.1.28**), a contradiction. Therefore $\chi = \chi'$, hence $\chi\beta + \rho = \chi\beta + \rho'$, finally $\rho = \rho'$ by (3.1.33). □

Proposition 3.1.10 *Let $\beta \neq 0$ be an ordinal and α a limit ordinal. Then*

(3.1.53) $$\alpha\beta = \lim_{\gamma<\alpha} \gamma\beta.$$

PROOF: We prove first that the successor ξ of the set $\{\gamma\beta\}_{\gamma<\alpha}$, which exists by Corollary 3.1.3, coincides with $\alpha\beta$. But $\gamma < \alpha$ implies $\gamma\beta \leq \alpha\beta$ by (3.1.43**), while $\gamma\beta = \alpha\beta$ would imply $\gamma = \alpha$ by (3.1.46), a contradiction. So $\gamma\beta < \alpha\beta$ for every $\gamma < \alpha$, therefore $\xi \leq \alpha\beta$ and it remains to prove the converse inequality. Now $\xi = \chi\beta + \rho$ with $\rho < \beta$, by Proposition 3.1.9. As $\chi\beta \leq \xi$ by (3.1.28**), the inequality $\alpha\beta \leq \xi$ will be established if we only prove that $\alpha\beta \leq \chi\beta$. But the definition of ξ implies that

$$\gamma\beta < \xi = \chi\beta + \rho < \chi\beta + \beta = (\chi+1)\beta \quad (\forall \gamma < \alpha)$$

by (3.1.29*) and (3.1.40), therefore $\gamma < \chi + 1$ by (3.1.44**). Since the latter inequality holds for all $\gamma < \alpha$, it follows that $\alpha \leq \chi + 1$ by Proposition 3.1.1, therefore $\alpha < \chi + 1$ because α is of the second kind by Proposition 3.1.2. Now $\alpha \leq \chi$ by Corollary 3.1.5, hence $\alpha\beta \leq \chi\beta$ by (3.1.42**).

So $\alpha\beta$ is the successor of $\{\gamma\beta\}_{\gamma<\alpha}$ and in view of Definition 3.1.6 it suffices to show that this set has no greatest element. Otherwise $\gamma\beta \leq \gamma_0\beta$ for all $\gamma < \alpha$ and some $\gamma_0 < \alpha$. If $\gamma\beta < \gamma_0\beta$ then $\gamma < \gamma_0$ by (3.1.44*), while if $\gamma\beta = \gamma_0\beta$ then $\gamma = \gamma_0$ by (3.1.46). Therefore $\gamma \leq \gamma_0$, which would mean that γ_0 is the greatest element of $\mathbf{Ord}(\alpha)$, in contradiction with Proposition 3.1.2. □

The next theorem is perhaps the most important from the point of view of applications.

Let $\pi(x)$ be a property in the sense discussed in Section 1 1 after ZF2. Then

Theorem 3.1.4 I. *Let* $X = \{x_\xi\}_{\xi<\alpha}$ *be an arbitrary set indexed by a segment of ordinals* (cf. Corollary 3.1.2). *Let* $\pi(x)$ *be a property defined for all* $x_\xi \in X$. *Assume that*

I.1 $\pi(x_0)$ *holds ;*

I.2 $\beta + 1 < \alpha$ *and* $\pi(x_\beta)$ *imply* $\pi(x_{\beta+1})$ *;*

I.3. *if* $\gamma < \alpha$ *is a limit ordinal and* $\pi(x_\beta)$ *holds for all* $\beta < \gamma$, *then* $\pi(x_\gamma)$ *also holds.*

Then $\pi(x_\xi)$ *holds for all* $x_\xi \in X$.

II. *Let* α *be an ordinal and* $\pi(\xi)$ *a property defined for all ordinals* $\xi \geq \alpha$. *Assume that*

II.1. $\pi(\alpha)$ *holds ;*

II.2 $\alpha \leq \beta$ *and* $\pi(\beta)$ *imply* $\pi(\beta + 1)$ *;*

II.3 *if* $\gamma > \alpha$ *is a limit ordinal and* $\pi(\beta)$ *holds for all* β *with* $\alpha \leq \beta < \gamma$, *then* $\pi(\gamma)$ *also holds.*

Then $\pi(\xi)$ *holds for all ordinals* $\xi \geq \alpha$.

Comments 1) Each of the above properties I and II may be called the *principle of transfinite induction.*

2) In particular I applies to sets of ordinals.

PROOF: I. Let $S = \{\xi \in \mathbf{Ord}(\alpha) \mid \pi(x_\xi) \text{ holds}\}$. In view of Theorem 2.5.1, it suffices to take an ordinal $\xi \in \mathbf{Ord}(\alpha)$ such that $\mathbf{Ord}(\xi) \subseteq S$ and prove that $\xi \in S$. If $\xi = 0$ this is true by I.1. If $\xi = \beta + 1 < \alpha$, then $\beta \in \mathbf{Ord}(\xi)$, hence $\beta \in S$, therefore $\xi \in S$ by I.2. Finally if $\xi < \alpha$ is a limit ordinal, then as all $\beta < \xi$ belong to S, it follows that $\xi \in S$ by I.3. Thus $S = \mathbf{Ord}(\alpha)$, as desired.

II. Since $\pi(\alpha)$ holds by II.1, it suffices to take an ordinal $\zeta > \alpha$ and prove that $\pi(\xi)$ holds for all $\xi \in [\alpha,\zeta\,)$ (see Proposition 3.1.7), because $\pi(\zeta)$ is then verified by II.2 if $\zeta = \beta + 1$ (since $\alpha \leq \beta$ by Corollary 3.1.5) and by II.3 if ζ is a limit ordinal.

Indeed, Proposition 3.1.7 shows that $[\alpha,\zeta\,)$ can be indexed in the form $x_\eta = \alpha + \eta$, where $\eta < \delta$ with $\alpha + \delta = \zeta$. It remains to prove that property I is fulfilled for the set $X = \{x_\eta\}_{\eta<\delta}$. Indeed, $\pi(x_0) = \pi(\alpha)$ holds. Property II.2 becomes $\pi(x_\beta) \Longrightarrow \pi(x_{\beta+1})$, which is stronger than I.2. Finally, a property stronger than I.3 holds: if ξ is a limit ordinal and $\pi(x_\eta)$ holds for all $\eta < \xi$, then $\pi(x_\xi)$ also holds. Indeed, $\alpha + \xi$ is a limit ordinal by Proposition 3.1.6, then $\pi(\alpha + \eta)$ holds for all $\alpha + \eta < \alpha + \xi$, therefore $\pi(x_\xi) = \pi(\alpha + \xi)$ also holds by II.3. □

Corollary 3.1.6 *The principle of complete induction holds for* $\mathbb{N}$.

PROOF: Apply Theorem 3.1.4.I to $\mathbb{N} = \{n\}_{n<\omega}$: here $\alpha := \omega$ and condition I.3 is vacuously fulfilled. □

Theorem 3.1.5 I. *Let* $X = \{x_\xi\}_{\xi<\alpha}$ *and* Y *be two sets. A function* $f : X \longrightarrow Y$ *is defined whenever one has the following three things:*

(i) $f(x_0) \in Y$ *;*

(ii) *a rule which uniquely defines* $f(x_{\beta+1}) \in Y$ *whenever* $\beta + 1 < \alpha$ *and* $f(x_\beta)$ *is defined ;*

(iii) *a rule which uniquely defines* $f(x_\gamma) \in Y$ *whenever* $\gamma < \alpha$ *is a limit ordinal and* $f(x_\beta) \in Y$ *is defined for all* $\beta < \gamma$.

II. *Let* α *be an ordinal. With each ordinal* $\xi \geq \alpha$ *is associated a uniquely determined element* $f(\xi)$ *whenever one has the following three things:*

(i) $f(\alpha)$ *;*

(ii) *a rule which uniquely defines* $f(\beta + 1)$ *whenever* $\alpha \leq \beta$ *and* $f(\beta)$ *is defined ;*

(iii) *a rule which uniquely defines* $f(\gamma)$ *whenever* $\gamma > \alpha$ *is a limit ordinal and* $f(\beta)$ *is defined for all* β *with* $\alpha \leq \beta < \gamma$.

Comments 1) In II the nature of the elements $f(x)$ is not defined.

2) Each of the above two constructions is called a *definition by transfinite induction.*

PROOF: To prove II (to prove I) apply Theorem 3.1.4 to the property "$\pi(x) :=$ $f(x)$ exists (exists in Y) and is uniquely determined". □

As an application, let us try to define a "good" product $P = \prod_{\xi<\zeta} \alpha_\xi$ of a family $\{\alpha_\xi\}_{\xi<\zeta}$ of ordinals in such a way as to extend Definition 3.1.8. Clearly we should set $P = 0$ if at least one of the α_ξs is 0. Otherwise we can use a definition by transfinite induction:

(i) if $\zeta = 0$ then $P = 1$:

(ii) $\prod_{\xi<\zeta+1} \alpha_\xi = (\prod_{\xi<\zeta} \alpha_\xi)\alpha_{\zeta+1}$;

(iii) if ζ is a limit ordinal, then P is the successor of the products $\prod_{\xi<\beta} \alpha_\xi$ for $\beta < \zeta$.

In mathematical practice it is often very convenient to replace transfinite induction by an application of the Zorn lemma; cf. Theorem 2.4.1.

Finite products of ordinals can be simply defined by mathematical induction. It is important to notice that *for finite ordinals m,n, the power* m^n *coincides with the usual power of natural numbers.*

We close this section with a sketch of the "beginning" of the class **Ord** of ordinals; however we do not forget that this beginning has no end !

We have already seen that the first ordinals are (identified with) the natural numbers $0, 1, 2, \ldots, n, \ldots$, and that their successor is the first infinite ordinal ω; cf. (3.1.20). The following ordinals are $\omega+1, \omega+2, \ldots, \omega+n, \ldots$, by Proposition 3.1.3. Clearly the successor of all these ordinals coincides with the successor of the set $\{\omega + n\}_{n<\omega}$, which is $\omega + \omega$ by Proposition 3.1.6, that is 2ω by (3.1.49). Now the process continues: the following ordinals are $2\omega + 1, \ldots, 2\omega + n, \ldots$; their successor is $2\omega + \omega = 2\omega + 1 \cdot \omega = (2 + 1)\omega = 3\omega$, etc. We thus obtain a set of ordinals of the form $m\omega + n$, where $m, n <\omega$. As $m\omega + n < (m + 1)\omega$ for every $m, n <\omega$, it follows that the successor of the ordinals $m\omega + n$ is the same as the successor of the ordinals $m\omega$; but the latter is $\omega \cdot \omega = \omega^2$ by Proposition

3.1.10. We now obtain the ordinals $\omega^2 + m\omega + n$, where $m, n < \omega$. Using again Proposition 3.1.6, we find that the successor of the latter ordinals – and in fact of all the ordinals constructed so far – is $\omega^2 + \omega^2 = 2\omega^2$. The iteration of the above construction yields the set of all "polynomial" ordinals

$$m_0\omega^n + m_1\omega^{n-1} + \cdots + m_{n-2}\omega^2 + m_{n-1}\omega + m_n \ , \tag{3.1.54}$$

where $m_0, m_1, \ldots, m_n$ and n run over $\mathbb{N}$; the order is lexicographic with respect to the powers of ω.

Exercise 3.1.8 Show that the successor of the ordinals (3.1.54) is ω^ω.

Remark 3.1.3 Every non-empty class $\mathbf{K}$ of ordinals has first element: if $\gamma \in \mathbf{K}$, the first element of the set $\{\xi \in \mathbf{K} \mid \xi \leq \gamma\} \subseteq \mathbf{Ord}(\gamma + 1)$ is easily seen to be the first element of $\mathbf{K}$.

3.2 Cardinals

In this section we indicate the construction of cardinals as special ordinals, a procedure which may be regarded as a concrete construction of cardinals within the axiomatic theory of ordinals, as developed in §1. Since our eBook is oriented towards order-theoretical aspects, we confine this section to the connection between ordinals and cardinals. Thus many important problems related to cardinals will be disregarded, e.g., the arithmetic of ordinals, the continuum hypothesis, etc.

The concept of a cardinal number is the analogue, for arbitrary sets, of the concept of ordinal number for well-ordered sets.

Definition 3.2.1 The *cardinal number* or the *cardinality* of a set A is an object $|A|$ associated with A in such a way that (i) $|A|$ is uniquely determined by A, and (ii) $|A| = |B|$ if and only if there is a bijection between A and B. The cardinal $|A|$ is said to be *finite* (*infinite*) if A is finite (infinite).

The above definition does not specify the nature of the object $|A|$ associated with the set A. As a matter of fact, there is a natural doubt concerning the consistency of Definition 3.2.1: do cardinal numbers actually exist? Two possible answers would be: 1) take as $|A|$ the class of all sets B in bijection with A, or 2) take as $|A|$ an element of this class. The discussion of these two solutions is quite analogous to the discussion of the corresponding solutions for ordinals in the beginning of §1. A third solution constitutes the object of this section.

Definition 3.2.2 The *concrete cardinal* $\overline{\overline{A}}$ of a set A is the least ordinal among the ordinals of the wosets with underlying sets in bijection with A.

Theorem 3.2.1 *The concrete cardinals are cardinals in the sense of Definition* 3.2.1.

PROOF: Wosets with underlying sets in bijection with A do exist by the Zermelo theorem. The least ordinal $\overline{\overline{A}}$ of these wosets exists by Remark 3.1.3 and is uniquely determined by A. To check (ii), consider two sets A, B. If $\overline{\overline{A}} = \overline{\overline{B}}$, then there is an isomorphism f between two wosets with underlying sets A and B; but f is a fortiori a bijection between A and B. Conversely, if there is a bijection $f : A \longrightarrow B$, then $\overline{\overline{B}}$ is the ordinal of a woset with underlying set B in bijection with A, hence $\overline{\overline{A}} \leq \overline{\overline{B}}$ and similarly $\overline{\overline{B}} \leq \overline{\overline{A}}$, therefore $\overline{\overline{A}} = \overline{\overline{B}}$. □

Remark 3.2.1 The notation $\overline{\overline{A}}$ is defined for every set A, while $\overline{A}$ is meaningful only for a woset $(A, \leq)$, abusively denoted by its underlying set A.

Remark 3.2.2 $\overline{\overline{A}} \leq \overline{A}$ for every woset A.

At this point we can do the junction with the conventional theory of cardinal numbers. The idea is that the cardinality should extend the familiar concept of number of elements of a finite set; this has led to Definition 3.2.1. A companion idea is the introduction of an order relation between cardinals in such a way that $|A| \leq |B|$ should express the fact that the set A has at most as many elements as the set B. This has led to the definition: $|A| \leq |B|$ iff there is an injection $f : A \longrightarrow B$. The next proposition shows that concrete cardinals actually satisfy this requirement.

Proposition 3.2.1 *For every two sets A,B, $\overline{\overline{A}} \leq \overline{\overline{B}}$ if and only if there is an injection from A to B.*

PROOF: Let $(A, \leq)$ and $(B, \leq)$ be wosets such such that $\overline{(A, \leq)} = \overline{\overline{A}}$ and $\overline{(B, \leq)} = \overline{\overline{B}}$. If $\overline{\overline{A}} \leq \overline{\overline{B}}$, then either $\overline{\overline{A}} = \overline{\overline{B}}$, in which case $(A, \leq) \approx (B, \leq)$, hence there is a bijection $f : A \longrightarrow B$, or $\overline{(A, \leq)} < \overline{(B, \leq)}$, in which case there is an isomorphism $f : (A, \leq) \longrightarrow (B(b), \leq)$, where $b \in B$. In this case $f : A \longrightarrow B(b)$ is a bijection, hence $f : A \longrightarrow B$ is injective.

Conversely, suppose $f : A \longrightarrow B$ is an injection. Then we obtain a bijection $f : A \longrightarrow f(A)$ where $f(A) \subseteq B$. Let $\leq$ be a relation on B which makes B into a woset of type $\overline{\overline{B}}$. Proposition 2.5.12 shows that taking the restriction of $\leq$ to $f(A)$ one obtains a woset $(f(A), \leq)$ whose order type α satisfies $\alpha \leq \overline{\overline{B}}$. Setting $x \leq_A y \Longleftrightarrow f(x) \leq f(y)$ for any $x, y \in A$ we obtain a woset $(A, \leq_A)$ isomorphic to $(f(A), \leq)$, hence $\overline{(A, \leq_A)} = \alpha$. It follows that $\overline{\overline{A}} \leq \alpha \leq \overline{\overline{B}}$. □

Theorem 3.2.1 and Proposition 3.2.1 show that the following relationships hold between the "abstract" cardinals $|A|$ in Definition 3.2.1 and the concrete cardinals $\overline{\overline{A}}$ in Definition 3.2.2:

$$|A| = |B| \Longleftrightarrow \overline{\overline{A}} = \overline{\overline{B}} \;, \tag{3.2.1}$$

$$|A| \leq |B| \Longleftrightarrow \overline{\overline{A}} \leq \overline{\overline{B}} \;, \tag{3.2.2}$$

therefore, as concerns properties relative to order, there is no loss of generality in working with concrete cardinals in the rôle of cardinals $|A|$. We will do so in the following.

Example 3.2.1 Every finite ordinal n is a finite cardinal, and conversely.

Example 3.2.2 The cardinal of a countable set is denoted by $\aleph_0$ (aleph zero); it is the first infinite cardinal. The concrete cardinal of a countable set is the ordinal ω.

Example 3.2.3 The antisymmetry of the order relation of cardinals, that is, $|A| \leq |B| \,\&\, |B| \leq |A| \Longrightarrow |A| = |B|$, is nothing but the Cantor-Bernstein theorem.

As a matter of fact, we have obtained a new proof of the Cantor-Bernstein theorem.

Example 3.2.4 Another well-known set-theoretical property can be paraphrased in the form $|A| < |2^A|$.

Theorem 3.2.2 *The relations $\leq$ and $<$ in Proposition 3.2.1 are a total order and the corresponding strict total order. Besides, every set of cardinals is well-ordered under $\leq$.*

PROOF: Immediate from Theorem 3.1.1 and Definition 3.2.2. □

Exercise 3.2.1 Prove that every non-empty class of ordinals has least element.

Corollary 3.2.1 (the principle of transfinite induction for cardinals) *Let $\pi(\mu)$ be a property defined for all cardinals and such that for every cardinal μ, the validity of $\pi(\nu)$ for all cardinals $\nu < \mu$ implies that $\pi(\mu)$ holds. Then $\pi(\mu)$ is valid for all cardinals μ.*

PROOF: To show that $\pi(\mu)$ holds for an arbitrary but fixed cardinal μ, apply Theorem 2.5.1 to the set $\mathbf{Card}(\mu]$ of all cardinals $\nu \leq \mu$ and the subset S of those cardinals $\nu \leq \mu$ which satisfy $\pi(\nu)$. □

Exercise 3.2.2 State and prove a principle of transfinite induction for all cardinals $\geq \mu_0$, where μ_o is a fixed cardinal.

Now we deal with successors of cardinals.

Lemma 3.2.1 *If $(A, \leq)$ and $(2^A, \leq')$ are wosets, then $(A, \leq) < (2^A, \leq')$.*

PROOF: Otherwise $(2^A, \leq') \leq (A, \leq)$, that is, there is an isomorphism f between $(2^A, \leq')$ and a segment $(S, \leq)$ of $(A, \leq)$. But $S \subseteq A$ and f would be a bijection between 2^A and S, which is impossible. □

Theorem 3.2.3 *Every set of cardinals has a successor in the class* **Card** *of cardinals.*

PROOF: Let $\mathbf{K}$ be a set of cardinals. Viewed as a set of ordinals, the set $\mathbf{K}$ has a successor $\alpha = \overline{(A, \leq)}$ by Theorem 3.1.3. So $\mathbf{K} < \alpha$, but $\alpha < \overline{\overline{2^A}}$ by Lemma 3.2.1, therefore $\mathbf{K} < \overline{\overline{2^A}}$. The least cardinal κ such that $\mathbf{K} < \kappa$, which exists by Remark 3.1.3, is the sought successor. □

Remark 3.2.3 The successor α of K in **Ord** and the cardinal successor κ of K satisfy $\alpha \leq \kappa$, because κ is one of the strict upper bounds of K.

Exercise 3.2.3 Prove that $\alpha < \kappa$.

It follows from Theorem 3.2.3, in the same way as the Burali-Forti antinomy was inferred from Theorem 3.1.3, that there is no set of all cardinals; in other words, *the class* **Card** *of all cardinals is a proper class.*

We now generalize Example 3.2.2:

Definition 3.2.3 The cardinal $\aleph_\alpha$ (*aleph alpha*) associated with an ordinal α is given by the following rules:

(A0) $\aleph_0$ is the cardinal of countable sets;

(A1) if $\alpha > 0$, then $\aleph_\alpha$ is the successor in **Card** of the set $\{\aleph_\xi \mid \xi < \alpha\}$.

Theorem 3.2.4 (i) *For every ordinal α, the cardinal $\aleph_\alpha$ exists and is uniquely determined.*

(ii) *Conversely, every infinite cardinal κ can be written in the form $\kappa = \aleph_\xi$ for some uniquely determined ordinal ξ.*

(iii) *For every two ordinals α, β,*

$$\alpha < \beta \Longleftrightarrow \aleph_\alpha < \aleph_\beta \ . \tag{3.2.3}$$

(iv) *For every ordinal α,*

$$\alpha = \overline{\mathbf{Card}(\aleph_\alpha)} \ , \tag{3.2.4}$$

where **Card**(κ) *denotes the set of all cardinals* $< \kappa$.

Comment Property (iv) is the analogue of Theorem 3.1.2.

PROOF: (i) The construction by transfinite induction is justified by Theorem 3.1.5.II with $\alpha = 0$. Indeed, (A0) is II(i), (A1) is a stronger form of II(iii), and II(ii) also holds, because if $\aleph_\beta$ is determined, then either $\beta = 0$, in which case $\aleph_{\beta+1}$ is the cardinal successor $\aleph_1$ of $\aleph_0$, or $\beta > 0$, in which case the set $\{\aleph_\xi \mid \xi < \beta\}$ exists because $\aleph_\beta$ is its cardinal successor, so that $\aleph_{\beta+1}$ is determined as the cardinal successor of the set $\{\aleph_\gamma \mid \gamma \leq \beta\}$. The existence of the successor is ensured by Theorem 3.2.3, and the uniqueness of the successor is well known.

(iii) If $\alpha < \beta$ note first that $\beta \neq 0$. Then $\aleph_\alpha \in \{\aleph_\xi \mid \xi < \beta\}$ and $\aleph_\beta$ is the successor of the latter set, therefore $\aleph_\alpha < \aleph_\beta$. Conversely, the latter inequality implies $\alpha < \beta$, otherwise $\beta \leq \alpha$ and the already proved part of (iii) would imply $\aleph_\beta \leq \aleph_\alpha$, a contradiction.

(ii) The uniqueness of a representation $\kappa = \aleph_\xi$ follows by (iii). To prove the existence, it suffices to take an arbitrary but fixed cardinal $\kappa > \aleph_0$ such that all the cardinals χ with $\aleph_0 < \chi < \kappa$ are alephs and prove that κ is an aleph, too; cf. Corollary 3.2.1 and Exercise 3.2.2.

Let $\Omega = \{\beta \in \mathbf{Ord} \mid \aleph_\beta < \kappa\}$ and denote by α the successor of Ω; cf. Theorem 3.1.3. If we succeed to prove that $\Omega = \mathbf{Ord}(\alpha)$, then, taking into account the inductive hypothesis, it will follow in turn that

$$\{\chi \in \mathbf{Card} \mid \aleph_o < \chi < \kappa\} = \{\aleph_\beta \mid \beta \in \Omega\} = \{\aleph_\beta \mid \beta < \alpha\} \ ,$$

$$\mathbf{Card}(\kappa) = \mathbb{N} + \{\aleph_\beta \mid \beta < \alpha\}$$

hence the successor κ of $\mathbf{Card}(\kappa)$ will coincide with the successor of $\{\aleph_\beta \mid \beta < \alpha\}$, which is $\aleph_a$ by Definition 3.2.3.

But$\Omega < \alpha$, i.e., $\Omega \subseteq \mathbf{Ord}(\alpha)$, and it remains to prove that $\mathbf{Ord}(\alpha) \subseteq \Omega$. The latter inclusion will be established if for every $\beta < \alpha$ we can find $\gamma \in \Omega$ such that $\beta \leq \gamma$, because in view of (iii) this will imply $\aleph_\beta \leq \aleph_\gamma < \kappa$, hence $\aleph_\beta < \kappa$, that is, $\beta \in \Omega$.

Suppose, by way of contradiction, the existence of an ordinal $\beta < \alpha$ such that $\gamma < \beta$ for all $\gamma \in \Omega$. Since α is the successor of Ω, the latter inequalities would imply that $\alpha \leq \beta$, a contradiction.

(iv) For every $\gamma \in \mathbf{Ord}(\alpha)$, property (iii) implies $\aleph_\gamma \in \mathbf{Card}(\aleph_\alpha)$, therefore the association $\gamma \mapsto \aleph_\gamma$ defines a mapping $\aleph : \mathbf{Ord}(\alpha) \longrightarrow \mathbf{Card}(\aleph_\alpha)$. This function is strictly increasing, again by (iii), and surjective by (ii) and (iii). Therefore the mapping $\aleph$ is an isomorphism by Proposition 2.3.4, so that using Theorem 3.1.2 we get

$$\alpha = \overline{\mathbf{Ord}(\alpha)} = \overline{\mathbf{Card}(\aleph_a)}\ .$$

□

3.3 Von Neumann's Construction of Ordinals

Whereas in §§1–2 we have followed an axiomatic line to the effect that we have assumed the ordinal $\overline{W}$ of any woset W exists and $\overline{W_1} = \overline{W_2}$ if and only if $W_1 \approx W_2$, in this section we follow von Neumann [1923] in proving the existence of ordinals by an actual construction. Unlike the solution sketched in the beginning of §1, von Neumann's ordinals are built up within the framework of set theory without classes. Another striking point is that although Definition 3.3.1 below concerns a recursive property, no induction is used in the subsequent proof.

We have already mentioned, in the beginning of §1, the idea of taking as ordinal for each class of isomorphic wosets, a distinguished woset of that class; the trouble was that this would have required a kind of a very strong axiom of choice. What von Neumann did was to pick up a representative from each class by an actual construction which avoids that "axiom of choice". Since ordinals have to be realized as wosets, the order relation $\leq$ in Section 2.5 makes sense and the idea is that the ordinal selected from each class of isomorphic wosets will be the set of all ordinals (= wosets) $\leq$ the wosets of that class.

So the ordinal of the class reduced to the empty woset $\varnothing$ will be $\varnothing$ itself, i.e., $0 = \varnothing$. The ordinal of the class of all one-element wosets will be $\{0\}$, i.e., $1 = \{0\}$. The ordinal of the class of all two-element wosets will be $\{0, 1\}$, i.e., $2 = \{0, 1\}$; etc. Now the aim is to extend this process to infinite wosets (in which case two wosets in bijection need not be isomorphic), so that any ordinal α will be the woset of all ordinals $\beta < \alpha$. We have to prove that this is possible in such a way that any woset W should possess an ordinal $\overline{W}$ and the fundamental property

$$\overline{W_1} = \overline{W_2} \Longleftrightarrow W_1 \approx W_2 \tag{3.3.1}$$

should hold.

Definition 3.3.1 Let W be a well-ordered set and

$$f : W \longrightarrow f(W) \tag{3.3.2}$$

a surjective mapping having the domain W. Then f is called a *Z-mapping*[1] if

$$f(x) = \{f(y) \mid y \in W(x)\} \tag{3.3.3}$$

for every $x \in W$.

We shall prove that every well-ordered set W possesses a unique Z-mapping f, from which it will follow that the set

$$\overline{W} = f(W) = \{f(x) \mid x \in W\} \tag{3.3.4}$$

exists, is uniquely determined by W and fulfills (3.3.1). In other words, the set $\overline{W}$ will be the *ordinal number* associated by von Neumann with the woset W.

Lemma 3.3.1 *The empty mapping $\perp : \varnothing \longrightarrow \varnothing$ is the unique Z-mapping of the woset $\varnothing$, while $\overline{\varnothing} = \varnothing$.*

PROOF: We already know (cf. Remark 1.2.2) that the empty mapping $\perp : \varnothing \longrightarrow \varnothing$ is the unique surjective mapping $f : \varnothing \longrightarrow f(\varnothing)$ for which the domain is empty. Now the property in Definition 3.3.1 can also ve written

$$(\forall x)\ x \in W \Longrightarrow f(x) = \{f(y) \mid y \in W(x)\}$$

and is satisfied for $W := \varnothing$, because the premise $x \in \varnothing$ is false. Thus $\perp$ is a Z-mapping and its codomain is $\overline{\varnothing} = \varnothing$. □

Lemma 3.3.2 *Every woset W has at most one Z-mapping.*

PROOF: For $W := \varnothing$ the property has already been established in Lemma 3.3.1. Let $W \neq \varnothing$ and suppose f and g are two distinct Z-mappings of W. Then $A = \{x \in W \mid f(x) \neq g(x)\}$ is a non-empty set and if a denotes the first element of A, it follows that the (possibly empty) set $W(a)$ is included in $W \setminus A$, hence

$$f(a) = \{f(x) \mid x \in W(a)\} = \{g(x) \mid x \in W(a)\} = g(a)\ ,$$

therefore $a \notin A$, a contradiction. □

Lemma 3.3.3 *If a woset $W \neq \varnothing$ has the Z-mapping f, then every proper segment $W(x)$ of W possesses the Z-mapping $f|_{W(x)}$.*

[1] The original German term is *Zählung*, which means the act of counting (*zählbare Menge* = countable set); however the set W need not be countable.

PROOF: Take $x \in W$ and set $g = f|_{W(x)}$. If $W(x) = \varnothing$, then g is obviously the empty mapping $\perp$, which is a Z-mapping by Lemma 3.3.1. If $W(x) \neq \varnothing$, then for every $y \in W(x)$ we have (even if $W(y) = \varnothing$),

$$g(y) = f(y) = \{f(z) \mid z \in W(y)\} = \{g(z) \mid z \in (W(x))(y)\} .$$

□

The converse of Lemma 3.3.3 is also valid:

Lemma 3.3.4 *If every proper segment of a woset $W \neq \varnothing$ has a Z-mapping, then W itself possesses a Z-mapping.*

PROOF: For every $x \in W$, let g_x be the Z-mapping of $W(x)$. Lemma 3.3.2 ensures that g_x is uniquely determined by x, therefore formula

$$f(x) = \{g_x(y) \mid y \in W(x)\} \quad (\forall x \in W) \tag{3.3.5}$$

defines a mapping $f : W \longrightarrow f(W)$.

To prove that f is a Z-mapping of W, we first show that

$$f|_{W(x)} = g_x \quad (\forall x \in W) . \tag{3.3.6}$$

For both sides of (3.3.6) reduce to the empty mapping if $W(x) = \varnothing$, while if $W(x) \neq \varnothing$ then for every $y \in W(x)$ it follows that $g_x|_{W(y)} = g_x|_{(W(x))(y)}$ is a Z-mapping of $W(y)$ by Lemma 3.3.3, therefore $g_x|_{W(y)} = g_y$ by Lemma 3.3.2, hence

$$f(y) = \{g_y(z) \mid z \in W(y)\} = \{g_x(z) \mid z \in (W(x))(y)\} = g_x(y) .$$

Finally (3.3.5) and (3.3.6) imply (3.3.3) □

Proposition 3.3.1 *Every woset has a Z-mapping and only one.*

PROOF: Since uniqueness is already established in Lemma 3.3.2, it remains to prove the existence of a Z-mapping for an arbitrary woset W. This is true for $W := \varnothing$ by Lemma 3.3.1. Now assume that a woset $W \neq \varnothing$ has no Z-mapping. Then Lemma 3.3.4 implies that the set

$$A = \{x \in W \mid W(x) \text{ has no Z} - \text{mapping}\}$$

is not empty; let a be the first element of A. If $W(a) = \varnothing$, then it possesses a Z-mapping by Lemma 3.3.1. If $W(a) \neq \varnothing$, then for every $y \in W(a)$ the proper segment $W(y) = (W(a))(y)$ has a Z-mapping, therefore $W(a)$ possesses a Z-mapping by Lemma 3.3.4. Thus in both cases $W(a)$ has a Z-mapping, therefore $a \notin A$, a contradiction. □

Lemma 3.3.5 *The Z-mapping f of any woset W fulfills $f(x) \notin f(x)$ for every $x \in W$.*

Warning More generally, the property $S \notin S$ is established for any set S within the framework of the axiomatic system of Zermelo-Fraenkel as a consequence of ZF8, while $f(x)$ is a set by (3.3.3) and ZF7. However we are going to prove $f(x) \notin f(x)$ directly, that is, without making use of axiomatic set theory.

PROOF: Suppose that $f(x) \in f(x)$ for some woset W, some $x \in W$ and the Z-mapping f of W. Then the set $A = \{x \in W \mid f(x) \in f(x)\}$ is not empty; let a be the first element of A. It follows that

$$f(a) \in f(a) = \{f(y) \mid y \in W(a)\} ,$$

hence $f(a) = f(y)$ for some $y \in W(a)$, thus implying $f(y) \in f(y)$, hence $y \in A$ although $y < a$, a contradiction. □

Proposition 3.3.2 *The Z-mapping $f : W \longrightarrow \overline{W}$ of a woset W establishes an order isomorphism between W and the set $\overline{W}$ ordered by set inclusion.*

PROOF: The mapping f is surjective by the very definition (3.3.4) of $\overline{W}$. To prove that

$$(\forall x, y \in W) \;\; x < y \Longrightarrow f(x) \subset f(y) \tag{3.3.7}$$

(cf. Lemma 2.3.1), take $x, y \in W$ with $x < y$. Then $W(x) \subset W(y)$, hence

$$f(x) = \{f(z) \mid z \in W(x)\} \subseteq \{f(z) \mid z \in W(y)\} = f(y) \tag{3.3.8}$$

and the inclusion is strict because $x \in W(y)$, hence $f(x) \in f(y)$ by (3.3.8), whereas $f(x) \notin f(x)$ by Lemma 3.3.5, therefore $f(x) \in f(y) \setminus f(x)$. □

Corollary 3.3.1 *The set $\overline{W}$ is well-ordered by set inclusion.*

PROOF: The property of being well-ordered is clearly preserved by an isomorphism. □

We are now in a position to prove the main result of this section, namely that (3.3.4) is a good definition for ordinal numbers.

Theorem 3.3.1 *For each well-ordered set W, let*

$$\overline{W} = f(W) = \{f(x) \mid x \in W\} , \tag{3.3.4}$$

where f is the Z-mapping of W. Then:

(i) *$\overline{W}$ exists and is uniquely determined by W.*

(ii) *For every two wosets W_1 and W_2,*

$$\overline{W_1} = \overline{W_2} \Longleftrightarrow W_1 \approx W_2 . \tag{3.3.1}$$

PROOF: Property (i) being a direct consequence of Proposition 3.3.1, it remains to prove (ii).

Let W_1, W_2 be two wosets and f_1, f_2 the Z-mappings of W_1 and W_2, respectively.

Suppose $\overline{W_1} = \overline{W_2}$; then Proposition 3.3.2 implies that $f_2^{-1} f_1$ is an isomorphism between W_1 and W_2.

Conversely, assume the existence of an isomorphism $h : W_1 \longrightarrow W_2$. If we succeed to prove that $f_2h : W_1 \longrightarrow \overline{W_2}$ is a Z-mapping, it will follow that $f_2h = f_1$ by Lemma 3.3.2, and this will imply

$$\overline{W_1} = f_1(W_1) = f_2h(W_1) = f_2(W_2) = \overline{W_2} \ .$$

But f_2h is surjective because so are h and f_2; then for every $x \in W_1$, the hypotheses and Proposition 2.3.4 imply

$$(f_2h)(x) = f_2(h(x)) = \{f_2(y) \mid y \in W_2(h(x))\} = \{f_2(y) \mid y \in W_2 \,\&\, y < h(x)\}$$

$$= \{f_2(h(z)) \mid z \in W_1 \,\&\, h(z) < h(x)\} = \{(f_2h)(z) \mid z \in W_1 \,\&\, z < x\} \ ,$$

therefore $(f_2h)(x) = \{(f_2h)(z) \mid z \in W_1(x)\}$. □

At this moment we know that the whole theory of ordinals studied in §1 applies to von Neumann's "concrete" ordinals. The latter have certain supplementary properties which cannot be established within the axiomatic theory.

The most important such property is that von Neumann's ordinals verify the condition mentioned in the beginning of this section:

Theorem 3.3.2 *Every ordinal α is the set of all ordinals $\beta < \alpha$.*

Comments 1) Some authors include this property in the very definition of ordinals. Yet it must be proved that the definition is consistent.
2) Written in the form

$$\alpha = \mathbf{Ord}(\alpha) \ , \qquad (3.3.9)$$

this result appears as a strengthening of Theorem 3.1.1.
Sketch of PROOF : As shown by Pierce [1968, Introduction, §4], this happens if the ordinals satisfy the following conditions:

(i) every ordinal α is a set of sets;

(ii) if $\beta \in \alpha$, then $\beta \subset \alpha$;

(iii) for every $\varnothing \neq A \subseteq \alpha$, there is $\beta_0 \in A$ such that $\beta_0 \in \beta$ for every $\beta \in A \setminus \beta_0$.

It is easy to see that conditions (i)-(iii) are fulfilled by von Neumann's ordinals. Indeed: (i) The ordinal (3.3.4) is the set consisting of the sets (3.3.3). (ii) Let $\alpha = f(W)$. If $W = \varnothing$, then Lemma 3.3.1 shows that $\alpha = \overline{\varnothing} = \{\varnothing\}$, hence $\beta = \varnothing \subset \overline{\varnothing}$. If $W \neq \varnothing$, then $\beta = f(x)$ for some $x \in W$; but $f(x) \subseteq f(W)$ by (3.3.3) and (3.3.4). In fact $f(x) \in f(W) \setminus f(x)$ by Lemma 3.3.5, therefore $f(x) \subset f(W)$. (iii) Suppose $\varnothing \neq A \subseteq f(W)$. Then each $a \in A$ has a unique representation $a = f(x_a)$ with $x_a \in W$. Set $B = \{x_a \mid a \in A\}$. Then $\varnothing \neq B \subseteq W$ and let x_{a_0} be the first element of B. It follows that $a_0 \in A$ and for each $a \in A \setminus \{a_0\}$ we have $x_{a_0} < x_a$, hence $a_0 = f(x_{a_0}) \in f(x_a) = a$ by (3.3.3). □

Corollary 3.3.2 *The successor of an ordinal α is the set of ordinals $\{0, 1, \ldots, \alpha\}$, while the successor of a set of ordinals Ω without greatest element is Ω itself.*

Chapter 4

Lattices

After the essentials of the theory of transfinite numbers, investigated in the previous chapter, we now present a few topics from another specialization of the concept of order: lattice theory. Our aim is to provide lattice-theoretical tools broadly used in various fields of mathematics and in particular some lattice-theoretical background necessary to universal algebra.

Our way from posets to lattices passes through the intermediate level of semilattices; in §1 the relationship between the order-theoretical and the algebraic definition of these concepts is described in the categorical language. Several characterizations of the important subclasses of modular, distributive and Boolean lattices are given in §2. Complete lattices are dealt with in §3, with emphasis on complete Boolean algebras, which appear in several mathematical contexts; compactly generated lattices are also introduced. The next section is devoted to closure operators, an order-theoretical tool which is applied in almost all branches of mathematics to construct the substructures of a set endowed with a certain structure. Galois connections, intimately related to closure operators, are dealt with in the last section.

There is a huge literature on lattices and their applications. For lattices in general we refer the reader e.g. to Birkhoff [1967], Blyth [2005], Crawley and Dilworth [1973], Davey and Priestley [1990], Dubreil-Jacotin, Lesieur and Croisot [1953], Grätzer [1978a], Roman [2008], Padmanabhan and Rudeanu [2008], Szász [1963]. For Boolean algebras see e.g. Dwinger [1961], Halmos [1963], Monk and Bonnet [1989], Rudeanu [1973], [2001], Sikorski [1964], Vladimirov [1969]. For Boolean algebras in other fields see e.g. Halmos [1950] for measure theory, Nöbeling [1954], Ponasse and Carrega [1979], Vaidyanathaswamy [1960] for topology, and Kappos [1969], Onicescu and Cuculescu [1976] for probability theory. Last but not least, a branch of algebra which owes much to lattice theory is universal algebra. The reader is referred e.g. to Burris and Sankappanavar [1981], Grätzer [1978b], and Pierce [1968].

Keywords: Semilattice, lattice, sublattice, Hasse diagram, modular lattice, distributive lattice, Boolean algebra, duality, Boolean ring, complete lattice, complete homomorphism, compact element, closure operator, Moore family,

Sergiu Rudeanu

Galois connection.

4.1 Fundamentals of Lattices and Semilattices

Lattices can be viewed as posets for which every two elements have a least upper bound and a greatest lower bound. It turns out that taking the l.u.b. and the g.l.b. of two elements in a lattice defines two binary operations on the given lattice and this yields an equivalent definition of lattices as algebraic structures. In categorical terms, there is an isomorphism between the category of posets with l.u.b and g.l.b and the category of the corresponding algebraic structures. This will be shown in the present section, starting with the preliminary important case of semilattices.

Definition 4.1.1 A *meet semilattice* (*join semilattice*) is a poset $S, \leq)$ such that every two elements $x, y \in S$ have a greatest lower bound $x \wedge y$ (least upper bound $x \vee y$).

Note that this is a particular case of Definition 2.2.2(iv).

Example 4.1.1 The posets V and Y in Example 2.1.8 are meet semilattices.

Example 4.1.2 The poset in Example 2.2.8′ is a join semilattice.

Example 4.1.3 If X is an arbitrary set and $S \neq \varnothing$ is a meet semilattice (join semilattice), then the poset S^X of all mappings $f : X \longrightarrow S$ (cf. Example 2.1.11) is a meet semilattice (join semilattice).

Example 4.1.4 If S_1 and S_2 are meet semilattices (join semilattices), then their direct product (cf. Example 2.1.10) is also a meet semilattice (join semilattice).

Example 4.1.5 If I is a meet semilattice (join semilattice) and $\{S_i\}_{i \in I}$ is a family of meet semilattices with greatest element1 (join semilattices with least element 0), then the sum $\sum_{i \in I} S_i$ (cf. Example 2.1.10) is also a meet semilattice (join semilattice).

See also Examples 4.1.11-4.1.20.

Remark 4.1.1 The dual of a meet semilattice (join semilattice) is a join semilattice (meet semilattice). Therefore the following *duality principle* holds for semilattices: from each theorem concerning meet semilattices one obtains by duality a theorem concerning join semilattices, and conversely. On the other hand, the duality principle holds neither within the category of meet semilattices, nor within the category of join semilattices.

As a matter of fact, we have not yet defined the category of meet semilattices (join semilattices); but the above Remark 4.1.1 is true for any "reasonable" definition of these categories, i.e., provided the morphisms of one category should be the duals of the morphisms of the other category. Another possibility would be to take as morphisms for both categories the isotone mappings, thus obtaining two full subcategories of **Pos**. However it turns out that the most interesting category having as objects all meet semilattices (join semilattices) is obtained by taking a poorer class of morphisms. This follows from an algebraization of semilattices which we are now going to describe and which will justify the adequacy of Definition 4.1.4 below.

Note that we have already informally used the term "binary operation", which appears in Definition 4.1.2 below. A *binary operation* on a set X is a function $f : X^2 \longrightarrow X$; the elements $f(x,y)$ are usually written $x\, f\, y$.

Definition 4.1.2 The *meet* operation (*join* operation) of a meet semilattice (join semilattice) is the binary operation $\wedge : S^2 \longrightarrow S$ such that

$$x \wedge y = \text{g.l.b. of } x \text{ and } y \tag{4.1.1$'$}$$

for every $x, y \in S$ (the binary operation $\vee : S^2 \longrightarrow S$ such that

$$x \vee y = \text{l.u.b. of } x \text{ and } y \tag{4.1.1$''$}$$

for every $x, y \in S$). More generally, by $x_1 \wedge \ldots \wedge x_n$ (by $x_1 \vee \ldots \vee x_n$) we denote the g.l.b. (the l.u.b.) of the set $\{x_1, \ldots, x_n\}$.

In view of Definition 2.2.2 and Theorem 2.2.1, for every $x, y \in S$, if S is a meet semilattice, then $x \wedge y$ is the unique element such that

$$x \wedge y \leq \{x, y\}\ , \tag{4.1.2$'$}$$

$$(\forall t \in S)\ \ (t \leq \{x,y\} \Longrightarrow t \leq x \wedge y) \tag{4.1.3$'$}$$

while if S is a join semilattice, then $x \vee y$ is the unique element such that

$$\{x, y\} \leq x \vee y\ , \tag{4.1.2$''$}$$

$$(\forall t \in S)\ \ (\{x,y\} \leq t \Longrightarrow x \vee y \leq t)\ . \tag{4.1.3$''$}$$

This shows in particular that $\wedge$ (that $\vee$) is actually a binary operation on the meet semilattice (join semilattice) S. Moreover, we have:

Proposition 4.1.1 *If S is a meet semilattice (join semilattice), the pair $(S, \wedge)$ (the pair $(S, \vee)$) is an idempotent commutative semigroup,* that is, *for every $x, y, z \in S$:*

$$(x \wedge y) \wedge z = x \wedge (y \wedge z)\ \ \text{(associativity)}, \tag{4.1.4$'$}$$

$$x \wedge y = y \wedge x\ \ \text{(commutativity)}, \tag{4.1.5$'$}$$

$$x \wedge x = x\ \ \text{(idempotency)}, \tag{4.1.6$'$}$$

and, moreover,

(4.1.7′) $$x \le y \Longleftrightarrow x \wedge y = x$$

(that is, *for every* $x, y, z \in S$,

(4.1.4″) $$(x \vee y) \vee z = x \vee (y \vee z) \text{ (associativity)},$$

(4.1.5″) $$x \vee y = y \vee x \text{ (commutativity)},$$

(4.1.6″) $$x \vee x = x \text{ (idempotency)}),$$

and, moreover,

(4.1.7″) $$x \le y \Longleftrightarrow x \vee y = y \text{)}.$$

PROOF: In view of the duality principle it suffices to establish properties (4.1.4′)-(4.1.7′). Clearly $x \le y$ if and only if the g.l.b. of $\{x, y\}$ is x; this is precisely (4.1.7′). The property $x \le x$ and the trivial implication $t \le x \Longrightarrow t \le x$ show that x is the g.l.b. of the set $\{x\} = \{x, x\}$, that is, (4.1.6′). Then $\{x, y\} = \{y, x\}$ implies (4.1.5′). Finally, setting $(x \wedge y) \wedge z = s$, we get $s \le z$ and $s \le x \wedge y \le \{x, y\}$ by (4.1.2′), hence $s \le \{x, y, z\}$. This implies $s \le \{y, z\}$, hence $s \le y \wedge z$ by (4.1.3′), therefore $s \le x \wedge (y \wedge z)$ again by (4.1.3′). Similarly $x \wedge (y \wedge z) \le (x \wedge y) \wedge z$. □

The following converse of Proposition 4.1.1 is also valid:

Proposition 4.1.2 *Let* $(S, \circ)$ *be an idempotent commutative semigroup and define two binary relations* $x \le'$ *and* $\le''$ *on* S *by*

(4.1.8′) $$x \le' y \Longleftrightarrow x \circ y = x\ ,$$

(4.1.8″) $$x \le'' y \Longleftrightarrow x \circ y = y\ .$$

Then:

(i) $(S, \le')$ *is a meet semilattice having* $\circ$ *as meet operation.*

(ii) $(S, \le'')$ *is a join semilattice having* $\circ$ *as join operation.*

(iii) *The relations* $\le'$ *and* $\le''$ *are dual to each other.*

PROOF: (i) The idempotency $x \circ x = x$ implies $x \le' x$. If $x \le' y$ and $y \le' z$ then

$$x \circ z = (x \circ y) \circ z = x \circ (y \circ z) = x \circ y = x\ ,$$

hence $x \le' z$. If $x \le' y$ and $y \le' x$, then $x = x \circ y = y \circ x = y$. Thus $(S, \le')$ is a poset and it remains to show that $x \circ y =$ g.l.b. of $\{x, y\}$. Indeed,

$$(x \circ y) \circ x = x \circ (x \circ y) = (x \circ x) \circ y = x \circ y\ ,$$

hence $x \circ y \le' x$; then $(x \circ y) \circ y = x \circ (y \circ y) = x \circ y$, hence $x \circ y \le' y$. Finally $z \le' \{x, y\}$ implies

$$z \circ (x \circ y) = (z \circ x) \circ y = z \circ y = z\ ,$$

that is, $z \leq' x \circ y$.

(ii) Similar proof

(iii) $x \leq' y \Longleftrightarrow x \circ y = x \Longleftrightarrow y \circ x = x \Longleftrightarrow y \leq'' x$. □

Propositions 4.1.1 and 4.1.2 suggest that semilattices can also be defined as idempotent commutative semigroups. This idea can be given a more precise and stronger meaning in terms of category theory, as we are now going to do.

Definition 4.1.3 A function $f : (S, \leq) \longrightarrow (S', \leq)$ between two meet semilattices (join semilattices) is called a *meet homomorphism* (*join homomorphism*) provided it *preserves finite meets* (*finite joins*), i.e.,

$$f(x \wedge y) = f(x) \wedge f(y) \tag{4.1.9'}$$

for every $x, y \in S$ (i.e.,

$$f(x \vee y) = f(x) \vee f(y) \tag{4.1.9''}$$

for every $x, y \in S$).

Properly speaking, the property of preserving finite meets is

$$f(x_1 \wedge \ldots \wedge x_n) = f(x_1) \wedge \ldots \wedge f(x_n)$$

for every $n = 2, 3, \ldots$ and every $x_1, \ldots, x_n \in S$; this is obviously equivalent to (4.1.9$'$). Similar comment for (4.1.9$''$).

Proposition 4.1.3 *Every meet homomorphism (join homomorphism) is an order homomorphism; the converse is not true.*

PROOF: If f is a meet homomorphism and $x \leq y$, then $f(x) = f(x \wedge y) = f(x) \wedge f(y)$, hence $f(x) \leq f(y)$; dual proof for join homomorphisms. The converse is not true, as shown, e.g., in the example below. □

Example 4.1.6 Take the meet semilattices V and Y in Fig. 4.1; cf. Example 4.1.1. Define $f : V \longrightarrow Y$ by $f(x) = x'$ for all $x \in V$. Then f is isotone but $f(a \wedge b) = f(0) = 0'$, whereas $f(a) \wedge f(b) = a' \wedge b' = c'$.

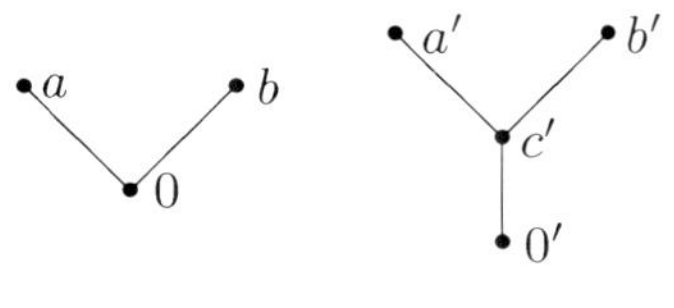

Fig. 4.1

Definition 4.1.4 Let **MSem** (let **JSem**) denote the category of all meet semilattices (join semilattices) and all meet homomorphisms (join homomorphisms) between them. Let **Sem** denote the category of all idempotent commutative semigroups and all semigroup homomorphisms between them

Now the main result is

Theorem 4.1.1 *The categories* **MSem, JSem** *and* **Sem** *are isomorphic.*

Comment This theorem means that Propositions 4.1.1 and 4.1.2 establish a bijection between meet or join semilattices and idempotent commutative semigroups, while the morphisms are the same for each category.

PROOF: In view of duality, it suffices to establish the isomorphism between **MSem** and **Sem**.

Note first that meet homomorphisms coincide with the morphisms of **Sem**, by (4.1.9′).

For every meet semilattice $(S, \leq)$, let $F'(S, \leq) = (S, \wedge)$, where $\wedge : S^2 \longrightarrow S$ is defined by (4.1.1′); for every meet homomorphism f, let $F'f = f$. Then Proposition 4.1.1 immediately implies that $F' :$ **MSem** $\longrightarrow$ **Sem** is a functor.

For every idempotent commutative semigroup $(S, \circ)$, let $G'(S, \circ) = (S, \leq')$, where $\leq'$ is the binary relation on S defined by (4.1.8′); for every morphism f of **Sem**, let $G'f = f$. Then Proposition 4.1.2 immediately implies that $G' :$ **Sem** $\longrightarrow$ **MSem** is a functor.

It remains to prove that $G' \circ F'$ and $F' \circ G'$ are the identity functors. This is trivial for the morphism functions and we shall prove it for the object function.

Let $(S, \leq)$ be a meet semilattice, $F'(S, \leq) = (S, \wedge)$ and $G'(S, \wedge) = (S, \leq')$; we have to prove that $\leq'$ coincides with $\leq$. But

$$x \leq' y \Longleftrightarrow x \wedge y = x \Longleftrightarrow \text{g.l.b.}\{x, y\} = x \Longleftrightarrow x \leq y \, .$$

Finally let $(S, \circ)$ be an object of **Sem**, $G'(S, \circ) = (S, \leq')$ and $F'(S, \leq') = (S, \wedge)$; we have to prove that $\wedge$ coincides with $\circ$. But $x \wedge y =$g.l.b.$\{x, y\}$ by the definition of F', while $x \circ y =$g.l.b.$\{x, y\}$ by Proposition 4.1.2, therefore $x \wedge y = x \circ y$ by the uniqueness of the g.l.b. □

Warning The equalities $F'f = f$ and $G'g = g$ used in the above proof are not quite correct. As a matter of fact, if $f : (S, \leq) \longrightarrow (S', \leq')$ is a morphism of **MSem**, then we denote by $F'f$ a thing different from this morphism, namely the morphism $f : (S, \wedge) \longrightarrow (S', \wedge')$ of **Sem**. However these two morphisms correspond to the same mapping $f : S \longrightarrow S'$ between the adjacent sets, and this essential fact is retained in formula $F'f = f$. Such simplifications of notation will be tacitly admitted in the sequel (this resembles the simplification which consists in writing $(S, \leq)$, $(S', \leq)$, etc.).

In view of Theorem 4.1.1, we shall identify each meet semilattice $(S, \leq)$ with the corresponding semigroup $(S, \wedge)$ and each join semilattice $(S, \leq)$ with the corresponding semigroup $(S, \vee)$. Besides, $(S, \circ)$ may be $(S, \wedge)$ or $(S, \vee)$. Moreover, we shall use:

Definition 4.1.Sem The term *semilattice* is a common name for meet semilattices, join semilattices, and idempotent commutative semigroups. The term *semilattice homomorphism* is a common name for the morphisms of the categories **MSem, JSem** and **Sem**.

From a practical point of view, we are entitled to use simultaneously properties (4.1.2)-(4.1.7) established above, as well as those which will be demonstrated below.

Proposition 4.1.4 *If S is a meet semilattice (a join semilattice), then for every $x, y, z, x', y' \in S$,*

$$x \leq y \Longrightarrow x \wedge z \leq y \wedge z \, , \tag{4.1.10'}$$

$$x \leq y \;\&\; x' \leq y' \Longrightarrow x \wedge x' \leq y \wedge y' \, , \tag{4.1.11'}$$

$$x \# y \Longleftrightarrow x \neq x \wedge y \neq y \, , \tag{4.1.12'}$$

(then for every $x, y, z, x', y' \in S$,

$$x \leq y \Longrightarrow x \vee z \leq y \vee z \, , \tag{4.1.10''}$$

$$x \leq y \;\&\; x' \leq y' \Longrightarrow x \vee x' \leq y \vee y' \, , \tag{4.1.11''}$$

$$x \# y \Longleftrightarrow x \neq x \vee y \neq y \;). \tag{4.1.12''}$$

PROOF: It suffices to prove (4.1.10′)-(4.1.12′). If $x \leq y$, then (4.1.4′)-(4.1.7′) imply

$$(x \wedge y) \wedge (y \wedge z) = ((x \wedge z) \wedge y) \wedge z = (x \wedge (z \wedge y)) \wedge z = (x \wedge (y \wedge z)) \wedge z$$
$$= ((x \wedge y) \wedge z) \wedge z = x \wedge (z \wedge z) = x \wedge z \, ,$$

hence $x \wedge z \leq y \wedge z$ by (4.1.7′). If $x \leq y$ and $x' \leq y'$, then (4.1.10′) implies $x \wedge x' \leq y \wedge x'$ and $y \wedge x' \leq y \wedge y'$, therefore $x \wedge x' \leq y \wedge y'$. Finally (4.1.12′) follows immediately from Definition 2.1.2 and (4.1.7′). □

Exercise 4.1.1 (cf. Definition 4.1.2). Let $x_1, \ldots, x_n$ be elements of a meet semilattice (join semilattice) S. Then $x_1 \wedge \ldots \wedge x_n$ (then $x_1 \vee \ldots \vee x_n$) is the unique element such that, for every $t \in S$,

$$t \leq \{x_1, \ldots, x_n\} \Longleftrightarrow t \leq x_1 \wedge \ldots \wedge x_n \tag{4.1.13'}$$

(for every $t \in S$,

$$\{x_1, \ldots, x_n\} \leq t \Longleftrightarrow x_1 \vee \ldots \vee x_n \leq t \,). \tag{4.1.13''}$$

Exercise 4.1.2 The identities

$$(x \wedge y) \wedge z = x \wedge (y \wedge z) = x \wedge y \wedge z \tag{4.1.14'}$$

hold in any meet semilattice; dual property for join semilattices (cf. Definition 4.1.2).

Exercise 4.1.3 Carry out the details in obtaining (4.1.7″) and (4.1.10″) by duality from (4.1.7′) and (4.1.10′).

Proposition 4.1.5 *In each of the categories* **MSem**, **JSem** *and* **Sem**:

1^0 *Monomorphisms coincide with injective morphisms.*

2^0 *The following conditions are equivalent for a mapping $f : S \longrightarrow S'$ between two semilattices:*

(i) *f is an isomorphism;*

(ii) *f is a bijective morphism;*

(iii) *f is an order isomorphism.*

Comment The concept of order isomorphism is the one in Definition 2.1.6 (cf. Definition 2.1.5). The concept of isomorphism in **MSem** or in **JSem** follows from Definition 4.1.3. An order isomorphism between two objects $(S_1, \circ_1)$ and $(S_2, \circ_2)$ of **Sem** holds if and only if there is an order isomorphism between their isomorphic images from Theorem 4.1.1, that is, $G'(S_1, \circ_1) = (S_1, \leq_1')$ and $G'(S_2, \circ_2) = (S_2, \leq_2')$, or equivalently, $G''(S_1, \circ_1) = (S_1, \leq_1'')$ and $G''(S_2, \circ_2) = (S_2, \leq_2'')$.

PROOF: In view of Theorem 4.1.1 and the above comment, it suffices to demonstrate the proposition e.g. for **MSem**.

1^0 Modify the proof of Proposition 2.1.4(i) as follows.

Let P, Q be semilattices and $f : P \longrightarrow Q$ a meet homomorphism such that $f(a) = f(a')$ for some $a, a' \in P$ with $a \neq a'$. Note first that $a \not\leq a'$ or $a' \not\leq a$, so that having in view (4.1.7′), we may suppose without loss of generality that $a \wedge a' < a$.

The toset $\{0, 1\}$ with $0 < 1$ is in particular a meet semilattice. Define $u, v : \{0, 1\} \longrightarrow P$ by $u(0) = u(1) = v(1) = a$ and $v(0) = a \wedge a'$. Then $u \neq v$ and since u is a constant mapping while $v(0) < v(1)$, it follows readily that u and v are meet homomorphisms. Finally $fu(1) = fv(1)$ and

$$fv(0) = f(a \wedge a') = f(a) \wedge f(a') = f(a) = fu(0)$$

2^0 (i)$\Longrightarrow$(ii): True in any category of sets and mappings; cf. Remark 1.3.3.

(ii)$\Longrightarrow$(iii): f is an order homomorphism by Definition 2.1.5 and Proposition 4.1.3. It remains to prove that f^{-1} is also isotone. Indeed, suppose that $x', y' \in S'$ fulfill $x' \leq y'$ and set $f^{-1}(x') = x, f^{-1}(y') = y$; we have to show that $x \leq y$. But from

$$f(x) = x' = x' \wedge y' = f(x) \wedge f(y) = f(x \wedge y)$$

we infer $x = x \wedge y$, that is, $x \leq y$.

(iii)$\Longrightarrow$(i): Take $x, y \in S$ and prove that $f(x \wedge y)$ is the g.l.b. of $\{f(x), f(y)\}$. But $x \wedge y \leq x$ implies $f(x \wedge y) \leq f(x)$ and similarly $f(x \wedge y) \leq f(y)$. Now take an element $z' \in S'$ such that $z' \leq \{f(x), f(y)\}$ and prove that $z' \leq f(x \wedge y)$. Indeed, $z' = f(z)$ for some $z \in S$ and the isotony of f^{-1} implies $z \leq \{x, y\}$, hence $z \leq x \wedge y$, therefore $z' = f(z) \leq f(x \wedge y)$. So $f(x \wedge y) = f(x) \wedge f(y)$ and interchanging the rôles of f and f^{-1}, it follows that f^{-1} is also a meet homomorphism. □

We now introduce

Definition 4.1.5 A *subsemilattice* of a semilattice $(S, \circ)$ is a subset $T \subseteq S$ such that $x, y \in T \Longrightarrow x \circ y \in T$.

Example 4.1.7 Let S be a semilattice. Then S itself is a subsemilattice; we refer to the other subsemilattices as the *proper* subsemilattices of S. If $a, b \in S$, then $\{a, b, a \wedge b\}$ is a subsemilattice. Every singleton is a subsemilattice.

Example 4.1.8 The subsemilattices of the semilattice in Example 2.2.8′ are the subsets of $\mathbb{Z}$ with or without a.

Example 4.1.9 In Example 4.1.4, if T_1 and T_2 are subsemilattices of S_1 and S_2, respectively, then $T_1 \times T_2$ is a subsemilattice of the direct product $S_1 \times S_2$.

Example 4.1.10 In Example 4.1.5, if T_i are subsemilattices of S_i for $i \in I$, then $\sum_{i\in I} T_i$ is a subsemilattice of the direct sum $\sum_{i\in I} S_i$.

Proposition 4.1.6 (i) *A subset T of a semilattice $(S, \circ)$ is a subsemilattice of $(S, \circ)$ if and only if T is a semilattice $(T, \circ)$ under the restriction of the operation $\circ$ of S to the elements of T.*

(ii) *If $f : (S, \circ) \longrightarrow (S', \circ')$ is a homomorphism, then $f(S)$ is a subsemilattice of S'.*

(iii) *If, moreover, f is injective, then S is isomorphic to $f(S)$.*

PROOF: (i) Obvious: in particular the axioms of semilattices hold in T.

(ii) Take $f(x), f(y) \in f(S)$; then $f(x) \circ' f(y) = f(x \circ y) \in f(S)$.

(iii) The corestriction $f : S \longrightarrow f(S)$ is a bijective morphism, hence an isomorphism by Proposition 4.1.5.2^0. □

Corollary 4.1.1 *The concept of a subsemilattice coincides, up to an isomorphism, with the concept o a subobject in the category of semilattices* (cf. Definition 1.3.5).

PROOF: If T is a subsemilattice of S, then the inclusion mapping $i : T \longrightarrow S$ is a monomorphism because it is injective. Conversely, if $f : T \longrightarrow S$ is a monomorphism, then T is isomorphic to the subsemilattice $f(T)$ by Propositions 4.1.5 and 4.1.6. □

In Proposition 4.1.6(i), the condition that the meet operation of T should be the restriction of the meet operation of S is essential. More precisely:

Remark 4.1.2 Given a meet semilattice (join semilattice) $(S, \leq)$, a subset $T \subseteq S$ such that T itself is a meet semilattice (join semilattice) with respect to the restriction of the order of S to the elements of T, need not be a subsemilattice of $(S, \wedge)$ (of $(S, \vee)$). For instance, the V-shaped subset $T = \{a', b', 0'\}$ of the Y-shaped meet semilattice $S = \{a', b', c', 0'\}$ from Example 4.1.6 is a meet semilattice with respect to the order $\leq$ induced by S, but not a subsemilattice of S because $a' \wedge b' = c' \notin T$.

The above results clearly demonstrate the convenience of working with the categories **MSem, JSem** and **Sem** of semilattices and semilattice homomorphisms. However it turns out to be worth noting the behaviour of semilattices with respect to mappings that are just isotone.

Proposition 4.1.7 *Let S and S' be meet semilattices (join semilattices) and $f : S \longrightarrow S'$ an isotone mapping. Then:*

(i) *The relation*

$$f(x \wedge y) \leq f(x) \wedge f(y) \tag{4.1.15'}$$

(the relation

$$(4.1.15'') \qquad f(x) \vee f(y) \leq f(x \vee y)\ )$$

holds for every $x, y \in S$.

(ii) *The set* $f(S)$ *is a semilattice under the operation* $\bullet$ *defined by*

$$(4.1.16) \qquad f(x) \bullet f(y) = f(x \circ y)\ ,$$

where $\circ$ *denotes the meet (join) operation of* S.

PROOF: (i) From $x \wedge y \leq x$ we get $f(x \wedge y) \leq f(x)$ and similarly $f(x \wedge y) \leq f(y)$, thus implying (4.1.15′). Property (4.1.15″) follows by duality.

(ii) The associativity, commutativity and idempotency of the operation $\bullet$ follow from the corresponding properties of $\circ$, for instance,

$$f(x \bullet y) = f(x \circ y) = f(y \circ x) = f(y) \bullet f(x)\ .$$

□

There are actually cases in which the relations (4.1.15) are fulfilled as strict inequalities, so that the operation $\bullet$ defined by (4.1.16) is not the restriction of $\circ$ and $f(S)$ is not a subsemilattice of S'. For instance, for the mapping $f : V \longrightarrow Y$ in Example 4.1.6 we have $a' \bullet b' = f(a \wedge b) = 0'$, whereas $a' \wedge b' = c'$.

Incidentally, in the above example $f(V)$ is isomorphic to V, but this need not hold in general; se e.g. the proof of Corollary 2.1.1.

Exercise 4.1.4 Study the full subcategory of **Pos** consisting in all meet semilattices (join semilattices) and all order homomorphisms between them.

We now resume the above study within the more restrictive but important framework of lattices.

Definition 4.1.6 A *lattice* is a poset $(L, \leq)$ which is both a meet semilattice and a join semilattice with respect to $\leq$.

In other words, a lattice is characterized by the fact that every two elements x, y have a g.l.b. $x \wedge y$ and a l.u.b. $x \vee y$.

Example 4.1.11 The empty set $\varnothing$ is a lattice (cf. Example 2.1.1).

Example 4.1.12 Every totally ordered set is a lattice, with $x \wedge y = \min(x, y)$ and $x \vee y = \max(x, y)$.

Example 4.1.13 The set $\mathbb{N} = \{0, 1, 2, \ldots\}$ of natural numbers, endowed with the divisibility relation $\mid$, is a lattice; here the lattice-theoretical operations $\wedge$ and $\vee$ are g.c.d. and l.c.m., respectively (cf. Example 2.1.5).

Example 4.1.14 The poset of all functions $f : \mathbb{R} \longrightarrow \mathbb{R}$, endowed with the pointwise order, is a lattice (cf. Example 2.1.7).

Example 4.1.15 If X is an arbitrary set and $L \neq \varnothing$ is a lattice, then the poset L^X of all mappings $f : X \longrightarrow L$ is a lattice (cf. Examples 2.1.7 and 4.1.3).

Example 4.1.16 If L_1 and L_2 are lattices, then so is their direct product (cf. Examples 4.1.4 and 2.1.7).

Example 4.1.17 If I is a lattice and $\{L_i\}_{i\in I}$ a family of lattices with 0 and 1, then the sum $\sum_{i\in I} L_i$ is also a lattice (cf. Examples 4.1.5 and 2.1.10).

Example 4.1.18 The set 2^X of all subsets of an arbitrary set, equipped with set inclusion, is a lattice. Here the lattice-theoretical operations of meet and join are the set-theoretical operations of intersection and union, respectively (cf. Example 2.1.7).

Example 4.1.19 The set of all subgroups (normal subgroups) of a group G is a lattice with respect to set inclusion. Here meet coincides with set-theoretical intersection, while the join of two subgroups (normal subgroups) H_1 and H_2 is the subgroup (normal subgroup) generated by the set $H_1 \cup H_2$.

Numerous variants of this example can be easily constructed for other algebraic structures; cf. §4.

Example 4.1.20 The three-dimensional space $\mathbb{R}^3$, the planes, the straight lines, the points and the empty set $\varnothing$ form a lattice under set inclusion. For instance, if x is a line and y a point not on x, then $x \wedge y = \varnothing$, while $x \vee y = \mathbb{R}^3$.

Proposition 4.1.8 *If $(L, \leq)$ is a lattice, then the system $(L, \wedge, \vee)$, where $\wedge, \vee : L^2 \longrightarrow L$ are defined by*

$$x \wedge y = \text{g.l.b.}\{x, y\},\ x \vee y = \text{l.u.b.}\{x, y\}\ , \tag{4.1.1}$$

satisfies

$$(x \wedge y) \wedge z = x \wedge (y \wedge z),\ (x \vee y) \vee z = x \vee (y \vee z)\ , \tag{4.1.4}$$

$$x \wedge y = y \wedge x,\ x \vee y = y \vee x\ , \tag{4.1.5}$$

$$x \wedge x = x,\ x \vee x = x\ , \tag{4.1.6}$$

$$x \leq y \Longleftrightarrow x \wedge y = x \Longleftrightarrow x \vee y = y\ , \tag{4.1.7}$$

and

$$x \wedge (x \vee y) = x \vee (x \wedge y) = x\ \ \text{(absorption)}\ . \tag{4.1.17}$$

PROOF: Immediate from Proposition 4.1.1, plus a mere translation of $x \wedge y \leq x \leq x \vee y$. □

The following converse of Proposition 4.1.8 is also valid:

Proposition 4.1.9 *Let the system $(L, \wedge, \vee)$, where $\wedge, \vee : L^2 \longrightarrow L$, fulfill* (4.1.4), (4.1.5) *and* (4.1.17). *Then:*

(i) *Properties* (4.1.6) *and*

(4.1.18) $$x \wedge y = x \Longleftrightarrow x \vee y = y$$

also hold.

(ii) *The binary relation* $\leq$ *on* L *defined by*

(4.1.19) $$x \leq y \Longleftrightarrow x \wedge y = x \Longleftrightarrow x \vee y = y$$

makes L *into a lattice* $(L, \leq)$ *having* $\wedge$ *and* $\vee$ *as meet and join operation.*

PROOF: (i) Use (4.1.17) twice, taking first an arbitrary y, then $y := x \wedge y$; this yields $x \wedge x = x \wedge (x \vee (x \wedge y)) = x$ and similarly $x \vee x = x$. Further, if $x \wedge y = x$, then $x \vee y = (x \wedge y) \vee y = y \vee (y \wedge x) = y$ by (4.1.5) and (4.1.17); the other implication (4.1.18) is established similarly.

(ii) The definition (4.1.19) is coherent by (4.1.18). Besides, by applying Proposition 4.1.2, (i) and (ii), to $(L, \wedge)$ and to $(L, \vee)$, respectively, it follows that $(L, \leq)$ is a meet semilattice with $\wedge$ as meet operation and also a join semilattice with $\vee$ as join operation. □

Summarizing, if $(L, \leq)$ is a lattice, then Proposition 4.1.8 constructs an algebra $(L, \wedge, \vee)$, where the operations are commutative, associative, idempotent and satisfy the absorption laws. Conversely, if $(L, \wedge, \vee)$ satisfies commutativity, associativity and absorption, then idempotency is also satisfied and Proposition 4.1.9 constructs a lattice $(L, \leq)$ in which g.l.b.$\{x, y\}$ and l.u.b.$\{x, y\}$ are precisely $x \wedge y$ and $x \vee y$, respectively. Moreover, this correspondence is a bijection; we postpone the proof to the next theorem.

We have thus extended the construction already done for semilattices and, just as in the case of semilattices, the bijection can be extended to an isomorphism of categories, as we are going to show.

Definition 4.1.7 A mapping $f : L \longrightarrow L'$ between two lattices is called a *lattice homomorphism* if it is both a meet homomorphism and a join homomorphism.

This concept is actually stronger than those of meet homomorphism and join homomorphism, as shown e.g. by the following extension of Example 4.1.6:

Example 4.1.21 Take the two lattices in Fig. 4.2 and the function f defined by $f(x) = x'$ for all $x \in \{0, a, b, 1\}$. Then f is a join homomorphism but not a meet homomorphism, because $f(a \wedge b) = 0'$ whereas $f(a) \wedge f(b) = c'$. Therefore f is not a lattice homomorphism.

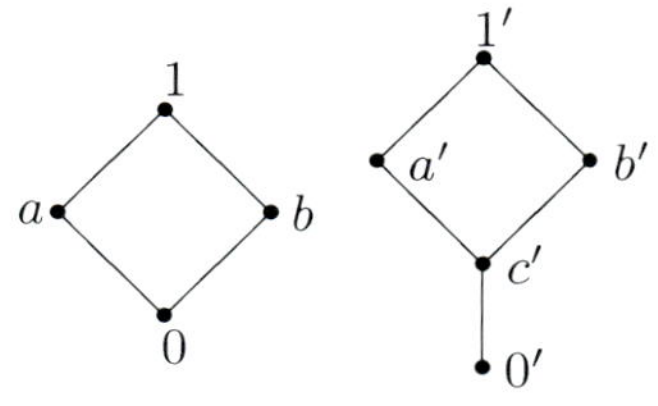

Fig. 4.2

Definition 4.1.8 Denote by **Lat** the category of all lattices and all lattice homomorhisms between them. Denote by **Lat**$'$ the category of all algebras $(L, \wedge, \vee)$ which satisfy associativity (4.1.4), commutativity (4.1.5) and absorption (4.1.17), the morphisms being those functions between them that preserve $\wedge$ and $\vee$, i.e., satisfying (4.1.9$'$) and (4.1.9$''$).

Warning The definition of **Lat**$'$ is symmetric with respect to $\wedge$ and $\vee$. However we have in mind that the objects of **Lat**$'$ are associated with lattices $(L, \leq)$ by (4.1.1), that is, $x \wedge y$ and $x \vee y$ are g.l.b. and l.u.b., respectively. The order in which one indicates the two binary operations is, of course, a matter of convention. In this eBook we adopt the convention of writing $\wedge$ (meet) first.

Theorem 4.1.2 *The categories* **Lat** *and* **Lat**$'$ *are isomorphic.*

Comment This theorem means that Propositions 1.4.8 and 1.4.9 establish a bijection between lattices and algebras $(L, \wedge, \vee)$ satisfying associativity, commutativity and absorption, while the morphisms are the same in both categories.
PROOF: For every $(L, \leq) \in$ **Lat** define $F(L, \leq) = (L, \wedge, \vee)$, where $x \wedge y =$ g.l.b.$\{x, y\}$ and $x \vee y =$ l.u.b.$\{x, y\}$; then $F(L, \leq) \in$ **Lat**$'$ by Proposition 4.1.8.

If $(L', \leq') \in$ **Lat** and $F(L', \leq') = (L', \wedge', \vee')$, then every morphism $f \in$ **Lat**(L, L') satisfies $f(x \wedge y) = f(x') \wedge' f(y')$ and $f(x \vee y) = f(x') \vee' f(y')$ by Definition 4.1.3, hence $f : F(L, \leq) \to F(L', \leq')$ is a morphism of **Lat**$'$.

Therefore, setting also $Ff = f$, we get a functor $F :$ **Lat** $\to$ **Lat**$'$.

For every $(L, \wedge, \vee) \in$ **Lat**$'$ define $G(L, \wedge, \vee) = (L, \leq)$, where $x \leq y \Longleftrightarrow x \wedge y = x \Longleftrightarrow x \vee y = y$. The definition is meaningful and we have $G(L, \wedge, \vee) \in$ **Lat** by Proposition 4.1.9.

If $(L', \wedge', \vee') \in$ **Lat**$'$ and $G(L', \wedge', \vee') = (L', \leq')$, then every morphism $f \in$ **Lat**$'(L, L')$ satisfies $f(x \wedge y) = f(x') \wedge' f(y')$ and $f(x \vee y) = f(x') \vee' f(y')$, hence $f : G(L, \leq) \to G(L', \leq')$ is a lattice homomorphism, again by Proposition 4.1.9.

Therefore, setting also $Gf = f$, we get a functor $G :$ **Lat**$'$ $\to$ **Lat**.

It remains to prove that relations $G \circ F = I_{\mathbf{Lat}}$ and $F \circ G = I_{\mathbf{Lat}'}$ hold on objects.

If $(L, \leq) \in$ **Lat**, then setting $F(L, \leq) = (L, \wedge, \vee)$, we have $GF(L, \leq) = G(L, \wedge, \vee) = (L, \leq')$, where $x \leq' y \Longleftrightarrow x \wedge y = x \Longleftrightarrow$ g.l.b.$\{x, y\} = x \Longleftrightarrow x \leq y$, therefore $GF(L, \leq) = (L, \leq)$.

Now we use the functors $F' :$ **MSem** $\to$ **Sem** and $G' :$ **Sem** $\to$ **MSem** constructed in the proof of Theorem 4.1.1. If $(L, \wedge, \vee) \in$ **Lat**$'$, set $G(L, \wedge, \vee) = (L, \leq)$ and $FG(L, \wedge, \vee) = F(L, \leq) = (L, \wedge', \vee')$. Then $F(L, \leq) \in$ **Sem** and $x \wedge' y =$ g.l.b.$\{x, y\}$, hence $(L, \wedge') = F'(L, \leq)$. Also, $(L, \wedge) \in$ **MSem** and $x \leq y \Longleftrightarrow x \wedge y = x$, hence $(L, \leq) = G'(L, \wedge)$. Therefore $(L, \wedge') = F'G'(L, \wedge) = (L, \wedge)$ and similarly $(L, \vee') = (L, \vee)$, that is, $FG(L, \wedge, \vee) = (L, \wedge, \vee)$. □

Warning See the warning to Theorem 4.1.1.

Now we adopt the following analogue of Definition 4.1.Sem:

Definition 4.1.Lat The term *lattice* is a common name for lattices in the sense of Definition 4.1.6 and algebras $(L, \wedge, \vee)$ satisfying (4.1.4), (4.1.5) and (4.1.7). The term *lattice homomorphism* designates any of their morphisms.

From a practical point of view, we are entitled to use simultaneously all the properties of lattices (and semilattices!), both order-theoretical and algebraic. In this context it is worth mentioning that lattices were first introduced by Dedekind as algebras $(L, \wedge, \vee)$; later on, Ore gave their equivalent definition as special posets $(L, \leq)$.

Remark 4.1.3 Let $(L, \leq)$ be a lattice and let $\wedge$ and $\vee$ be its meet and join operation, that is, $F(L, \leq) = (L, \wedge, \vee)$; then $F(L, \geq) = (L, \vee, \wedge)$. Therefore the *duality principle* holds for lattices: from any theorem concerning lattices, a dual theorem is obtained by interchanging $\leq$ with $\geq$ and $\wedge$ with $\vee$ (from a practical point of view it is often preferred to replace $x \leq y$ by $y \leq x$ rather than to replace $\leq$ by $\geq$). $(L, \wedge, \vee)$ and $(L, \vee, \wedge)$ are said to be *dual* lattices.

The first part of Remark 4.1.3 can be reformulated as follows. Let $(L, \circ, *)$, where $\circ, * : L^2 \longrightarrow L$, fulfill (4.1.4), (4.1.5) and (4.1.17). Then we can interpret either $\circ$ as the meet and $*$ as the join of a lattice, or $*$ as the meet and $\circ$ as the join of a lattice; the two lattices obtained in this way are $G(L, \circ, *)$ and its dual.

The next two Propositions extend certain properties from semilattices to lattices. This is possible due to the fact that lattice homomorphisms coincide with functions that are both meet homomorphisms and join homomorphisms.

Proposition 4.1.10 *In the category* **Lat***:*

1^0 Monomorphisms coincide with injective morphisms.

2^0 The following conditions are equivalent for a mapping $f : L \longrightarrow L'$ between two lattices L and L';

(i) *f is an isomorphism;*

(ii) *f is a bijective morphism;*

(iii) *f is an order isomorphism.*

PROOF: Immediate from Proposition 4.1.5. □

Now we introduce:

Definition 4.1.9 A *sublattice* of a lattice $(L, \wedge, \vee)$ is a subset $K \subseteq L$ such that $x, y \in K \Longrightarrow x \wedge y, x \vee y \in K$.

Proposition 4.1.11 (i) *A subset K of a lattice L is a sublattice of $(L, \wedge, \vee)$ if and only if K is itself a lattice $(K, \wedge, \vee)$ under the restrictions of the operations $\wedge, \vee$ of L to the elements of K.*

(ii) *If $f : L \longrightarrow L'$ is a lattice homomorphism, then $f(L)$ is a sublattice of L'.*

(iii) *If, moreover, f is injective, then L is isomorphic to $f(L)$.*

PROOF: Immediate from Proposition 4.1.6. □

Corollary 4.1.2 *The concept of a sublattice coincides, up to an isomorphism, with the concept of a subobject in the category* **Lat***;* cf. Definition 1.3.5.

PROOF: As for Corollary 4.1.1, but using Propositions 4.1.10 and 4.1.11. □

Remark 4.1.4 Given a lattice $(L, \leq)$, a subset $K \subseteq L$ such that K is itself a lattice with respect to the restriction of $\leq$ to the elements of K, need not be a sublattice of $(L, \wedge, \vee)$. For instance, see the subset $\{0', a', b', 1'\}$ of the lattice $\{0', a', b', c', 1'\}$ in Example 4.1.21, Fig. 4.2.

Remark 4.1.5 In view of Theorems 4.1.1 and 4.1.2 and of Propositions 4.1.5.2^0 and 4.1.10.2^0, whenever we work with (some of) the categories **Pos**, **MSem**, **JSem**, **Sem** and **Lat**, we can unambiguously use the term *isomorphism* without further specification.

Exercise 4.1.5 Study the following categories: 1) all lattices and all meet homomorphisms; 2) all lattices and all join homomorphisms; 3) all lattices and all order homomorphisms.

Exercise 4.1.6 Show that the following concepts coincide in the category **Tos**: 1) order homomorphism; 2) meet homomorphism; 3) join homomorphism; 4) lattice homomorphism.

Exercise 4.1.7 Prove that $\{(4.1.4), (4.1.5), (4.1.17)\}$ is an independent system of axioms for lattices. See also Padmanabhan and Rudeanu [2008].

We conclude with a few words about the category **ALat**, obtained from the category **Lat** much in the same way as the category **APos** has been obtained from the category **Pos**. More precisely, the objects of **ALat** are all lattices, while the morphisms from L to L' are the functions $f : L \longrightarrow L'$ which *interchange* $\wedge$ *and* $\vee$, i.e., such that

$$f(x \wedge y) = f(x) \vee f(y), \quad f(x \vee y) = f(x) \wedge f(y) \tag{4.1.9*}$$

for every $x, y \in L$. The morphisms (isomorphisms) of the category **ALat** are referred to as *dual lattice homomorphisms* (*dual lattice isomorphisms*); they are particular cases of antitone mappings. The properties of the category **Alat** are parallel to those of **Lat** and left to the reader.

4.2 Modular, Distributive and Boolean Lattices

Lattice theory pays special attention to the classes of modular, distributive and Boolean lattices, which appear in other fields of mathematics as well. In particular Boolean lattices, alias Boolean algebras, have numerous applications both at the mathematical and the practical level. In this section we give several characterizations and simple properties of these types of lattices.

The starting point is the remark that if x, y, z are elements of an arbitrary lattice, then

$$(x \wedge y) \vee (y \wedge z) \vee (x \wedge z) \leq (x \vee y) \wedge (y \vee z) \wedge (x \vee z) \tag{4.2.1}$$

and

(4.2.2) $$x \leq z \Longrightarrow x \vee (y \wedge z) \leq (x \vee y) \wedge z$$

because each term from the left side of $\leq$ is included in each term of the right side. The properties obtained from (4.2.1) and (4.2.2) by replacing the sign $\leq$ by $=$ characterize two important proper subclasses of **Lat**.

We begin this study with (4.2.2).

Theorem 4.2.1 *The following conditions are equivalent for a lattice L:*

(4.2.3) $$x \leq z \Longrightarrow x \vee (y \wedge z) = (x \vee y) \wedge z \; ;$$

(4.2.4) $$x \leq z \Longrightarrow (x \wedge y) \vee (y \wedge z) \vee (x \wedge z) = (x \vee y) \wedge (y \vee z) \wedge (x \vee z) \; ;$$

(4.2.5) $$x \leq z \;\&\; (\, \exists t \;\; x \wedge t = z \wedge t \;\&\; x \vee t = z \vee t \,) \Longrightarrow x = z \; ;$$

(4.2.6) *L includes no sublattice isomorphic to the lattice in* Fig. 4.3 .

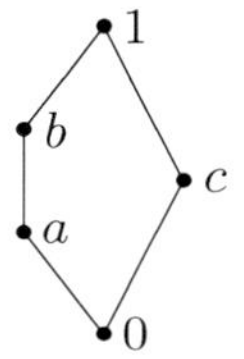

Fig. 4.3

PROOF: (4.2.3)$\Longrightarrow$(4.2.4): If $x \leq z$ then

$$(x \wedge y) \vee (y \wedge z) \vee (x \wedge z) = (x \wedge y) \vee (y \wedge z) \vee x = x \vee (y \wedge z)$$

$$= (x \vee y) \wedge z = (x \vee y) \wedge z \wedge (y \vee x) = (x \vee y) \wedge (x \vee z) \wedge (y \vee z) \, .$$

(4.2.4)$\Longrightarrow$(4.2.5): Assume (4.2.4) and the hypotheses of (4.2.5). Then

$$z = z \wedge (z \vee t) \wedge (z \vee t) = (x \vee z) \wedge (z \vee t) \wedge (x \vee t)$$

$$= (x \wedge z) \vee (z \wedge t) \vee (x \wedge t) = x \vee (x \wedge t) = x \, .$$

(4.2.5)$\Longrightarrow$(4.2.6): It suffices to prove that the lattice in Fig. 4.3 does not verify (4.2.5). Indeed, $a < b$ and $(a \wedge c) \vee (c \wedge b) \vee (a \wedge b) = 0 \vee 0 \vee a = a$, but $(a \vee c) \wedge (c \vee b) \wedge (a \vee b) = 1 \wedge 1 \wedge b = b$.

(4.2.6)$\Longrightarrow$(4.2.3): Assuming that (4.2.3) is not satisfied for some $p, q, r \in L$, we shall prove that the elements $0 = q \wedge r$, $a = p \vee (q \wedge r)$, $b = (p \vee q) \wedge r$, $c = q$ and $1 = p \vee q$ form a sublattice isomorphic to that in Fig. 4.3.

Since (4.2.3) is anyway fulfilled for $x = z$, the assumption we have made amounts to $p < r$ and $p \vee (q \wedge r) < (p \vee q) \wedge r$. Thus

(4.2.7) $$0 \leq a < b \leq 1 \;\text{ and }\; 0 \leq c \leq 1 \, .$$

From (4.2.7) we infer $0 \leq a \wedge c \leq b \wedge c = (p \vee q) \wedge r \wedge q = q \wedge r = 0$, therefore

(4.2.8) $$0 = a \wedge c = b \wedge c\,.$$

Also, $1 = p \vee q \vee (q \wedge r) = a \vee c \leq b \vee c \leq (p \vee q) \vee q = 1$, hence

(4.2.9) $$1 = a \vee c = b \vee c\,.$$

Now we prove that $a\#c$ and $b\#c$. Indeed, using (4.2.7) and $p < r$, the opposite assumption leads to a contradiction:

$$\neg(a\#c) \text{ or } \neg(b\#c) \iff a \leq c \text{ or } c \leq a \text{ or } b \leq c \text{ or } c \leq b \iff b \leq c \text{ or } c \leq a$$

$$\iff (p \vee q) \wedge r \leq q \text{ or } q \leq p \vee (q \wedge r) \Longrightarrow p \leq q \text{ or } q \leq r$$

$$\Longrightarrow b = q \wedge r \leq a \text{ or } a = p \vee q \geq b \Longrightarrow b \leq a\,.$$

It remains to prove that all the inequalities in (4.2.7) are strict. Assuming the contrary, the previous results yield the following contradictions: $0 = a \Longrightarrow a \leq c$, $0 = c \Longrightarrow c \leq a$, $b = 1 \Longrightarrow c \leq b$, $c = 1 \Longrightarrow b \leq c$. □

Definition 4.2.1 A lattice satisfying (any of) the equivalent conditions in Theorem 4.2.1 is said to be a *modular lattice.*

Modular lattices are also called Dedekind lattices because they were introduced by Dedekind [1900], who proved that the normal subgroups of a group (cf. Example 4.1.16) satisfy (4.2.3) and began the study of lattices with this property. There are many significant examples of modular lattices which belong to the subclasses of distributive lattices and Boolean lattices, to be also studied in this section.

At this point it is convenient to introduce an informal definition: a statement which coincides with its dual is said to be *self-dual*[1] .

Remark 4.2.1 Conditions (4.2.3)-(4.2.6) are self-dual (to see this for (4.2.3) and (4.2.4) interchange x and z in the dual statement and use commutativity). Therefore the *duality principle* holds for the class of modular lattices.

Remark 4.2.2 A lattice is modular if and only if it satisfies the identity

(4.2.10) $$x \vee (y \wedge (x \vee z)) = (x \vee y) \wedge (x \vee z)\,,$$

which is obtained by taking $x \vee z$ in the rôle of z instead of assuming $x \leq z$ in (4.2.3). Thus the class of modular lattices is a *variety* or an *equational class,* which means that modular lattices can be defined by a system of identities (namely, the axioms of lattices plus (4.2.10)).

Exercise 4.2.1 Every sublattice of a modular lattice, every homomorphic image of a modular lattice and every direct product of arbitrarily many modular lattices are also modular lattices. (Hint. As a matter of fact, this is a particular case of a well-known characterization of varieties in universal algebra).

Exercise 4.2.2 If the lattice $L \neq \varnothing$ is modular, so is the lattice L^X of all

[1] The term "statement" remains undefined.

mappings $f : X \longrightarrow L$ (cf. Example 4.1.15).

Exercise 4.2.3 The lattice in Example 4.1.20 is not modular.

Exercise 4.2.4 If p, q are elements of a poset P and $p < q$, define the *interval* $[p, q] = \{x \in P \mid p \leq x \leq q\}$. Now let a, b be elements of a lattice L; for every $x \in [\, a \wedge b,\, b\,]$ set $\varphi(x) = a \vee x$ and for every $y \in [\, a,\, a \vee b\,]$ set $\psi(y) = b \wedge y$. Show that $\varphi : [\, a \wedge b,\, b,\,] \longrightarrow [\, a,\, a \vee b\,]$ and $\psi : [\, a,\, a \vee b\,] \longrightarrow [\, a \wedge b,\, b\,]$ are isotone mappings. If L is modular, then φ and $\psi = \varphi^{-1}$ are lattice isomorphisms. (Hint. Use Remark 4.1.4 and (4.2.3)).

It turns out that the equivalent conditions in Theorem 4.2.1 can be sharpened in such a way as to obtain a new collection of equivalent conditions defining an important proper subclass of the class of modular lattices. As for Theorem 4.2.1, we first remark that relations

(4.2.1′) $$x \wedge (y \vee z) \geq (x \wedge y) \vee (x \wedge z) \ ,$$

(4.2.11″) $$x \vee (y \wedge z) \leq (x \vee y) \wedge (x \vee z) \ ,$$

hold identically in any lattice. We now prove:

Theorem 4.2.2 *The following conditions are equivalent for a lattice L:*

(4.2.12′) $$x \wedge (y \vee z) = (x \wedge y) \vee (x \wedge z) \ ;$$

(4.2.12″) $$x \vee (y \wedge z) = (x \vee y) \wedge (x \vee z) \ ;$$

(4.2.13) $$(x \wedge y) \vee (y \wedge z) \vee (x \wedge z) = (x \vee y) \wedge (y \vee z) \wedge (x \vee x) \ ;$$

(4.2.14) $$(\ \exists t\ x \wedge t = z \wedge t \ \&\ x \vee t = z \vee t \Longrightarrow x = z \ ;$$

(4.2.15) *L includes no sublattice isomorphic to the lattice in* Fig.4.3 *or to the lattice in* Fig.4.4 .

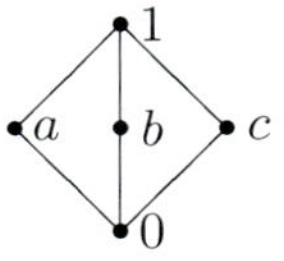

Fig. 4.4

PROOF: (4.2.12′)$\Longrightarrow$(4.2.14): If $\wedge t = z \wedge t$ and $x \vee t = z \vee t$, then

$$x = x \wedge (x \vee t) = x \wedge (z \vee t) = (x \wedge z) \vee (x \wedge t)$$

$$= (z \wedge x) \vee (z \wedge t) = z \wedge (x \vee t) = z \wedge (z \vee t) = z \ .$$

(4.2.14)$\Longrightarrow$(4.2.15): It suffices to prove that the lattices in Fig. 4.3 and Fig. 4.4 do not verify (4.2.14). We already know from Theorem 4.2.1 that the lattice in Fig. 4.3 does not fulfill condition (4.2.5), which is weaker than (4.2.14). For the lattice in Fig. 4.4, we have $a \wedge b = c \wedge b$ and $a \vee b = c \vee b$, yet $a \neq c$.

$\neg$(4.2.13) $\Longrightarrow \neg$(4.2.15): If the lattice L is modular, then (4.2.15) fails by Theorem 4.2.1. Now we suppose L is modular and prove that L includes a sublattice isomorphic to that in Fig. 4.4.

Take $p, q, r \in L$ such that $(p \wedge q) \vee (q \wedge r) \vee (p \wedge r) < (p \vee q) \wedge (q \vee r) \wedge (p \vee r)$. Define $0 = (p \wedge q) \vee (q \wedge r) \vee (p \wedge r)$, $1 = (p \vee q) \wedge (q \vee r) \wedge (p \vee r)$, $a = (q \wedge r) \vee (p \wedge (q \vee r))$, $b = (p \wedge r) \vee (q \wedge (p \vee r))$, $c = (p \vee q) \wedge (r \vee (p \wedge q))$ and prove that $\{0, a, b, c, 1\}$ is the sought lattice. We have $0 < 1$ and the symmetry of the above formulas will spare 2/3 of the subsequent proofs.

From $(p \wedge q) \vee (q \wedge r) \leq p \wedge (q \vee r)$ we get $0 \leq a$. The modular law implies $a = ((q \wedge r) \vee p) \wedge (q \vee r)$ and since $p \vee (q \wedge r) \leq (p \vee q) \wedge (p \vee r)$, we obtain $a \leq 1$. Therefore

$$0 \leq a, b, c \leq 1 \ .$$

Now $a \vee b \leq (q \wedge r) \vee (p \wedge r) \leq 0$, hence $a \vee b = 0$. But we have also $b = (p \vee r) \wedge (q \vee (p \wedge r))$, hence $a \wedge b \geq (q \vee r) \wedge (p \vee r) \geq 1$, which implies $a \wedge b = 1$. Therefore

$$a \wedge b = b \wedge c = c \wedge a = 1 \ \ \& \ \ a \vee b = b \vee c = c \vee a = 0 \ ,$$

so that it remains to prove that $a \# b$ and $b \# c$ and $a \# c$. Indeed, suppose e.g. that $a \leq b$. Then $a = a \wedge b = 0 = a \wedge c \leq c$, hence $b = a \vee b = 1 = a \vee c = c$, therefore $1 = 1 \wedge 1 = b \wedge c = 0$, a contradiction.

(4.2.13)$\Longrightarrow$(4.2.12$'$): Since property (4.2.13) is stronger than (4.2.5), Theorem 4.2.1 shows that L is modular. Then, using modularity with $x := (x \wedge y) \vee (x \wedge z)$, $y := y \wedge z$, $z := x$, we get

$$x \wedge (y \vee z) = x \wedge (x \vee y) \wedge (x \vee z) \wedge (y \vee z) = x \wedge ((x \wedge y) \vee (x \wedge z) \vee (y \wedge z))$$

$$= ((x \wedge y) \vee (x \wedge z)) \vee (y \wedge z \wedge x) = (x \wedge y) \vee (x \wedge z) \ .$$

Thus properties (4.2.12$'$), (4.2.14), (4.2.15) and (4.2.13) are equivalent. Therefore (4.2.12$''$) and (4.2.13) are also equivalent by duality. □

Definition 4.2.2 A lattice satisfying (any of) the equivalent conditions in Theorem 4.2.2 is said to be a *distributive lattice*.

Example 4.2.1 The empty set $\varnothing$ is a distributive lattice (cf. Example 4.1.11).

Example 4.2.2 Every totally ordered set is a distributive lattice (cf. Example 4.1.12).

Example 4.2.3 The set $\mathbb{N} = \{0, 1, 2, \ldots\}$ of natural numbers, endowed with the divisibility relation $|$, is a distributive lattice (cf. Example 4.1.13). This may be a tool for proving certain number-theoretical properties. For instance, the distributivity of the greatest common divisor over the least common multiple and the divisibility of the l.c.m. over the g.c.d. require only one arithmetical proof, as identities (4.1.12$'$) and (4.1.12$''$) are equivalent in any lattice.

Example 4.2.4 The lattice of all functions $f : \mathbb{R} \longrightarrow \mathbb{R}$ is distributive (cf. Example 4.1.14).

Example 4.2.5 If X is an arbitrary set and $L \neq \varnothing$ is a distributive lattice, then the lattice L^X is a distributive lattice (cf. Example 4.1.15).

Example 4.2.6 In particular the lattice $(2^X, \cup, \cap)$ of all subsets of a set X, is distributive (cf. Example 4.1.18).

Example 4.2.7 It can be proved that the modular lattice of all normal subgroups of a group is not distributive (cf. Example 4.1.19).

Other significant examples of distributive lattices belong to the proper subclass of Boolean algebras, which is the next topic in this section.

The following basic properties of distributive lattices are at hand.

Remark 4.2.3 Every distributive lattice is modular. A modular lattice is not distributive if and only if it includes a sublattice isomorphic to that in Fig. 4.4.

Remark 4.2.4 The class of distributive lattices is equational and the principle of duality holds in this class. For distributive lattices can be defined by a self-dual system of equations for lattices plus the pair of dual axioms (4.2.12′), (4.2.12″).

Exercise 4.2.5 Every sublattice of a distributive lattice, every homomorphic image of a distributive lattice and every direct product of arbitrarily many distributive lattices are also distributive lattices. (Hint. See Exercise 4.2.1).

Now we are going to prepare the introduction of the subclass of Boolean algebras.

Definition 4.2.3 In the sequel we denote by 0 (by 1) the least (greatest) element of a lattice, if any.

Remark 4.2.5 Every finite lattice has 0 and 1.

Remark 4.2.6 The dual of a lattice with 0 and 1 is also a lattice with 1 and 0. Therefore the duality principle holds for lattices with 0 and 1; in applying it one interchanges $\leq$ with $\geq$, $\wedge$ with $\vee$, and 0 with 1.

Proposition 4.2.1 *The following properties hold in a lattice with* 0 *and* 1*:*

(4.2.16) $$x \wedge 0 = 0, \; x \vee 1 = 1 \, ,$$

(4.2.17) $$x \wedge 1 = x, \; x \vee 0 = x,$$

(4.2.18) $$x \wedge y = 1 \Longleftrightarrow x = y = 1, \;\; x \vee y = 0 \Longleftrightarrow x = y = 0 \, .$$

PROOF: Immediate from $0 \le x \le 1$ and (4.1.19). □

Exercise 4.2.6 The class of lattices $(L, \wedge, \vee, 0, 1)$ with 0 and 1 is equational, hence it is closed with respect to sublattices, homomorphic images and direct products of arbitrarily many factors. Therefore so is the class of distributive lattices with 0 and 1. (Hint. See Remark 4.2.4 and Exercise 4.2.5).

Definition 4.2.4 Let L be a lattice with 0 and 1. The *complement* of an element $x \in L$ is an element $x' \in L$, if any, such that

(4.2.19) $$x \wedge x' = 0,\ x \vee x' = 1\ .$$

Every element having a complement is said to be *complemented* and if every element of a lattice L is complemented then L is said to be *complemented.*

Remark 4.2.7 Let L be a lattice with 0 and 1. Let further x be a complemented element and x' a complement of x; then x is a complement of x' and so x' is complemented. In particular

(4.2.20) $$0' = 1,\ 1' = 0\ .$$

An element x of a lattice with 0 and 1 may have several complements, a single complement or none. Thus, for instance, in Fig. 4.3 the element c is the single complement of a and of b, while the complements of c are a and b. In Fig. 4.4 each of the elements a, b, c has the two other elements as complements. It follows immediately from (4.2.17) and (4.2.19) that in every lattice with 0 and 1 the unique complement of 0 (of 1) is 1 (is 0). In a chain with 0 and 1, every element distinct from 0 and 1 has no complement (because (4.2.19) implies $x = 0$ or $x = 1$ according as $x \le x'$ or $x' \le x$).

Several supplementary properties of complements are valid in the case of distributive lattices.

Proposition 4.2.2 *Let L be a distributive lattice with 0 and 1. Then the complement of an element $x \in L$ either does not exist or is unique.*

PROOF: If x' and y are complements of x, then

$$y = y \wedge 1 = y \wedge (x \vee x') = (y \wedge x) \vee (y \wedge x') = 0 \vee (y \wedge x')$$
$$= (x \wedge x') \vee (y \wedge x') = (x \vee y) \wedge x' = 1 \wedge x' = x'\ .$$

□

Proposition 4.2.3 *Assume the elements x,y of a lattice L with 0 and 1 are complemented. Then $x \wedge y$ and $x \vee y$ are also complemented and*

(4.2.21) $$(x \wedge y)' = x' \vee y',\ (x \vee y)' = x' \wedge y' \text{ (De Morgan's laws)}\ ,$$

(4.2.22) $$x'' = x \text{ (principle of double negation)}\ ,$$

(4.2.23) $$x \wedge (x' \vee y) = x \wedge y,\ x \vee (x' \wedge y) = x \vee y,$$

(4.2.24) $$x \le y \Longleftrightarrow x \wedge y' = 0 \Longleftrightarrow x \vee y' = 1 \, ,$$

(4.2.25) $$x = y \Longleftrightarrow (x \wedge y') \vee (x' \wedge y) = 0 \Longleftrightarrow (x' \vee y) \wedge (x \vee y') = 1 \, ,$$

(4.2.26) $$x \le y \Longleftrightarrow y' \le x' \, .$$

PROOF: In view of the duality principle, it suffices to demonstrate half of the above properties.

If we succeed in showing that

$$(x \wedge y) \wedge (x' \vee y') = 0, \ (x \wedge y) \vee (x' \vee y') = 1 \, ,$$

this will mean that $x' \vee y'$ is a complement of $x \wedge y$, hence it is the unique complement of $x \wedge y$ by Proposition 4.2.2; in other words, this will mean that $x \wedge y$ is complemented and $(x \wedge y)' = x' \vee y'$. But

$$(x \wedge y) \wedge (x' \vee y') = (x \wedge y \wedge x') \vee (x \wedge y \wedge y') = 0 \vee 0 = 0,$$

$$(x \wedge y) \vee (x' \vee y') = (x \vee x' \vee y') \wedge (y \vee x' \vee y') = 1 \wedge 1 = 1 \, .$$

Similarly, since x is a complement of x' by Remark 4.2.7, this means that $(x')' = x$, that is, (4.2.22).

The equalities (4.2.23) are immediate.

If $x \le y$, then (4.1.10′) implies $x \wedge y' \le y \wedge y' = 0$, hence $x \wedge y' = 0$. Conversely, if $x \wedge y' = 0$, then $x \wedge y = (x \wedge y) \vee (x \wedge y') = x \wedge (y \vee y') = x$, therefore $x \le y$.

Properties (4.2.25) follow immediately from (4.2.24). Finally it follows from (4.2.24) and (4.2.22) that $x \le y \Longleftrightarrow x \wedge y' = 0 \Longleftrightarrow y' \wedge x'' = 0 \Longleftrightarrow y' \le x'$.

□

Exercise 4.2.7 Let L be a distributive lattice with 0 and 1. If $a, b, c \in L$ are complemented, so is $(a \wedge b) \vee (a' \wedge c)$; namely,

(4.2.27) $$((a \wedge b) \vee (a' \wedge c))' = (a \wedge b') \vee (a' \wedge c') \, .$$

After this preparation, we are ready to introduce *Boolean algebras*: they are complemented distributive lattices.

Yet this short definition needs a few explanations. The first remark is that the existence of 0 and 1 is implicit in the property of being complemented. But an important specification concerns what is known as the *signature* of the lattice. We have defined lattices as algebras $(L, \wedge, \vee)$. When we refer to a lattice with 0 and 1 we may understand either a lattice $(L, \wedge, \vee)$ in which there exist elements $0, 1 \in L$ satisfying (4.2.16) [or (4.2.17)], or we may understand an algebra $(L, \wedge, \vee, 0, 1)$ in which $(L, \wedge, \vee)$ is a lattice and 0,1 are *zero-ary operations*, that is, constants of L, which satisfy (4.2.16) [or (4.2.17)]. So, there are two ways of understanding "complemented lattice"; they differ by the *signature* $\wedge, \vee$ versus the *signature* $\wedge, \vee, 0, 1$. Accordingly, we have two possible interpretations for the term "complemented distributive lattice". However the uniqueness of the complement, stated in Proposition 4.2.2, enables one to introduce complementation as a unary operation which maps each x to x'. So the above informal definition of Boolean algebras generates two closely related explicit concepts:

Definition 4.2.5 A *Boolean algebra* is an algebra $(B, \wedge, \vee, ', 0, 1)$ in which $(B, \wedge, \vee, 0, 1)$ is a distributive lattice with 0 and 1, and the unary operation $' : B \longrightarrow B$ satisfies (4.2.19). A *Boolean lattice* is a complemented distributive lattice $(B, \wedge, \vee, 0, 1)$ with 0 and 1.

To really understand the difference between Boolean algebras and Boolean lattices, consider the simplest example: the lattice $B = \{0, a, b, 1\}$ in Fig. 2.2. This distributive lattice with 0 and 1 is complemented: namely, $a' = b$ and $b' = a$. Thus $(B, \wedge, \vee, 0, 1)$ is a Boolean lattice. The chain $S = \{0, a, 1\}$ is a 0-1 sublattice of the lattice B, but the 0-1 distributive lattice $(S, \wedge, \vee, 0, 1)$ is not complemented. So, the class of Boolean lattices is not closed under taking subalgebras, therefore it is not equational: it cannot be defined by a system of identities (cf. Remark 4.2.2 and Exercise 4.2.5). On the contrary, Remark 4.2.9 below shows that the class of Boolean algebras is equational. One might say that except this matter, Boolean algebras and Boolean lattices are the same.

To sum up, Boolean algebras satisfy properties (4.1.4)-(4.1.7) of lattices, properties (4.1.12)-(4.1.14) of distributive lattices, properties (4.2.16)-(4.2.18) of 0 and 1, and properties (4.2.19)-(4.2.26) of the complement. These are very useful computational rules in a Boolean algebras and may also be viewed as a highly redundant equational self-dual system of axioms for Boolean algebras. This implies the following two remarks.

Remark 4.2.8 The dual of a Boolean algebra is also a Boolean algebra. Therefore the *duality principle* holds for Boolean algebras; in applying it one interchanges $\leq$ with $\geq$, $\wedge$ with $\vee$, and 0 with 1, while axioms (4.2.19) show that $'$ remains unchanged.

Remark 4.2.9 The class of Boolean algebras is a variety. Therefore (anticipating the definitions of subalgebras and homomorphisms), every subalgebra of a Boolean algebra, every homomorphic image of a Boolean algebra, and every direct product of arbitrarily many Boolean algebras are also Boolean algebras (cf. Remark 4.2.4 and Exercise 4.2.5).

Now we need examples of Boolean algebras.

Example 4.2.8 Every singleton is a Boolean algebra, which we call the *improper* Boolean algebra, while the other Boolean algebras are *proper* Boolean algebras. Like the empty set for several classes of lattices (cf. e.g. Examples 4.1.11, 4.2.1 and also for the class of modular lattices), the improper Boolean algebra does not really interest us.

There are two standard examples of Boolean algebras.

Example 4.2.9 The *two-element Boolean algebra* is $2 = (\{0, 1\}, \wedge, \vee, ', 0, 1)$, where the operations $\wedge, \vee$ and $'$ are given by Tables 4.1 below:

$\wedge$	0	1
0	0	0
1	0	1

$\vee$	0	1
0	0	1
1	1	1

	$'$
0	1
1	0

Tables 4.1

We denote this algebra and the set $\{0, 1\}$ itself by 2. It is noteworthy that in fact Tables 4.1 describe the behaviour of 0 and 1 in any proper Boolean algebra. In other words, 2 is a subalgebra (cf. Definition 4.2.7 below) of every proper Boolean algebra.

The next example is also standard (more properly speaking, it is a class of standard examples of Boolean algebras).

Example 4.2.10 The lattice $(2^X, \cap, \cup)$ of all subsets of a set X (cf. Example 4.1.18) becomes a Boolean algebra $(2^X, \cap, \cup, \mathbf{C}, \varnothing, X)$ by adding the unary operation $\mathbf{C}$ of set-theoretical complement and the constants $\varnothing$ and X in the rôles of 0 and 1, respectively. No doubt, this is obvious to any reader familiar with set-theoretical operations. In particular $2^{\varnothing}$ is the improper Boolean algebra.

The next three examples generalize 2^X.

Example 4.2.11 A non-empty subset $\mathcal{F} \subseteq 2^X$ closed with respect to set-theoretical union, intersection and complement, i.e., such that $A, B \in \mathcal{F} \Longrightarrow A \cap B, A \cup B, \mathbf{C}A \in \mathcal{F}$, satisfies also $\varnothing, X \in \mathcal{F}$ and is a Boolean algebra. Such a Boolean algebra is called a *field of sets* and it will be proven in Chapter 5 that every Boolean algebra is isomorphic to a field of sets. A practical consequence of this result is that computations in a Boolean algebra obey exactly the same rules as computations with set-theoretical operations.

Example 4.2.12 A subset S of an infinite set X is said to be *cofinite* if its complement $X \setminus S$ with respect to X is finite. The finite and cofinite subsets of X form a field of sets; $\varnothing$ is regarded as a finite set.

Example 4.2.13 If B is a Boolean algebra an X is a set, the lattice B^X of all functions $f : X \longrightarrow B$ is also a Boolean algebra (cf. Example 4.2.5). In particular $B^{\varnothing}$ is the improper Boolean algebra.

Example 4.2.14 If B is a Boolean algebra and $a, b \in B$ with $a < b$, the interval $[a, b] = \{x \in B \mid a \leq x \leq b\}$ is a Boolean algebra with respect to the order $\leq$ induced by that of B. The meet and join in $[a, b]$ are the restrictions of the meet and join of B, so that $[a, b]$ is a sublattice of $(B, \wedge, \vee)$, the complement in $[a, b]$ is defined by $\grave{x} = a \vee (b \wedge x')$, the zero and one of $[a, b]$ are a and b.

Example 4.2.15 The complemented elements of a distributive lattice L with 0 and 1 form a 0-1-sublattice of L (i.e., the having same zero and one), and which is a Boolean lattice under the restrictions of the operations of L to the elements of B. It is known as the *Boolean algebra* $B(L)$ *of complemented elements of* L and plays an important rôle in the study of lattices related to logic.

Now we wish to introduce categories having as classes of objects the various types of lattices studied in this section; it remains to choose appropriate definitions for the morphisms. Whereas lattice homomorphisms are the natural morphisms to be taken between modular or distributive lattices, the situation is different for lattices with 0 and 1 and Boolean algebras. The point is that it is natural to require that a morphism $f : L \longrightarrow L'$ between two lattices with 0 and 1 preserve both the lattice operations and the distinguished elements 0 and 1, that is, the identities (4.1.9) and

$$f(0) = 0, \quad f(1) = 1 \tag{4.2.28}$$

should hold (where, of course, like in(4.1.9), the same symbol 0 (symbol 1) denotes the zero (the one) in both lattices L and L'). Relations (4.2.28)need not be fulfilled by an arbitrary lattice homomorphism between two lattices with 0 and 1: take e.g. a constant mapping.

Definition 4.2.6 We denote by **Mod** (by **Dis**) the category of all modular (distributive) lattices and all lattice homomorphisms between them. We denote by **Lat01** (by **Bol**) the category of all lattices with 0 and 1 (all Boolean algebras) and all 0-1-preserving lattice homomorphisms between them, that is, all mappings fulfilling (4.1.9) and (4.2.28).

So, **Mod** is a full subcategory of **Dis** and of **Lat**, while **Dis** is a full subcategory of **Lat**, and **Bol** is a full subcategory of **Lat01**.

Proposition 4.2.4 *In each of the categories* **Mod, Dis, Lat01** *and* **Bol**:

1^0 *Monomorphisms coincide with injective morphisms.*

2^0 *The following conditions are equivalent for a mapping $f : L \longrightarrow L'$ between two objects L and L':*

(i) *f is an isomorphism ;*

(ii) *f is a bijective morphism ;*

(iii) *f is an order isomorphism.*

PROOF: 1^0 The injective morphisms are monomorphisms by Remark 1.3.3. It remains to prove that in each of the above categories, non-injective morphisms are not monomorphisms.

The construction in the proof of Theorem 4.1.5 works also for the categories **Mod** and **Dis** because the chain $\{0, 1\}$ is a distributive lattice and the functions u and v being isotone, are in fact lattice homomorphisms.

For the category **Lat01** we work in the same spirit but the situation is a bit more complicated. Take $f \in \mathbf{Lat01}(L, L')$ with $f(a) = f(b)$ for some $a, b \in L$ with $a \neq b$; we may suppose $a \wedge b < a$.

If $0 < a \wedge b < a < 1$, take the chain $\{0, \frac{1}{3}, \frac{2}{3}, 1\}$ and define $u(0) = v(0) = 0$, $v(\frac{1}{3}) = a \wedge b$, $u(\frac{1}{3}) = u(\frac{2}{3}) = v(\frac{2}{3}) = a$, $u(1) = v(1) = 1$.

If $0 = a \wedge b < a < 1$, take the chain $\{0, \frac{1}{2}, 1\}$ and define $u(0) = v(0) = 0 = u(\frac{1}{2})$, $v(\frac{1}{2}) = a$, $u(1) = v(1) = 1$.

If $0 < a \wedge b < a = 1$, take the chain $\{0, \frac{1}{2}, 1\}$ and define $u(0) = v(0) = 0$, $v(\frac{1}{2}) = a \wedge b$, $u(1) = v(1) = 1 = u(\frac{1}{2})$.

If $0 = a \wedge b < a = 1$ then $b = 0$, hence $f(0) = f(1)$ and since f is a morphism of **Lat01**, this implies that $L' = \{e\}$ is a singleton. Take the chain $\{0, 1\}$ and define $u(x) = 0$, $v(x) = 1$.

In all of these cases the functions u and v are lattice homomorphisms because the source of these isotone mappings is a chain. The other verifications are left to the reader.

For the category **Bol** take $f \in \mathbf{Bol}(B, B_1)$ with $f(p) = f(q)$ for some $p, q \in B_1$ with $p \neq q$. Take the Boolean algebra $\{0, a, b, 1\}$ from Fig. 4.2 and define $u(0) = v(0) = 0 \in B_1$, $u(1) = v(1) = 1 \in B_1$, $u(a) = p$, $u(b) = p'$, $v(a) = q$, $v(b) = q'$.

Since u and v are isotone, to prove that they are lattice homomorphisms it suffices to check (4.1.9) for $x\#y$. Indeed, to verify e.g. (4.1.9′) we compute $u(a \wedge b) = u(0) = 0 = p \wedge p' = u(a) \wedge u(b)$; a similar proof holds for v.

Finally, $f \circ u$ and $f \circ v$ are morphisms of **Bol**, hence they coincide on $\{0, 1\}$. Besides, $(f \circ u)(a) = f(p) = f(q) = (f \circ v)(a)$ and similarly $(f \circ u)(b) = (f \circ v)(b)$. Yet $u \neq v$.

2^0 For the full subcategories **Mod** and **Dis** of **Lat** this follows from Proposition 4.1.10.2^0. For the category **Lat01** we offer the following proof.

(i)$\Longrightarrow$(ii): By Remark 1.3.3.

(ii)$\Longrightarrow$(iii): Since f is also a bijective morphism in **Lat**, this follows from Proposition 4.1.10.2^0.

(iii)$\Longrightarrow$(i): Proposition 2.1.5 shows that f is a bijection and $x \leq y \Longleftrightarrow f(x) \leq f(y)$. Then f^{-1} is a bijection and $f^{-1}(x') \leq f^{-1}(y') \Longleftrightarrow x' \leq y'$, hence f^{-1} is also an order isomorphism. Therefore f and f^{-1} are isomorphisms in **Lat** by Proposition 4.1.10.2^0. But clearly *every surjective isotone function between two objects of* **Lat01** *preserves* 0 *and* 1 (exercise!). Therefore f and f^{-1} are isomorphisms in **Lat01**.

Finally, **Bol** satisfies 2^0 because it is a full subcategory of **Lat01**, which verifies 2^0. □

Remark 4.2.10 Whenever we work simultaneously with (some of) the categories **Pos**, **Sem**, **Lat**, **Mod**, **Dis** and **Bol**, we can use unambiguously the term <u>isomorphism</u> without further specification.

Exercise 4.2.8 Characterize the subobjects in**Lat01**.

Now we concentrate on the category **Bol**. Here are several equivalent characterizations of its morphisms. The most important point in Proposition 4.2.5 below is the fact that besides the definition (4.1.9)&(4.2.28), a Boolean morphism satisfies also (4.2.29), that is, it preserves the complement. This meets the signature we have chosen for Boolean algebras $(B, \wedge, \vee, ', 0, 1)$.

Proposition 4.2.5 *The following conditions are equivalent for a mapping* $f : B \longrightarrow B_1$ *between two Boolean algebras:*

(i) f *is a morphism of* **Bol**, i.e., it fulfills

(4.1.9) $$f(x \wedge y) = f(x) \wedge f(y), \quad f(x \vee y) = f(x) \vee f(y) ,$$

(4.2.28) $$f(0) = 0, \quad f(1) = 1 ;$$

(ii) f *fulfills* (4.1.9′) *and*

(4.2.29) $$f(x') = (f(x))' ;$$

(iii) *f fulfills* (4.1.9″) *and* (4.2.29) ;

(iv) *f fulfills*

(4.2.30′) $$f(x' \wedge y') = (f(x))' \wedge (f(y))' ;$$

(v) *f fulfills*

(4.2.30″) $$f(x' \vee y') = (f(x))' \vee (f(y))' .$$

PROOF: (i)$\Longrightarrow$(ii): Property (4.2.29) follows from

$$f(x) \wedge f(x') = f(x \wedge x') = f(0) = 0 \;\&\; f(x) \vee f(x') = f(x \vee x') = f(1) = 1$$

via Proposition 4.2.2.

(ii)$\Longrightarrow$(iv): Immediate.

(iv)$\Longrightarrow$(i): Taking $x = y$ we get (4.2.29), which we use below:

$$f(x \wedge y) = f(x'' \wedge y'') = (f(x'))' \wedge (f(y'))' = f(x'') \wedge f(y'') = f(x) \wedge f(y) ,$$

$$f(x \vee y) = f((x' \wedge y')') = ((f(x))' \wedge (f(y))')' = f(x) \vee f(y) ,$$

then $f(0) = f(x \wedge x') = f(x) \wedge (f(x))' = 0$ and $f(1) = 1$ by duality.

Thus (i)$\Longleftrightarrow$(ii)$\Longleftrightarrow$(iv) and (i)$\Longleftrightarrow$(iii)$\Longleftrightarrow$(v) by duality. □

Exercise 4.2.9 Let $X \neq \varnothing$ be a fixed arbitrary set. The *characteristic function* c_Y of a subset $Y \subseteq X$ is the mapping $c_Y : X \longrightarrow 2$ defined by $c_Y(x) = 1$ if $x \in Y$, else $c_Y(x) = 0$. The mapping which associates with each $Y \subseteq X$ the function c_Y is an isomorphism between the field of all subsets of X and the Boolean algebra of all mappings $\varphi : X \longrightarrow 2$ (cf. Examples 4.2.10, 4.2.11 and 4.2.13).

As a matter of fact, this isomorphism justifies the common notation 2^X for the above two Boolean algebras.

Exercise 4.2.10 Let X be a set, $\mathcal{B}$ a field of subsets of X and $x_0 \in X$. The mapping f defined for every $Y \in \mathcal{B}$ by $f(Y) = 1$ if $x_0 \in Y$, else $f(Y) = 0$, is a homomorphism $f : \mathcal{B} \longrightarrow 2$ (cf. Examples 4.2.9 and 4.2.11).

Exercise 4.2.11 Let $X \neq \varnothing \neq Y, \varphi : X \longrightarrow Y$, $\mathcal{A}$ a field of subsets of X, $\mathcal{B}$ a field of subsets of Y. The mapping φ^{-1} defined for every $Z \in \mathcal{B}$ by $\varphi^{-1}(Z) = \{x \in X \mid \varphi(x) \in Z\}$, is a homomorphism $\varphi^{-1} : \mathcal{B} \longrightarrow \mathcal{A}$ (cf. Example 4.2.11).

Exercise 4.2.12 Given the Boolean algebras B and $[a, b]$ in Example 4.2.14, the mapping f defined by $f(x) = a \vee (b \wedge x)$ for every $x \in B$ is a homomorphism $f : B \longrightarrow [a, b]$.

Exercise 4.2.13 Given a family $f_i : B_i \longrightarrow C_i$ $(i \in I)$ of morphisms in **Bol**, define a morphism $f : \prod_{i \in I} B_i \longrightarrow \prod_{i \in I} C_i$ (cf. Remark 4.2.9).

We are now looking for a good definition of subalgebras of Boolean algebras. It should fit the signature we have chosen for Boolean algebras. The next remark reveals certain variants that would be not suitable.

Remark 4.2.11 Every sublattice of a Boolean algebra B is a distributive lattice by Remark 4.2.5, but need not be a Boolean algebra; take e.g. a totally ordered subset of B (cf. the discussion before Proposition 4.2.2). Even if a sublattice S of a Boolean algebra B is itself a Boolean algebra with respect to the order induced by that of B, the complement and the constants zero and one of S need not be the same as those of B : cf. Example 4.2.14.

Definition 4.2.7 A *Boolean subalgebra* of a Boolean algebra $(B, \wedge, \vee, ', 0, 1)$ is a subset $A \subseteq B$ such that $0, 1 \in A$ and if $a, b \in A$, then $a', a \wedge b, a \vee b \in A$.

Proposition 4.2.6 *The following conditions are equivalent for a subset A of a Boolean algebra B :*

(i) *A is a Boolean subalgebra of B ;*

(ii) *$A \neq \varnothing$ and if $a, b \in A$, then $a', a \wedge b \in A$;*

(iii) *$A \neq \varnothing$ and if $a, b \in A$, then $a', a \vee b \in A$;*

(iv) *$A \neq \varnothing$ and if $a, b \in A$, then $a' \wedge b' \in A$;*

(v) *$A \neq \varnothing$ and if $a, b \in A$, then $a' \vee b' \in A$;*

(vi) *A is itself a Boolean algebra $(A, \wedge, \vee, ', 0, 1)$ under the restrictions to A of the operations $\wedge, \vee, '$ of B and with the same zero and one as B.*

PROOF: (i)$\Longleftrightarrow$(vi) Obvious.

(i)$\Longrightarrow$(ii): Trivial.

(ii)$\Longrightarrow$(v) For $a' \vee b' = (a \wedge b)'$.

(v)$\Longrightarrow$(i) Take $a, b \in A$. Then $a' = a' \vee a' \in A$, $1 = a' \vee a'' \in A$, $0 = 1' \in A$, $a \wedge b = (a' \vee b')' \in A$, $a \vee b = a'' \vee b'' \in A$.

So (i)$\Longleftrightarrow$(ii)$\Longleftrightarrow$(v) and (i)$\Longleftrightarrow$(iii)$\Longleftrightarrow$(iv) by duality. □

Proposition 4.2.7 (i) *If $f : B \longrightarrow C$ is a morphism in* **Bol**, *then $f(B)$ is a Boolean subalgebra of C.*

(ii) *If, moreover, f is injective, then B is isomorphic to $f(B)$.*

PROOF: (i) From Propositions 4.2.6(iv) and 4.2.5(iv).

(ii) The corestriction $f : B \longrightarrow f(B)$ is a bijective morphism, hence an isomorphism by Proposition 4.2.4. □

We conclude this section by showing that Boolean algebras can be identified with a special type of rings.

Definition 4.2.8 By a *Boolean ring* is meant a ring in which every element is *idempotent*, i.e.,

$$x \cdot x = 1 \,. \qquad (4.2.31)$$

Proposition 4.2.8 *Every Boolean algebra $(B, \wedge, \vee, ', 0, 1)$ becomes a Boolean ring with unit $(B, +, \cdot, 0, 1)$ under the operations*

$$x + y = (x \wedge y') \vee (x' \wedge y) \,, \qquad (4.2.32)$$

(4.2.33) $$x \cdot y = x \wedge y \; ,$$

the zero and the unit of the ring being the elements 0 *and* 1 *of the Boolean algebra, respectively.*

Comments As usual, we may write xy instead of $x \cdot y$. The operation (4.2.32) is known as *symmetric difference.*

PROOF: The operation $+$ is commutative. Then (4.2.31) and (4.2.27) imply

$$(x+y)+z = (((x \wedge y') \vee (x' \wedge y)) \wedge z') \vee (((x \wedge y) \vee (x' \wedge y')) \wedge z)$$

$$= (x \wedge y' \wedge z') \vee (x' \wedge y \wedge z') \vee (x \wedge y \wedge z) \vee (x' \wedge y' \wedge z) \; ;$$

since this expression is symmetric in x, y, z, it follows that the element $x + (y + z) = (y + z) + x$ has the same expression. Clearly $x + 0 = x$ and $x + x = 0$, so that $(B, +, 0)$ is an Abelian group in which $-x = x$.

The multiplication (4.2.23) is commutative, associative, idempotent and

$$x(y+z) = (y+z)x = x \wedge ((y \wedge z') \vee (y' \wedge z)) = (x \wedge y \wedge z') \vee (x \wedge y' \wedge z)$$

$$= (x \wedge y \wedge (x' \vee z')) \vee (x \wedge z \wedge (x' \vee y')) = (x \wedge y) + (x \wedge z) = xy + xz \; .$$

Thus $(B, +, \cdot, , 0)$ is a Boolean ring and $x \cdot 1 = 1 \cdot x = x$. □

Example 4.2.16 Here is an example of a Boolean ring without unit. Let U be an infinite set and $\mathcal{P}_f(U)$ be the set of all finite subsets of U. In particular $\varnothing \in \mathcal{P}_f(U)$. (There is an axiomatic definition of finite sets, which says that a set S is finite if there is no bijection between S and a proper subset of S; this is fulfilled by $\varnothing$, because there is no proper subset of $\varnothing$). Clearly if X and Y are finite subsets of U, so is their intersection $X \cap Y$ and their symmetric difference

(4.2.32′) $$X + Y = (X \cap (U \setminus Y)) \cup ((U \setminus X) \cap Y) \; .$$

Since $(\mathcal{P}(U), \cap, \cup, \mathbf{C}, \varnothing, U)$ is a Boolean algebra (cf. Example 4.2.10), the properties used in the proof of Proposition 4.2.8 remain valid for $(\mathcal{P}_f(U), +, \cap, \varnothing)$, except $x \cdot 1 = 1 \cdot x = x$, because there is no finite subset 1 of U satisfying $X \cap 1 = X$, or, equivalently $X \subseteq 1$, for all finite $X \subset U$.

To prove the converse of Proposition 4.2.8, we first establish a lemma which shows that the properties of the Boolean rings obtained obtained in Proposition 4.2.8, except the existence of the unit 1, hold in fact in every Boolean ring.

Lemma 4.2.1 *Every Boolean ring* $(B, +, \cdot, 0)$ *is commutative and of characteristic* 2, i.e., for every $x, y \in B$,

(4.2.34) $$xy = yx \; ,$$

(4.2.35) $$x + x = 0 \; .$$

PROOF: From

$$x + y = (x + y)(x + y) = x^2 + xy + yx + y^2 = x + xy + yx + y$$

we infer $xy + yx = 0$. Taking $y := x$ we get (4.2.35), or, equivalently, $-x = x$, therefore the previous relation yields $yx = -xy = xy$. □

Proposition 4.2.9 *Every Boolean ring with unit* $(B, +, \cdot, 0, 1)$ *becomes a Boolean algebra* $(B, \wedge, \vee, 0, 1)$ *under the operations*

$$x \wedge y = x \cdot y \ , \tag{4.2.36}$$

$$x \vee y = x + y + xy \ , \tag{4.2.37}$$

$$x' = x + 1 \ , \tag{4.2.38}$$

the elements 0 *and* 1 *of the Boolean algebra being the zero and the unit of the ring, respectively.*

PROOF: $(B, \wedge)$ is a semilattice by hypothesis and lemma 4.2.1. The operation $\vee$ is clearly commutative and idempotent. Besides,

$$(x \vee y) \vee z = (x + y + xy) + z + (x + y + xy)z = x + y + z + xy + xz + yz + xyz \ ;$$

since this expression is symmetric in x, y, z, it follows that the element $x \vee (y \vee z) = (y \vee z) \vee x$ has the same expression, therefore $(B, \vee)$ is a semilattice. But

$$x \wedge (x \vee y) = x(x + y + xy) = x + xy + xy = x \ ,$$

$$x \vee (x \wedge y) = x + xy + x \cdot xy = x + xy + xy = x \ ,$$

so that $(B, \wedge, \vee)$ is a lattice. Furthermore,

$$x \wedge (y \vee x) = xy + xz + xyz = (x \wedge y) + (x \wedge z) + (x \wedge y)(x \wedge z) = (x \wedge y) \vee (x \wedge z) \ ,$$

showing the distributivity of $(B, \wedge, \vee)$. Finally,

$$x \wedge 0 = x \cdot 0 = 0, \ x \wedge 1 = x \cdot 1 = x \ ,$$

$$x \wedge x' = x(x + 1) = x + x = 0 \ ,$$

$$x \vee x' = x + (x + 1) + x(x + 1) = x + x + 1 + x + x = 1 \ .$$

□

Category theory is best qualified to express the relationship between Boolean algebras and Boolean rings with unit

Proposition 4.2.10 *The following conditions are equivalent for a mapping* $f : B \longrightarrow C$ *between two Boolean algebras:*

(i) *f is a morphism of* **Bol***;*

(ii) *f is a morphism of* **Rng1**, i.e., a ring homomorphism preserving the unit.

PROOF: (i)$\Longrightarrow$(ii) Proposition 4.2.8 and 4.2.5 imply

$$f(x+y) = (f(x) \wedge f(y)') \vee ((f(x)' \wedge f(y)) = f(x) + f(y) ,$$

$$f(xy) = f(x) \wedge f(y) = f(x)f(y) ,$$

$$f(1) = 1 .$$

(ii)$\Longrightarrow$(i) Proposition 4.2.9 implies $x' \wedge y' = (x+1)(y+1)$, hence

$$f(x' \wedge y') = (f(x) + f(1))(f(y) + f(1)) = (f(x))' \wedge f(y))' ,$$

therefore f is a morphism of **Bol** by Proposition 4.2.5. □

To state the main result, we need the following

Definition 4.2.9 Let **BRng1** be the category of all Boolean rings with unit and all 1-preserving ring homomorphisms between them.

Note that **BRng1** is a full subcategory of **Rng1**.

Theorem 4.2.3 *The categories* **Bol** *and* **BRng1** *are isomorphic.*

PROOF: For every object B of **Bol** (of **BRng1**), let FB (let GB) be the Boolean ring with unit (the Boolean algebra) constructed in Proposition 4.2.8 (Proposition 4.2.9); for every morphism f of **Bol** (of **BRng1**) let $Ff = f$ (let $Gf = f$). Then $F : \mathbf{Bol} \longrightarrow \mathbf{BRng1}$ and $G : \mathbf{BRng1} \longrightarrow \mathbf{Bol}$ are functors by Propositions 4.2.8–4.2.10. It remains to prove that GF and FG are the identity functors; this is trivial for the morphism functions and we shall prove it for the object functions.

If $(B, \wedge, \vee, ', 0, 1) \in \mathbf{Bol}$, then $FB = (B, +, \wedge, 0, 1)$ and $GFB = (B, \wedge, \sqcup, ^*, 0,$ $1)$; we have to prove that $\sqcup$ and * coincide with $\vee$ and $'$, respectively. Indeed,

$$x \sqcup y = x + y + xy = ((x+y) \wedge (xy)') \vee ((x+y)' \wedge (xy))$$

$$= (((x \wedge y') \vee (x' \wedge y)) \wedge (x' \vee y')) \vee (((x \wedge y) \vee (x' \wedge y')) \wedge (x \wedge y))$$

$$= (x \wedge y') \vee (x' \wedge y) \vee (x \wedge y) = (x \wedge y') \vee y = x \vee y ,$$

$$x^* = x + 1 = (x \wedge 1') \vee (x' \wedge 1) = 0 \vee x' = x' .$$

If $(B, +, \cdot, 0, 1) \in \mathbf{BRng1}$, then $GB = (B, \cdot, \vee, ', 0, 1)$ and $FGB = (B, \oplus, \cdot, 0,$ $1)$; we have to prove that $\oplus$ coincides with $+$. Indeed,

$$x \oplus y = xy' \vee x'y = x(y+1) + (x+1)y + xy' \cdot x'y = xy + x + xy + y + 0 = x + y .$$

□

Exercise 4.2.14 Embed every Boolean ring without unit into a Boolean ring with unit.

In conclusion we sketch a construction similar to the one which has produced the categories **Apos** and **ALat** from the categories **Pos** and **Lat**. To be specific, we introduce the categories **AMod, ADis, ALat01** and **ABol**, which have the same objects as the categories **Mod, Dis, Lat01** and **Bol**, respectively. The morphisms of the categories **AMod** and **ADis** are the dual homomorphisms between the objects, i.e., they satisfy (4.1.9*), while the morphisms of the categories **ALat01** and **ABol** are those dual homomorphisms which interchange 0 with 1, i.e., $f(1) = 0$ and $f(0) = 1$.

The properties of the new four categories are parallel to the ones studied in this section; the details are left to the reader.

4.3 Complete Lattices and Semilattices

Another specialization of lattices is obtained by requiring that every subset, not only the finite ones, have inf and sup. This is the concept of a complete lattice, which we study in the present section; we note the existence of complete infinite semilattices that are not complete lattices. We pay special attention to complete Boolean algebras, which appear in several mathematical contexts and have interesting supplementary properties. We also deal with compactly generated lattices on account of their importance in universal algebra.

Remark 4.3.1 Let A be an arbitrary subset of a poset P. Then $\bigwedge A$ (then $\bigvee A$), provided it exists, is the unique element of P such that

(4.3.1′) $$x \leq A \Longleftrightarrow x \leq \bigwedge A$$

(such that

(4.3.1″) $$A \leq x \Longleftrightarrow \bigvee A \leq x\ ) ;$$

(cf. Exercise 4.1.1).

In particular, since the properties $x \leq \varnothing \leq x$ hold for every x (the implications $t \in \varnothing \Longrightarrow x \leq t$ and $t \leq x$ are valid!), it follows from Remark 4.3.1 that

Remark 4.3.2 For every poset P, either P has no greatest element and $\bigwedge \varnothing$ does not exists (P has no least element and $\bigvee \varnothing$ does not exists), or $\bigwedge \varnothing = 1$ is the greatest element of P ($\bigvee \varnothing = 0$ is the least element of P).

Remark 4.3.3 If A is a non-empty subset of poset P, then $\bigwedge A \subseteq \bigvee A$ whenever $\bigwedge A$ and $\bigvee A$ exist.

Remark 4.3.4 If A and B are subsets of a poset P, then

(4.3.2′) $$A \subseteq B \Longrightarrow \bigwedge B \leq \bigwedge A$$

whenever $\bigwedge A$ and $\bigwedge B$ exist (then

(4.3.2″) $$A \subseteq B \Longrightarrow \bigvee A \leq \bigvee B$$

whenever $\bigvee A$ and $\bigvee B$ exist).

Definition 4.3.1 A poset $S \neq \varnothing$ is called a *complete meet semilattice* (*complete join semilattice*) if every non-empty subset of S has g.l.b. (has l.u.b.). The term *complete semilattice* is a common name for complete meet semilattices and complete join semilattices.

Trivially every complete semilattice (complete meet semilattice, complete join semilattice) is a semilattice (meet semilattice, join semilattice).

Example 4.3.1 Every finite non-empty semilattice is complete.

Example 4.3.2 Every well-ordered set is a complete meet semilattice, in which $\bigwedge X$ is the first element of X.

Remark 4.3.5 Every complete meet semilattice (complete join semilattice) S has least element $0 = \bigwedge S$ (has greatest element $1 = \bigvee S$).

Note that a complete meet semilattice is not required to contain $\bigwedge \varnothing$ and a complete join semilattice is not required to contain $\bigvee \varnothing$. That is why there exist complete meet semilattices without 1 and complete join semilattices without 0; see, for instance, Example 4.3.2 for the set $\mathbb{Z}$ and Example 4.3.6(b) below, in which there is no greatest element (here the number 1 is the least element).

Definition 4.3.2 A *complete lattice* is a poset $L \neq \varnothing$ in which every subset has g.l.b. and l.u.b.

In other words, a complete lattice is both a complete meet semilattice with $1 = \bigwedge \varnothing$ and a complete join semilattice with $0 = \bigvee \varnothing$ (cf. Remark 4.3.2). As a matter of fact, the last two conditions are equivalent: each of them ensures that the complete semilattice is in fact a complete lattice, as shown in Proposition 4.3.2 below.

Example 4.3.3 Every finite non-empty lattice is complete.

Example 4.3.4 Let $X \neq \varnothing$. If S is a complete semilattice (lattice), so is S^X (cf. Example 4.1.3).

Example 4.3.5 If S_1 and S_2 are complete semilattices (complete lattices), so is their direct product (cf. Example 4.1.4).

Example 4.3.6 (a) The lattice $(\mathbb{N}, |)$ in Example 4.1.13 is complete. (b) The pair $(\mathbb{N} \setminus \{0\}, |)$ is a complete meet semilattice with respect to the operation of taking the g.c.d.

Example 4.3.7 If I is a complete lattice and $\{L_i\}_{i \in I}$ a family of complete lattices, then their sum $\sum_{i \in I} L_i$ is a complete lattice (cf. Example 4.1.17). For instance, given a subset X and setting $J = \{i \in I \mid \exists x \ (i, x) \in X\}$, we have $\bigwedge X = (j_0, 1)$, where $j_0 = \bigwedge J$.

Example 4.3.8 The lattice of all (normal) subgroups of a group is complete (cf. Example 4.1.19). Also, the lattices of all subrings of a ring, of all ideals of a ring, of all submodules of a module, of all subspaces of a vector space, of all subspaces of a topological space, etc., are complete. The deep reason explaining this family of examples will be studied in the next section.

Exercise 4.3.1 The reader is urged to look for semilattices and lattices that are not complete among the examples in the previous two sections.

Remark 4.3.6 The dual of a complete meet (join) semilattice is a complete join (meet) semilattices; the dual of a complete lattice is a complete lattice. Therefore the *duality principle* holds for complete lattices; in applying it one interchanges $\leq$ with $\geq$, g.l.b. with l.u.b., and 0 with 1.

In a complete meet semilattice (complete join semilattice) there are many subsets which have l.u.b. (which have g.l.b.):

Proposition 4.3.1 *In a complete meet semilattice (complete join semilattice) S, every subset A which has an upper bound (lower bound) has also* l.u.b. (g.l.b.). Namely if $A^+ = \{x \in S \mid A \leq x\} \neq \varnothing$ (if $A^- = \{x \in S \mid x \leq A\} \neq \varnothing$), then $\bigvee A = \bigwedge A^+$ (then $\bigwedge A = \bigvee A^-$).

PROOF: In view of the duality principle it suffices to prove the first statement. If $A^+ \neq \varnothing$, there exists $b = \bigwedge A^+$.

If $A = \varnothing$ then $A^+ = S$, hence $b = 0 = \bigvee \varnothing = \bigvee A$ by Remarks 4.3.5 and 4.3.2. Now suppose $A \neq \varnothing$. Then for every $a \in A$ we have $a \leq u$ for every $u \in A^+$, hence $a \leq b$ for every $a \in A$. We have thus proved that $b \in A^+$, hence b is the least element of A^+, that is, $b =$l.u.b. of A. □

Now we provide several equivalent characterization of complete lattices.

Proposition 4.3.2 *The following conditions are equivalent for a poset L:*

(i) *L is a complete lattice ;*

(ii) *L has greatest element* 1 *and every non-empty subset of L has* g.l.b. ;

(iii) *L has least element* 0 *and every non-empty subset of L has* l.u.b. ;

(iv) *$L \neq \varnothing$ and every subset of L has* g.l.b. ;

(v) *$L \neq \varnothing$ and every subset of L has* l.u.b. .

PROOF: (i)$\Longrightarrow$(ii): Trivial.

(ii)$\Longrightarrow$(iv): The empty subset has also g.l.b.: $\bigvee \varnothing = 1$ by Remark 4.3.2.

(iv)$\Longrightarrow$(ii): We have $\bigwedge \varnothing = 1$

(ii)$\Longrightarrow$(i): L is a complete meet semilattice and 1 is an upper bound of every subset, therefore every subset has l.u.b. by Proposition 4.3.1.

It follows that (i)$\Longleftrightarrow$(ii)$\Longleftrightarrow$(iv), therefore (i)$\Longleftrightarrow$(iii)$\Longleftrightarrow$(v) by duality. □

The following laws of *infinite associativity* generalize identities (4.1.14) in Exercise 4.1.2:

Proposition 4.3.3 *The identity*

(4.3.3′) $$\bigwedge\{\bigwedge A_i \mid i \in I\} = \bigwedge \bigcup_{i\in I} A_i$$

(the identity

(4.3.3″) $$\bigvee\{\bigvee A_i \mid i \in I\} = \bigvee \bigcup_{i\in I} A_i \;)$$

holds for every non-empty family $(A_i)_{i\in I}$ *of subsets of a complete meet semilattice (complete join semilattice). In a complete lattice both relations* (4.3.4) *hold.*

PROOF: In view of duality it suffices to prove (4.3.3′). Set $\bigcup_{i\in I} = A$.

For each $i \in I$ we have $A_i \subseteq A$, hence $\bigwedge A \leq \bigwedge A_i$ by Remark 4.3.4. So $\bigwedge A$ is a lower bound of the set $\{\bigwedge A_i \mid i \in I\}$, therefore

(4.3.4.1) $$\bigwedge A \leq \bigwedge\{\bigwedge A_i \mid i \in I\}\ .$$

The converse inequality holds for $A = \varnothing$ because $\bigwedge \varnothing = 1$. To prove it for $A \neq \varnothing$ we note that for each $a \in A$ there exists $i \in I$ such that $a \in A_i$, hence for that i we have $\bigwedge A_i \leq a$, therefore $\bigwedge\{\bigwedge A_i \mid i \in I\} \leq a$. We have thus proved that $\bigwedge\{\bigwedge A_i \mid i \in I\}$ is a lower bound of A, therefore

(4.3.4.2) $$\bigwedge\{\bigwedge A_i \mid i \in I\} \leq \bigwedge A\ .$$

□

The following *Knaster-Tarski theorem* establishes an important property of complete lattices.

Proposition 4.3.4 *Every isotone mapping f of a complete lattice L into itself has fixed points,* i.e., $f(a) = a$ for some $a \in L$. *More precisely, the least (greatest) fixed point is the least (greatest) element of the set* $\{x \in L \mid f(x) \leq x\}$ *(of the set* $\{x \in L \mid x \leq f(x)\}$*).*

PROOF: In view of duality, we prove only the first statement. Let $A = \{x \in L \mid f(x) \leq x\}$ and put $\bigwedge A = a$. As $A \neq \varnothing$ because $1 \in A$, for every $x \in A$ we have $a \leq x$, hence $f(a) \leq f(x) \leq x$. Thus $f(a) \leq A$, therefore $f(a) \leq a$. The latter relation further implies $f(f(a)) \leq f(a)$, hence $f(a) \in A$, therefore $a \leq f(a)$. Thus $f(a) = a$ is a fixed point. Since every fixed point belongs to A, it follows that a is the least fixed point. □

The above property does not hold for complete semilattices, even under stronger assumptions for f, as shown e.g. in :

Example 4.3.9 The set $\mathbb{N}$ of natural numbers is a lattice and a complete meet semilattice under its usual ordering. The mapping $f : \mathbb{N} \longrightarrow \mathbb{N}$ defined by $f(n) = n + 1$ is a lattice homomorphism and even a complete meet homomorphism in the sense of Definition 4.3.3 below, but it has no fixed points.

As a matter of fact, the converse of Proposition 4.3.4, proved by Davis [1955], shows that *a lattice L is complete if and only if every isotone mapping* $f : L \longrightarrow L$ *has fixed points.*

Roughly speaking, complete semilattices and complete lattices can be regarded as algebraic structures with infinitary operations. To be specific, if S is a complete meet semilattice (complete join semilattice, complete lattice), then to each cardinal satisfying $2 \leq \alpha \leq \mid S \mid$ there corresponds the operations of taking the g.l.b. (the operation of taking the l.u.b., the operations of taking the g.l.b. and the l.u.b.) of the subsets $T \subseteq S$ with $\mid T \mid = \alpha$.

We now come to the problem of an appropriate definition for a category of complete semilattices (lattices). One possibility would be to take the full subcategory of **Sem** (of **Lat**) having all complete semilattices (complete lattices) as objects. Another possibility is to restrict the class of morphisms as well; the point is that it seems natural – by analogy with what we have done for the category **Lat01** – to require that the morphisms of the new subcategories preserve the supplementary operations defined on the objects, namely all g.l.b and all l.u.b., and in particular 0 and 1 . To do so, we introduce:

Definition 4.3.3 A mapping $f : S \longrightarrow S'$ between two complete meet semilattices (complete join semilattices, complete lattices) is called a *complete meet homomorphism* (*complete join homomorphism, complete homomorphism*), provided it preserves arbitrary non-empty meets, i.e.,

$$f(\bigwedge A) = \bigwedge f(A) \tag{4.3.5'}$$

for every non-empty subset $A \subseteq S$ (provided it preserves arbitrary non-empty joins, i.e.,

$$f(\bigvee A) = \bigvee f(A) \tag{4.3.5''}$$

for every non-empty subset $A \subseteq S$, provided f fulfills both relations (4.3.5) for every non-empty subset $A \subseteq S$). The term *complete semilattice homomorphism* is a common name for complete meet homomorphisms and complete join homomorphisms.

Warning If the function $f : S \longrightarrow S'$ between two complete meet semilattices (complete join semilattices) satisfies (4.3.5′) (satisfies (4.3.5″)) for $A \neq \varnothing$, then $f(0)$ is the least element of $f(S)$ (then $f(1)$ is the greatest element of $f(S)$) , not necessarily of S'. (Take e.g. the complete meet semilattice $(\mathbb{N}, \leq)$ and define $f(0) = 1$, $f(n) = n$ for $n \geq 1$, or the set $\{0,1\}$ with $0 < 1$, and define $f(0) = f(1) = 1$).

The next Proposition provides a simple and very natural characterization of complete homomorphisms.

Proposition 4.3.5 *A mapping $f : L \longrightarrow L'$ between two complete lattices is a complete homomorphism if and only if it satisfies both relations* (4.3.5) *for every subset $A \subseteq L$,* not just for *$A \neq \varnothing$.*

PROOF: By Remark 4.3.2 and $f(\varnothing) = \varnothing$. □

Clearly every complete semilattice homomorphism is a semilattice homomorphism, and every complete homomorphism is a lattice homomorphism preserving 0 and 1. The converse is not true:

Example 4.3.10 The set $\mathbf{Ord}(\omega+2) = \{0, 1, \ldots, n, \ldots, \omega, \omega+1\}$ is a complete lattice under the usual ordering. Define $f : \mathbf{Ord}(\omega + 2) \longrightarrow \mathbf{Ord}(\omega + 2)$ by $f(0) = 0$, $f(n) = n + 1$ for $n \in \mathbb{N} \setminus \{0\}$ and $f(\omega) = f(\omega + 1) = \omega + 1$. Then the 0-1 preserving isotone map f between two chains is a lattice homomorphism and even a complete meet homomorphism, but not a complete join homomorphism because $\bigvee f(\mathbb{N}) = \bigvee \mathbb{N} = \omega$, whereas $f(\bigvee \mathbb{N}) = f(\omega) = \omega + 1$.

We may be interested in the behaviour of an arbitrary isotone mapping with respect to arbitrary g.l.b. and l.u.b. in complete semilattices (complete lattices). Thus e.g. the relation $\bigvee f(\mathbb{N}) < f(\bigvee \mathbb{N})$ in Example 4.3.10 is ot incidental, as shown in:

Proposition 4.3.6 *If $f : S \longrightarrow S'$ is an isotone mapping between two complete meet semilattices (complete join semilattices, complete lattices) S and S', then*

(4.3.6′) $$f(\bigwedge A) \leq \bigwedge f(A)$$

for every non-empty subset $A \subseteq S$ (then

(4.3.6″) $$\bigvee f(A) \leq f(\bigvee A)$$

for every non-empty subset $A \subseteq S$, then both inequalities (4.3.6) *hold for every subset $A \subseteq S$).*

Comment The fact that $\bigvee f(\mathbb{N}) < f(\bigvee \mathbb{N})$ in Example 4.3.10 also shows that Proposition 4.3.6 cannot be sharpened even if isotone mappings are replaced by lattice homomorphisms.

PROOF: In the case of complete semilattices, for every $a \in A$ we have $\bigwedge A \leq a$, hence $f(\bigwedge A) \leq f(a)$, which implies (4.3.6′). In the case of complete lattices we also have $\bigwedge f(\varnothing) = \bigwedge \varnothing = 1$. The proof is completed by duality. □

After this preparation we are ready to introduce:

Definition 4.3.4 We denote by **CMSem** (by **CJSem**, by **CLat**) the subcategory of **MSem** (of **JSem**, of **Lat**) having as objects all complete meet semilattices (complete join semilattices, complete lattices) and as morphisms all complete meet homomorphisms (complete join homomorphisms, complete homomorphisms) between them.

Proposition 4.3.7 *In each of the categories* **CMSem, CJSem,** *and* **CLat***:*

1^0 *Monomorphisms coincide with injective morphisms.*

2^0 *The following conditions are equivalent for a mapping $f : S \longrightarrow S'$ between two objects:*

(i) *f is an isomorphism;*

(ii) *f is a bijective morphism;*

(iii) *f is an order isomorphism.*

PROOF: Properties 1^0 and 2^0 hold in **Lat01** by Proposition 4.2.4, hence they are valid in the subcategory **CLat**, therefore they are also true in **CMSem** and **CJSem**. □

As concerns the subobjects in our categories, let us introduce:

Definition 4.3.5 By a *complete meet (join) subsemilattice* of a complete meet (join) semilattice S is meant a subset $C \subseteq S$ such that $0 \in C$ and $\varnothing \neq A \subseteq C \Longrightarrow \bigwedge A \in C$ (such that $1 \in C$ and $\varnothing \neq A \subseteq C \Longrightarrow \bigvee A \in C$). A *complete sublattice* of a complete lattice L is a subset $C \subseteq L$ which is both a complete meet subsemilattice of L and a complete join subsemilattice of L.

Remark 4.3.7 1^0 If a non-empty subset C of a complete meet (complete join) semilattice S fulfills only $\varnothing \neq A \subseteq C \Longrightarrow \bigwedge A \in C$ (only $\varnothing \neq A \subseteq C \Longrightarrow \bigvee A \in C$), then C is a complete meet(complete join) semilattice with respect to the restriction of the partial order of S and it has least element $\bigwedge C$ (it has greatest element $\bigvee C$).

2^0 In other words, a *complete sublattice* of a complete lattice L is a subset $C \subseteq L$ such that $\bigwedge A \in C$ and $\bigvee A \in C$ whenever $A \subseteq C$.

The next Proposition provides an important characterization of complete subsemilattices and complete sublattices.

Proposition 4.3.8 (i) *A subset* $C \neq \varnothing$ *of an object*[2] $S \in\mid \mathbf{CMSem} \mid$ ($S \in\mid \mathbf{CJSem} \mid, S \in\mid \mathbf{CLat} \mid$) *is a complete meet subsemilattice (complete join subsemilattice, complete sublattice) of* S *if and only if* $C \in\mid \mathbf{CMSem} \mid$ ($C \in\mid \mathbf{CJSem} \mid$, $C \in\mid \mathbf{CLat} \mid$) *if it is endowed with the operations of taking g.l.b.s (l.u.b.'s, g.l.b.s.'s and l.u.b.'s) of arbitrary non-empty subsets of* C.

(ii) *If* S, S' *are complete meet semilattices (complete join semilattices, complete lattices) and* $f : S \longrightarrow S'$ *is a complete meet homomorphism satisfying* $f(0) = 0$ *(complete join homomorphism satisfying* $f(1) = 1$, *complete homomorphism), then* $f(S)$ *is a complete meet subsemilattice (complete join subsemilattice, complete sublattice) of* S'.

(iii) *If, moreover,* f *is injective, then* S *is isomorphic to* $f(S)$.

PROOF: (i) Immediate.

(ii) For instance, in the case of **CMSem** we have $0 = f(0) \in f(S)$ and if $\varnothing \neq f(A) \subseteq f(S)$ then $\bigwedge f(A) = f(\bigwedge A) \in f(S)$.

(iii) By Proposition 4.3.6. □

Remark 4.3.8 So far we have encountered several definitions o substructures: Definition 4.1.5 of subsemilattices, Definition 4.1.9 of sublattices, Definition 4.2.7 of Boolean subalgebras, and now Definition 4.3.5. Each of them appears to define a *subobject* (cf. Definition 1.3.5) in the corresponding subcategory. Let us prove this e.g. for Definition 4.3.5. Note that the inclusion mapping $i : C \longrightarrow S$ is an injective morphism from $C \in\mid \mathbf{CMSem} \mid$ ($C \in\mid \mathbf{CJSem} \mid, C \in\mid \mathbf{CLat} \mid$) constructed in Proposition 4.3.8 to $S \in |\mathbf{CMSem}|, (S \in |\mathbf{CJSem}|, S \in |\mathbf{CLat}|)$, therefore it is a monomorphism $i : C \longrightarrow S$ by Proposition 4.3.7.

Note that in Proposition 4.3.8.(i) the completeness of the subsemilattice (sublattice) C is essential. More precisely, the equivalence may be lost if C is assumed to be only a meet subsemilattice (join subsemilattice, sublattice), even if other conditions are imposed on C. This is illustrated in:

[2]More properly said, of the underlying set of an object.

Example 4.3.11 Let $S = \{0, 1, \ldots, n, \ldots, \omega, \omega + 1\}$ and $C = \{0, 1, \ldots, n, \ldots, \omega\}$. Then S and C are complete lattices with respect to the usual ordering of ordinals, C is a sublattice of S, and even more, a complete meet subsemilattice of S, but not a complete joined subsemilattice of S, because the l.u.b. of the subset $\mathbb{N}$ of natural numbers is ω in S, but $\omega + 1$ in C. Hence the construction provided in Proposition 4.3.8.(i) does not work for this set C.

The relationship between the two l.u.b.'s in the above example is not incidental, as shown in (4.3.7″) below.

Proposition 4.3.9 *Let $(S, \leq)$ be a complete meet semilattice (complete join semilattice, complete lattice) and $C \subseteq S$ a complete meet semilattice (complete join semilattice, complete lattice) with respect to the restriction of $\leq$ to the elements of C. If $\bigwedge_C$ (if $\bigvee_C$) denotes the* g.l.b. *(the* l.u.b.*) in C, then:*

(i) *The inequality*

$$\bigwedge_C A \leq \bigwedge A \tag{4.3.7'}$$

holds for every $\varnothing \neq A \subseteq C$ (the inequality

$$\bigvee A \leq \bigvee_C A \tag{4.3.7''}$$

holds for every $\varnothing \neq A \subseteq C$, both inequalities (4.3.7) *hold for every $A \subseteq C$).*

(ii) *The equality $\bigwedge_C A = \bigwedge A$ $(\bigvee_C A = \bigvee A)$ holds if and only if $\bigwedge A \in C$ $(\bigvee A \in C)$.*

PROOF: (i) For every $a \in A$, we have $\bigwedge_C A \leq a$ in C, hence $\bigwedge_C A \leq a$ in S. This implies (4.3.7′), and similarly (4.3.7″). In the case of complete lattices, the inequalities (4.3.7) are extended as equalities for $A = \varnothing$, because $\bigwedge \varnothing = 1$ and $\bigvee \varnothing = 0$.

(ii) If $\bigwedge A \in C$, then $\bigwedge A$ is a lower bound of A in C, therefore $\bigwedge A \leq \bigwedge_C A$ in C, hence in S. This, together with (4.3.7′), yield $\bigwedge A = \bigwedge_C A$. The converse is trivial. □

We now turn to the important subclass of complete Boolean algebras.

Definition 4.3.6 By a *complete Boolean algebra* we mean a Boolean algebra $(B, \wedge, \vee, ', 0, 1)$ such that the lattice $(B, \wedge, \vee, 0, 1)$ is complete.

Example 4.3.12 Every finite Boolean algebra is complete.

Example 4.3.13 The Boolean algebra 2^X of all subsets of a set X is complete (cf. Examples 4.2.10 and 4.3.4).

Example 4.3.14 More generally, if B is a complete Boolean algebra , then the Boolean algebra B^X of all functions $f : X \longrightarrow B$ is complete (cf. Examples 4.2.13 and 4.3.4).

Example 4.3.15 If B is a complete Boolean algebra and $a, b \in B$ satisfy $a \leq b$, then the interval $[a, b]$ is a complete Boolean algebra (cf. Example 4.2.14).

Example 4.3.16 Every Cartesian product of complete Boolean algebras is a complete Boolean algebra (cf. an obvious generalization of Example 4.3.5).

Remarks 4.2.8 and 4.3.6 imply

Remark 4.3.9 The dual of a complete Boolean algebra is a complete Boolean algebra. Therefore the *duality principle* holds for complete Boolean algebras; in applying it we interchange $\leq$ with $\geq$, arbitrary g.l.b.'s with arbitrary l.u.b's, 0 with 1, while $'$ is left unchanged.

The distributivity and the *De Morgan laws* are generalized as follows to infinitary operations:

Proposition 4.3.10 *For every element x and every non-empty subset A of a complete Boolean algebra B,*

$$x \wedge \bigvee A = \bigvee(x \wedge A),\ x \vee \bigwedge A = \bigwedge(x \vee A)\ , \tag{4.3.8}$$

$$(\bigwedge A)' = \bigvee A',\ (\bigvee A)' = \bigwedge A'\ , \tag{4.3.9}$$

where we have set $x \circ A = \{x \circ a \mid a \in A\}$ for $\circ \in \{\wedge\ , \vee\}$ and $A' = \{a' \mid a \in A\}$.

PROOF: To prove that $x \wedge \bigvee A$ is the l.u.b. of $x \wedge A$, we show that $x \wedge A \leq x \wedge \bigvee A$ and $x \wedge A \leq y \Longrightarrow x \wedge \bigvee A \leq y$.

Indeed, $a \in A$ implies $a \leq \bigvee A$, hence $x \wedge a \leq x \wedge \bigvee A$. Further, suppose $x \wedge A \leq y$. Then for every $a \in A$ we have $x \wedge a \leq y$, whence by using (4.2.24) twice we get $x \wedge a \wedge y' = 0$, then $a \leq (x \wedge y')' = x' \vee y$. Therefore $\bigvee A \leq x' \vee y$, hence $x \wedge \bigvee A \leq x \wedge (x' \vee y) = x \wedge y \leq y$.

To prove that $(\bigwedge A)'$ is the l.u.b. of A', we first note that from $\bigwedge A \leq A$ we get $A' \leq (\bigwedge A)'$ by (4.2.26). It remains to prove that $A' \leq y \Longrightarrow (\bigwedge A)' \leq y$. Indeed,

$$a' \leq y\ (\forall\, a \in A) \Longrightarrow y' \leq a'' = a\ (\forall\, a \in A) \Longrightarrow y' \leq \bigwedge A \Longrightarrow (\bigwedge A)' \leq y'' = y\ .$$

Corollary 4.3.1 *If $f : B \longrightarrow B_1$ is a complete homomorphism between two complete Boolean algebras B and B_1, then for every $A \subseteq B$ we have*

$$f((\bigwedge A)') = \bigvee f(A'),\ f((\bigvee A)') = \bigwedge f(A')\ . \tag{4.3.10}$$

□

Property (4.3.9) also yields the following:

Example 4.3.17 The complemented elements of a complete distributive lattice L form a complete sublattice of L and a complete Boolean algebra with respect to its operations (cf. Example 4.2.15).

As we have done for all the categories introduced so far in this chapter, we now want to define a category of complete Boolean algebras, whose morphisms should preserve all the operations of these algebras. Taking into account Proposition 4.2.5 and Corollary 4.3.1, we are led to the following:

Definition 4.3.7 By **CBol** we denote the category of all complete Boolean algebras and all complete homomorphisms between them.

So, **CBol** is a full subcategory of **CMSem**, of **JMSem** and of **CLat**.

Proposition 4.3.11 *In the category* **CBol***:*

1^0 *Monomorphisms coincide with injective morphisms.*

2^0 *The following conditions are equivalent for a mapping* $f : B \longrightarrow B_1$ *between two complete Boolean algebras:*

(i) f *is an isomorphism of* **CBol***;*

(ii) f *is a bijective morphism of* **CBol***;*

(iii) f *is an order isomorphism.*

PROOF: Immediate from Proposition 4.3.7.

Remark 4.3.10 Propositions 4.1.5, 4.1.10, 4.2.4, 4,3,7, and 4.3.11 show that in each of the categories studied in this chapter:

1^0 Monomorphisms coincide with injective morphisms.

2^0 The concepts of isomorphism, bijective morphism and order isomorphism coincide.

Thus Remark 4.2.10 extends as follows: in all the categories studied in this chapter the term <u>isomorphism</u> can unambiguously be used without further specification.

The deep reason for which isomorphisms coincide with bijective morphisms is revealed in universal algebra; see e.g. Section 6.4, page 220.

The next step refers to subobjects.

Definition 4.3.8 A subset C of a complete Boolean algebra B is called a *complete Boolean subalgebra* of B provided that $c \in C \implies c' \in C$ and $A \subseteq C \implies \bigwedge A \in C \;\&\; \bigvee A \in C$.

In other words, a complete Boolean subalgebra C of B is a complete lattice and a Boolean subalgebra with respect to the restrictions of the operations $'$, $\bigwedge$ and $\bigvee$ of B; in Particular $0, 1 \in C$. We can prove more:

Proposition 4.3.12 *The following conditions are equivalent for a subset* C *of a complete Boolean algebra* B*:*

(i) C *is a complete Boolean subalgebra of* B*;*

(ii) C *is a complete sublattice of* B *and a Boolean subalgebra of* B*;*

(iii) C *is a complete sublattice of* B *and a Boolean algebra;*

(iv) C *is a complete meet subsemilattice of* B*, closed with respect to complementation;*

(v) *C is a complete join subsemilattice of B, closed with respect to complementation.*

PROOF: (i)$\Longrightarrow$(ii): C is a complete sublattice by hypothesis. Taking, in turn, $A := \varnothing$ and $A := \{a, b\}$ in Definition 4.3.7, it follows that C is also a Boolean subalgebra.

(ii)$\Longrightarrow$(iii): By Proposition 4.2.6, (i)$\Longleftrightarrow$(vi).

(iii)$\Longrightarrow$(iv): We only have to prove that C is closed with respect to complementation. Since C is a Boolean algebra, with each element $c \in C$ is associated a complement $\overline{c}$ in the Boolean algebra C; it suffices to show that $\overline{c} = c'$. But C is a complete sublattice of B, therefore $0, 1 \in C$ and the lattice-theoretical operations of C are the restrictions of those of B. So $c \wedge \overline{c} = 0$ and $c \vee \overline{c} = 1$, consequently $\overline{c} = c'$ by Proposition 4.2.2.

(iv)$\Longrightarrow$(i): We have $\bigvee \varnothing = 0 \in C$, hence $1 = 0' \in C$, or $\bigwedge \varnothing \in C$. Besides, $\varnothing \neq A \subseteq C$ implies $\bigwedge A \in C$, therefore it remains to prove that $\bigvee A \in C$. But clearly $\varnothing \neq A' \subseteq C$, therefore $\bigwedge A' \in C$ and $\bigvee A = (\bigwedge A')' \in C$.

So (i)$\Longleftrightarrow$(ii)$\Longleftrightarrow$(iii)$\Longleftrightarrow$(iv), hence (i)$\Longleftrightarrow$(v) by duality. □

It might be expected that Remark 4.3.8 extends to complete Boolean subalgebras. Indeed, the inclusion mapping $i : C \longrightarrow B$ is a morphism in **CLat** by Proposition 4.3.12(ii), hence it is also a morphism in **CBol**, therefore it is a monomorphism by Remark 4.3.10.

Proposition 4.3.13 (i) *If $f : B \longrightarrow B_1$ is a morphism in* **CBol**, *then $f(B)$ is a complete Boolean subalgebra of B_1.*

(ii) *If, moreover, f is injective, then B is isomorphic to $f(B)$.*

PROOF: (i) $f(B)$ is a Boolean subalgebra by Proposition 4.2.7(i) and a complete sublattice by Proposition 4.3.8(ii).

(ii) By Proposition 4.2.7(ii) and Remark 4.3.10. □

Let us also mention the category **ACLat**, which has the same objects as the category **CLat**, while the morphisms are the *complete dual homomorphisms* between complete lattices, that is, functions $f : L \longrightarrow L'$ which interchanges arbitrary meets with arbitrary joins, i.e.,

$$f(\bigwedge A) = \bigvee f(A), \; f(\bigvee A) = \bigwedge f(A) \tag{4.3.5*}$$

for every $A \subseteq L$. In particular the isomorphisms of **ACLat** will be referred to as *complete dual isomorphisms.* The properties of the category **ACLat** are parallel to the ones of **CLat** and left to the reader (cf. the categories **APos, ALat, AMod, ADis, ALat01** and **ABol**).

We conclude this section with a few words about the so-called compactly generated lattices; our approach follows essentially Pierce [1968]. A preliminary concept is introduced in:

Definition 4.3.9 [3] An element c of a complete join semilattice S is said to be *compact* provided every *covering* of c includes a *finite covering.* This means

[3] The duals of Definitions 4.3.8-4.3.10 will not be used; so we drop the notation $'$ and $''$.

That for every $A \subseteq S$ such that $c \leq \bigvee A$ there exists a finite subset $A_0 \subseteq A$ such that $c \leq \bigvee A_0$.

For instance, the element 0 of a complete lattice is compact. See also Examples 4.3.18-4.3.20 below.

To provide an equivalent definition of compact elements, we also introduce

Definition 4.3.10 A subset A of a poset P is said to be *upward directed* or simply *directed* provided every two elements of A have an upper bound in A. A *directed covering* of an element is a directed set which covers that element.

Clearly in the above definition "every two elements" can be replaced by "every finite non-empty subset".

Proposition 4.3.14 *An element c of a complete join semilattice S is compact if and only if every directed covering of c contains an upper bound,* i.e., for every directed subset $A \subseteq S$ such that $c \leq \bigvee A$ there is an element $A \in A$ such that $c \leq a$.

PROOF: Suppose c is compact and let $c \leq \bigvee A$ be a directed covering of c. Then $c \leq \bigvee A_0$ for some finite subset $A_o \subseteq A$. Since A is directed, there is an element $a \in A$ such that $A_0 \leq a$. This implies $\bigvee A_0 \leq a$, therefore $c \leq a$.

Conversely, assume that every directed covering of c contains an upper bound of c and let $c \leq \bigvee A$ be an arbitrary covering of c. Denote by $\overline{A}$ the set of all finite joins of elements of A. Obviously $\overline{A}$ is a directed set and $A \subseteq \overline{A}$. It follows that $\bigvee A \leq \bigvee \overline{A}$, hence $c \leq \bigvee \overline{A}$, therefore $c \leq \overline{a}$ for some $\overline{a} \in \overline{A}$. But $\overline{a} = a_1 \vee \ldots \vee a_m$, where $a_1, \ldots, a_m \in A$. □

Definition 4.3.11 A *compactly generated lattice* (*semilattice*) is a complete lattice (complete join semilattice) in which every element is a join of compact elements.

Example 4.3.18 Every finite lattice (join semilattice) is compactly generated.

Example 4.3.19 The lattice in Example 4.1.20 is compactly generated; the compact elements are the points.

For further examples and properties see Lemma 4.4.2 and Theorems 5.1.2 and 5.2.1.

4.4 Closure Operators

There is a construction widely used in mathematics in various contexts, e.g.: the subgroup generated by a subset of a group is the intersection of all subgroups which include that subset, the submodule generated by a subset of a module is the intersection of all submodules which include that subset, the closure of a subset of a topological space is the intersection of all closed subsets which include that subset, etc. The substratum of this technique is revealed by the concept of a closure operator over the subsets of a set, and, more generally, over a complete lattice.

In this section, following a suggestion of Bourbaki [1963], §1, Exercise 13 a), we describe a generalization of the above idea, namely the properties of closure operators within the more general framework of partially ordered sets. Then we discuss the intermediate level of complete lattices and complete semilattices, before recapturing the particular case of Moore families of sets, usually dealt with in the literature.

To facilitate the understanding of the general theory, let us begin with one of the many typical examples. Take, for instance, a group G. For every subset X of G, there exist subgroups of G which include X, e.g., G itself. The intersection $\overline{X}$ of all these subgroups is clearly a subgroup which includes X; more exactly, $\overline{X}$ is the smallest subgroup which includes X. This follows immediately from the obvious property that the intersection of every family of subgroups is a subgroup as well. The following properties are easily established: $X \subseteq \overline{X}$, $\overline{\overline{X}} = \overline{X}$, $X \subseteq Y \Longrightarrow \overline{X} \subseteq \overline{Y}$. Moreover, a subset $X \subseteq G$ is a subgroup of G if and only if $X = \overline{X}$.

The abstract counterpart of this and many other examples (cf. Examples 4.3.8 and 4.4.1-4.4.8) is at hand. Let $\mathcal{C}$ be a family of subsets of a non-empty set T, such that $T \in \mathcal{C}$ and every intersection of sets from $\mathcal{C}$ belongs to $\mathcal{C}$.Then $\mathcal{C}$ is called a *closure system* or a *Moore family* of sets, and for every $X \in 2^T$, the intersection $\overline{X}$ of all sets of $\mathcal{C}$ including X is the smallest set from $\mathcal{C}$ which includes X. Besides, the function $^- : 2^T \longrightarrow 2^T$ satisfies $X \subseteq \overline{X}$, $\overline{\overline{X}} = \overline{X}$ and $X \subseteq Y \Longrightarrow \overline{X} \subseteq \overline{Y}$. Conversely, if a mapping $^-$ from 2^T into itself verifies the above three properties, then $^-$ is said to be a *closure operator* on 2^T and the set of *closed subsets* of T, i.e., the subsets $X = \overline{X}$, is a Moore family of sets. This establishes a bijection between the closure systems and the closure operators on 2^T. Moore families and closure operators are encountered throughout the whole of mathematics.

In this section we study the properties of closure operators within the successive frameworks of posets, complete semilattices and complete lattices.

Definition 4.4.1 [4] A *closure operator* on a non-empty poset P is a mapping $\gamma : P \longrightarrow P$ satisfying the following conditions: for every $x, y \in P$,

(4.4.1) $$x \leq \gamma(x) \qquad (\textit{extensivity}),$$

(4.4.2) $$\gamma^2(x) = \gamma(x) \qquad (\textit{idempotency}),$$

where $\gamma^2(x)$ means $\gamma(\gamma(x))$,

(4.4.3) $$x \leq y \Longrightarrow \gamma(x) \leq \gamma(y) \quad (\textit{isotony}).$$

---

[4] In this section we don't use the notation $'$ and $''$ to indicate dual statements; cf. footnote 3.

The elements $x = \gamma(x)$ are said to be γ-*closed* or simply *closed* when no confusion is possible. Also, each $\gamma(x)$ is called the γ-*closure*, or simply the *closure* of x. The set $\mathcal{C}(P)$ of all closure operators on P is endowed with the pointwise partial order defined by: for every $\gamma, \delta \in \mathcal{C}(P)$,

$$\gamma \leq \delta \Longleftrightarrow \gamma(x) \leq \delta(x) \ (\forall x \in P) . \tag{4.4.4}$$

Definition 4.4.2 A *closure system* or a *Moore family* on a non-empty poset P is a subset $C \subseteq P$ such that for every $x \in P$, the set

$$C[x) = \{c \in C \mid x \leq c\} \tag{4.4.5}$$

has least element. The set $\mathcal{M}(P)$ of all Moore families on P is endowed with set inclusion $\subseteq$ as partial order.

Remark 4.4.1 A closure system C on P is *cofinal*, i.e., $C[x) \neq \varnothing$ for every $x \in P$.

The relationship between closure operators and closure systems is described in:

Theorem 4.4.1 *The mapping which associates with each* $\gamma \in \mathcal{C}(P)$*, the set*

$$C = \{x \in P \mid x = \gamma(x)\} , \tag{4.4.6}$$

and the mapping which associates with each $C \in \mathcal{M}(P)$*, the function* $\gamma : P \longrightarrow P$ *defined by*

$$\gamma(x) = \text{the least element of } C[x) \ (\forall x \in P) , \tag{4.4.7}$$

establish a dual isomorphism between the posets $(\mathcal{C}(P), \leq)$ *and* $(\mathcal{M}(P), \subseteq)$.

PROOF: For the sake of convenience, set

$$\Phi(\gamma) = \{x \in P \mid x = \gamma(x)\} \text{ for } \gamma \in \mathcal{C}(P), \tag{4.4.6*}$$

$$\psi(C)(x) = \text{ the least element of } C[x), \text{ for } C \in \mathcal{M}(P). \tag{4.4.7*}$$

If $\gamma \in \mathcal{C}(P)$, we must prove that $\Phi(\gamma)$ is a Moore family. Let us prove that for every $x \in P$, $\Phi(\gamma)[x)$ has least element $\gamma(x)$. Indeed,

$$\Phi(\gamma)[x) = \{c \in \Phi(\gamma) \mid x \leq c\} = \{c \in P \mid x \leq c = \gamma(c)\} .$$

So $\gamma(x) \in \Phi(\gamma)[x)$ because $x \leq \gamma(x) = \gamma(\gamma(x))$, while $c \in \Phi(\gamma)[x)$ implies $\gamma(x) \leq \gamma(c) = c$.

If $C \in \mathcal{M}(P)$, we must prove that $\psi(C)$ is a closure operator. To do this, we come back in this paragraph to the notation $\psi(C) = \gamma$. Then (4.4.7*) implies $\gamma(x) \in C[x)$, hence $x \leq \gamma(x)$, which is (4.4.1). If $x \leq y$ then $C[y) \subseteq C[x)$, hence from $\gamma(y) \in C[y)$ we infer $\gamma(y) \in C[x)$; but $\gamma(x)$ is the least element of of $C[x)$, therefore $\gamma(x) \leq \gamma(y)$ and we have thus proved (4.4.3). In view of (4.4.1), for (4.4.2) it suffices to prove that $\gamma(\gamma(x)) \leq \gamma(x)$. Indeed, from $\gamma(x) \leq \gamma(x) \in C$

we infer $\gamma(x) \in C[\gamma(x))$, while $\gamma(\gamma(x))$ is the least element of $C[\gamma(x))$, therefore $\gamma(\gamma(x)) \leq \gamma(x)$.

Now we prove that $\psi(\Phi)$ and $\Phi(\psi)$ are identity mappings.

If $\gamma \in \mathcal{C}(P)$, we have seen that for every $x \in P, \gamma\ (x)$ is the least element of $\Phi(\gamma)[x)$, that is, $\gamma(x) = \psi(\Phi(\gamma))(x)$, therefore $\gamma = \psi(\Phi(\gamma))$.

If $C \in \mathcal{M}(P)$ we must prove that $\Phi(\psi(C)) = C$. Indeed, from (4.4.6*), (4.4.7*) and using $x \leq x$ we infer

$$x \in \Phi(\psi(C)) \Longleftrightarrow x = \psi(C)(x) \Longleftrightarrow x \text{ is the least element of } C[x) \Longleftrightarrow x \in C\ .$$

The last point is to prove that the functions Φ and ψ are antitone.

If $\gamma \leq \delta$ then $\Phi(\delta) \subseteq \Phi(\gamma)$ because $x \in \Phi(\delta)$ implies $x \leq \gamma(x) \leq \delta(x) = x$, hence $x = \gamma(x)$, therefore $x \in \Phi(\gamma)$.

If $C \subseteq D$ then $\psi(D) \subseteq \psi(C)$ because for every $x \in P$, the element $\psi(D)(x)$ is the least element of $D[x)$; but $\psi(C)(x) \in C[x) \subseteq D[x)$, therefore $\psi(D)(x) \leq \psi(C)(x)$. □

From a practical point of view, Theorem 4.4.1 enables us to use simultaneously the closure-operator language and the Moore-family language, the properties (4.4.1)-(4.4.7) being simultaneously valid.

Remark 4.4.2 The following paraphrase of (4.4.7) is worth noting: the closure of an element is the least closed element which includes it.

Property (4.4.2) implies:

Remark 4.4.3 An element $x \in P$ is γ-closed if and only if it is of the form $\gamma(y)$. Hence the set C of γ-closed elements of P is

(4.4.8) $$C = \{x \in P \mid x = \gamma(x)\} = \{\gamma(y) \mid y \in P\} = \gamma(P)\ .$$

Remark 4.4.4 If the poset P has greatest element 1, then $\gamma(1) = 1$ for every $\gamma \in \mathcal{C}(P)$. In particular this happens when P is a complete join semilattice.

Exercise 4.4.1 The set (4.4.6) of fixed points of a mapping $\gamma : P \longrightarrow P$ fulfills $C = \gamma(P)$ if and only if γ is idempotent.

Exercise 4.4.2 Find the least element of $\mathcal{C}(P)$ and the greatest element of $\mathcal{M}(P)$.

Exercise 4.4.3 In a join semilattice a map $x \mapsto \bar{x}$ satisfies (4.4.1)and (4.4.2) if and only if

(4.4.9) $$x \vee \bar{\bar{x}} \leq \bar{x}\ ,$$

axiom (4.4.3) is equivalent to

(4.4.10) $$\bar{x} \vee \bar{y} \leq \overline{x \vee y}\ ,$$

while the system (4.4.1)-(4.4.3) is equivalent to

(4.4.11) $$x \vee \bar{x} \vee \bar{y} \vee \bar{\bar{x}} \leq \overline{x \vee y}\ .$$

Proposition 4.4.1 *A closure system C on a poset P verifies*

(4.4.12) $$X \subseteq C \text{ and } \bigwedge X \text{ exists} \Longrightarrow \bigwedge X \in C\ .$$

PROOF: Let $X \subseteq C$ and $c = \bigwedge X$. If $X = \varnothing$ then Remark 4.4.4 implies (4.4.12). Otherwise let γ be the closure operator associated with C. For every $x \in X$ we have $c \leq x \in C$, hence $\gamma(c) \leq \gamma(x) = x$ by (4.4.3) and (4.4.6). It follows that $\gamma(c) \leq c$ by the definition of the g.l.b., hence $c = \gamma(c)$, therefore $c \in C$. □

In the remainder of this section we carry out successive specializations of the poset P, until we recapture the conventional theory of closure operators on families of sets, which has numerous applications throughout the whole of mathematics.

Note the following immediate consequence of Proposition 4.4.1:

Corollary 4.4.1 *A closure system C on a meet semilattice (complete meet semilattice) $(S, \leq)$ is a subsemilattice of S (satisfies*

(4.4.13) $$\varnothing \neq X \subseteq C \Longrightarrow \bigwedge X \in C).$$

Exercise 4.4.4 In a complete meet semilattice S, every mapping $\gamma : S \longrightarrow S$ which satisfies (4.4.13) is extensive and idempotent.

As a matter of fact, the converse of Corollary 4.4.1 also holds:

Proposition 4.4.2 *The following conditions are equivalent for a subset C of a complete meet semilattice $(S, \leq)$:*

(i) *C is a closure system of S;*

(ii) *C is a cofinal complete meet semilattice with respect to the restriction of $\leq$ to the elements of C and for every non-empty subset $X \subseteq C$,*

(4.4.14) $$\bigwedge\nolimits_C X = \bigwedge X\ .$$

PROOF: (i)$\Longrightarrow$(ii): C is cofinal by Remark 4.4.1 and satisfies (4.4.13) by Corollary 4.4.1. Then for every non-empty subset $X \subseteq C$, since the g.l.b. of X is in C, it is the g.l.b. of X in C as well, i.e., (4.4.14) holds.

(ii)$\Longrightarrow$(i): Take $x \in S$. Then $\varnothing \neq C[x] \subseteq C$, hence there exists $a = \bigwedge C[x] = \bigwedge_C C[x]$, say a. It follows that $a \in C$. But x is a lower bound of $C[x]$, hence $x \leq a$, therefore $a \in C[x]$, consequently a is the least element of $C[x]$ by Lemma 2.2.3′. □

Corollary 4.4.2 *A closure operator γ on a complete meet semilattice $(S, \leq)$ satisfies, for every $x \in S$,*

(4.4.15) $$\gamma(x) = \bigwedge\{c \in S \mid x \leq c = \gamma(c)\}.$$

PROOF: Let C be the closure system associated with γ and a the right-hand side of (4.4.15). Then $a = \bigwedge C[x] = \bigwedge_C C[x]$ by (4.4.14), hence $a \in C$, while (4.4.5) implies $x \leq a$. Thus $a \in C[x]$, hence $a = \gamma(x)$ by Lemma 2.2.3′. □

In a complete join semilattice the converse of Proposition 4.4.1 also holds:

Proposition 4.4.3 *A subset C of a complete join semilattice S is a closure system of S if and only if C verifies* (4.4.12).

PROOF: It remains to prove that if (4.4.12) holds then C is a Moore family. Indeed, take $x \in S$. Since x is a lower bound of $C[x]$, Proposition 4.3.1 implies that $C[x]$ has g.l.b., say $\bigwedge C[x] = a$; then $x \leq a$. But (4.4.12) implies that $a \in C$, therefore $a \in C[x]$, hence a is the least element of $C[x]$ by Lemma 2.2.3′. □

Theorem 4.4.2 *If S is a complete join semilattice, then the dually isomorphic posets $(\mathcal{C}(P), \leq)$ and $(\mathcal{M}(P), \subseteq)$ are complete lattices, the complete meet operation in $(\mathcal{M}(P), \subseteq)$ being the complete set-theoretical intersection.*

PROOF: Since property (4.4.12) is clearly preserved by arbitrary intersections, it follows from Proposition 4.4.3 that the intersection $C = \bigcap_{i\in I} C_i$ of a non-empty family $(C_i)_{i\in I}$ of closure systems of S is again a closure system. Therefore C is the g.l.b. of $(C_i)_{i\in I}$ not only in $(2^S, \subseteq)$, but also in $(\mathcal{M}(S), \subseteq)$. Moreover, S is clearly a Moore family, therefore $(\mathcal{M}(S), \subseteq)$ is a complete lattice by Proposition 4.3.2. Now the conclusion for $(\mathcal{C}(S), \leq)$ follows by Theorem 4.4.1. □

At this point we concentrate on the important case of complete lattices. Trivially all previous results in this section are valid for complete lattices. The following theorem is specific.

Theorem 4.4.3 1^0 *The following conditions are equivalent for a subset C of a complete lattice $(L, \leq)$:*

(i) *C is a closure system on L;*

(ii) *C satisfies*

$$X \subseteq C \Longrightarrow \bigwedge X \in C; \tag{4.4.16}$$

(iii) *C is a complete lattice with respect to the restriction of $\leq$ to the elements of C, and satisfies*

$$X \subseteq C \Longrightarrow \bigwedge_C X = \bigwedge X. \tag{4.4.17}$$

2^0 *When this is the case,*

$$X \subseteq C \Longrightarrow \bigvee_C X = \bigwedge\{y \in C \mid X \leq y\} \stackrel{def}{=} \bigwedge\{y \in C \mid x \leq y \;\forall x \in X\}. \tag{4.4.18}$$

PROOF: 1^0 (i)$\Longleftrightarrow$(ii): By Proposition 4.4.3, in which (4.4.12) reduces to (4.4.16).

(i)$\Longrightarrow$(iii) and 2^0: As (i)$\Longleftrightarrow$(ii), it follows that (4.4.16) holds. In particular $1 = \bigwedge \varnothing \in C$. Therefore (4.4.18) holds by Proposition 4.3.1. We have thus proved that C is a complete lattice. Property (4.4.17) is a paraphrase of (4.4.16).

(iii)$\Longrightarrow$(ii): Immediate. □

Warning Although a closure system C on a complete lattice $(L \leq)$ is a complete lattice with respect to $\leq$, C need not be a complete sublattice. This happens if and only if the element 0 is closed, i.e., $0 \in C$.

Now let us say a bit more about the relationship between a closure operator γ and g.l.b.'s in L and in C.

Lemma 4.4.1 *Let C be a closure system on a complete lattice L and γ the associated closure operator. For every subset $X \subseteq L$,*

$$\gamma(\bigvee X) = \gamma(\bigvee \gamma(X)) = \bigvee_C \gamma(X) \; . \tag{4.4.19}$$

PROOF: We have $\gamma(\varnothing) = \varnothing$ and $\gamma(\bigvee \varnothing) = \gamma(0) =$ least element of C by (4.4.8). On the other hand, since $x \in \varnothing \Longrightarrow x \leq c$, formula (4.4.18) implies $\bigvee_C \varnothing = \bigwedge C =$ least element of C. Therefore (4.4.19) holds for $X = \varnothing$.

Now assume $X \neq \varnothing$. For every $x \in X$, we have $x \leq \gamma(x) \leq \bigvee \gamma(X)$, hence $\bigvee X \leq \bigvee \gamma(X)$, therefore

$$\gamma(\bigvee X) \leq \gamma(\bigvee \gamma(X)) \; .$$

Further, $\bigvee \gamma(X) \leq \bigvee_C \gamma(X)$ by Proposition 4.3.9 via Theorem 4.4.3, hence

$$\gamma(\bigvee \gamma(X)) \leq \gamma(\bigvee_C \gamma(X)) = \bigvee_C \gamma(X)$$

because $\bigvee_C \gamma(X) \in C$. Finally,

$$\bigvee_C \gamma(X) \leq \gamma(\bigvee X)$$

holds because $x \in X$ implies $x \leq \bigvee X$, hence $\gamma(x) \leq \gamma(\bigvee X) \in C$. □

Proposition 4.4.4 *Let C be a closure system on a complete lattice L and γ the associated closure operator. For every subset $X \subseteq C$,*

$$\bigvee_C X = \gamma(\bigvee X) \; , \tag{4.4.20}$$

or, equivalently,

$$\bigvee_C X = \bigwedge C[X] = \text{ least element of } C[X] \; , \tag{4.4.21}$$

where we have set

$$C[X] = \{c \in C \mid X \leq c\} \; . \tag{4.4.22}$$

Comment Property (4.4.21) via (4.4.20), is a generalization of (4.4.15).
PROOF: Relation (4.4.20) is a consequence of Lemma 4.4.1 via $\gamma(X) = X$. The first equality (4.4.21) holds by Proposition 4.3.1. The second equality follows from the remark that $c \in C[X]$ implies $\bigvee X \leq c \in C$, hence $\gamma(\bigvee X) \leq \gamma(c) = c$, so that $\gamma(\bigvee X)$ is indeed the least element of $C[X]$. □

We now apply the above results to establish a property which will be needed in the next section.

Lemma 4.4.2 *Let T be any set and $\gamma : 2^T \longrightarrow 2^T$ a closure operator. Suppose the associated closure system $\mathcal{C} \subseteq 2^T$ satisfies* (i) *$\mathcal{A} \subseteq \mathcal{C}$ implies $\bigcap \mathcal{A} \in \mathcal{C}$, and* (ii) *if $\mathcal{D} \subseteq \mathcal{C}$ is a directed set, then $\bigcup \mathcal{D} \in \mathcal{C}$. It follows that $(\mathcal{C}, \subseteq)$ is a compactly generated lattice, the elements $\gamma(\{x\})$, with $x \in T$ being compact.*

PROOF: Hypothesis (i) shows that $\bigcap \mathcal{A}$ is g.l.b.$\mathcal{A}$ in $(\mathcal{C}, \subseteq)$, hence $(\mathcal{C}, \subseteq)$ is a complete lattice by Proposition 4.3.2. Further, let $\gamma(\{x\}) \subseteq \bigvee_{\mathcal{C}} \mathcal{D}$, where $\mathcal{D} \subseteq \mathcal{C}$ is directed in $\mathcal{C}$. As $\bigvee_{\mathcal{C}} \mathcal{D} = \bigcup \mathcal{D}$ by (ii) *via* Proposition 4.3.9(ii), it follows that $\{x\} \subseteq \gamma(\{x\}) \subseteq \bigcup \mathcal{D}$, hence $x \in D$ for some $D \in \mathcal{D} \subseteq \mathcal{C}$, therefore $D \in \mathcal{C}[\{x\})$ by (4.4.5), consequently $\gamma(\{x\}) \subseteq D$ by Corollary 4.4.2. Thus $\gamma(\{x\})$ is compact by Proposition 4.3.14 and it remains to prove that any $X \in \mathcal{C}$ is a join of compacts. But $\{x\} \subseteq X$ implies $\gamma(\{x\}) \subseteq \gamma(X) = X$, therefore, using Proposition 4.3.9, we get

$$X = \bigcup\{\{x\} \mid x \in X\} \subseteq \bigcup\{\gamma(\{x\}) \mid x \in X\} \subseteq \bigvee_{\mathcal{C}}\{\gamma(\{x\}) \mid x \in X\} \subseteq X.$$

□

Exercise 4.4.5 If γ is a closure operator on a complete lattice L and $\varnothing \neq X \subseteq L$, then

$$\gamma(\bigwedge X) \leq \bigwedge \gamma(X) \leq \bigvee \gamma(X) \leq \gamma(\bigvee X)\,. \tag{4.4.23}$$

Exercise 4.4.6 A closure operator γ on a join semilattice fulfills the identities

$$\gamma(x \vee y) = \gamma(\gamma(x) \vee y) = \gamma(\gamma(x) \vee \gamma(y))\,. \tag{4.4.24}$$

Exercise 4.4.7 The concept dual to that of a closure operator is called *interior operator*: a mapping $\beta : P \longrightarrow P$ satisfying $\beta(x) \leq x, \beta^{\,2}(x) = \beta(x)$ and $x \leq y \Longrightarrow \beta(x) \leq \beta(y)$. Develop the theory of interior operators.

Exercise 4.4.8 (Bernard [1969], Lapscher [1972]). Let α be either a closure operator or an interior operator on a lattice L. Prove the identities

$$\alpha(x) = \alpha(x \wedge \alpha(x)) = \alpha(x \vee \alpha(x))$$

and show that the functions $\overline{\alpha}$ and $\underline{\alpha}$, defined by

$$\overline{\alpha}(x) = x \vee \alpha(x),\ \underline{\alpha}(x) = x \wedge \alpha(x)\ ,$$

are a closure operator and an interior operator, respectively.

Exercise 4.4.9 If α is a closure (an interior) operator on a Boolean algebra B, the mapping $\delta : B \longrightarrow B$ defined by

$$\delta(x) = (\alpha(x'))' \tag{4.4.25}$$

is an interior (a closure) operator on B.

In Lemma 4.4.2 and in most important applications, closure operators are defined on complete lattices of the form $\mathcal{L} = (2^T, \subseteq)$. In other words, the elements of the lattice L are the subsets of a certain "total" set T, the partial order is set inclusion $\subseteq$, while the (arbitrary – i.e., not necessarily finite) operations of meet and join are the (arbitrary) set-theoretical operations of intersection and union, respectively; cf. Examples 4.2.11, 4.3.13 and the introduction to this section).

Definition 4.4.3 By a *Moore family of sets* we mean a Moore family $\mathcal{C}$ on a complete lattice of the form $(2^T, \subseteq)$. Then $\mathcal{C}$ consists of elements called *$\mathcal{C}$-closed* sets or simply *closed* sets. The closure operator associated with $\mathcal{C}$ is sometimes denoted by a bar $^- : 2^T \longrightarrow 2^T$ and the image $\overline{X}$ of an arbitrary subset $X \subseteq T$ is sometimes called the *$\mathcal{C}$-closure* of X or simply the *closure* of X or the *$\mathcal{C}$-set generated by* X (the terminology depends on the context).

It is convenient to decompose condition (4.4.16) in Theorem 4.4.3.1^0 into the cases $X \neq \varnothing$ and $X = \varnothing$. We thus obtain:

Corollary 4.4.3 *Let T be a set and $\mathcal{C}$ a family of subsets of T, which we call closed subsets. Then $\mathcal{C}$ is a Moore family of sets on $(2^T, \subseteq)$ if and only if it satisfies the following two conditions:*

(i) *the intersection of every non-empty family of closed sets is a closed set, and*

(ii) *T is a closed set.*

Corollary 4.4.4 *Every Moore family of sets $\mathcal{C}$ on $(2^T, \subseteq)$ is a complete lattice with respect to $\subseteq$, in which the following properties hold:*

(4.4.26) $$\bigwedge_{\mathcal{C}} \mathcal{X} = \bigcap \mathcal{X} \quad (\forall \mathcal{X} \subseteq \mathcal{C}),$$

(4.4.27) $$\overline{X} = \bigcup \{Y \mid X \subseteq Y \in \mathcal{C}\} \quad (\forall X \subseteq T),$$

(4.4.28) $$\bigvee_{\mathcal{C}} \mathcal{X} = \overline{\bigcup \mathcal{X}} \quad (\forall \mathcal{X} \subseteq \mathcal{C});$$

in particular

(4.4.29) $$X \vee_{\mathcal{C}} Y = \overline{X \cup Y} \quad (\forall X, Y \in \mathcal{C}).$$

Corollary 4.4.5 *A Moore family of sets $\mathcal{C}$ on $(2^T, \subseteq)$ is a complete sublattice of $(2^T, \subseteq)$ if and only if*

(4.4.30) $$\mathcal{X} \subseteq \mathcal{C} \Longrightarrow \bigcup \mathcal{X} \in \mathcal{C}.$$

PROOF: By (4.4.28). □

Example 4.4.1 Recall that a binary relation ρ on a set E can be identified with its graph $\rho \subseteq E \times E$; the property $(x, y) \in \rho$ is written $x \, \rho \, y$. The set of all reflexive relations ρ on E, i.e., such that $x \, \rho \, x$ for every $x \in E$, is a Moore family on $2^{E \times E}$; the *reflexive closure* of an arbitrary relation $\rho \subseteq E \times E$ is the least reflexive relation including ρ, and is realized as the intersection of all reflexive relations that include ρ.

Example 4.4.2 The symmetric relations ρ on E, i.e., such that $x \, \rho \, y \Longrightarrow y \, \rho \, x$ for all $x, y \in E$, form a Moore family on $2^{E \times E}$; the *symmetric closure* of an arbitrary relation $\rho \subseteq E \times E$ is the least symmetric relation that includes ρ and is realized as the intersection of all symmetric relations that include ρ.

Example 4.4.3 The transitive relations ρ on E, i.e., such that $x \rho y$ & $y \rho z \Longrightarrow x \rho z$ for all $x, y, z \in E$, form a Moore family on $2^{E \times E}$; the *transitive closure* of an arbitrary relation $\rho \subseteq E \times E$ is the least transitive relation including ρ and is realized as the intersection of all transitive relations that include ρ.

Example 4.4.4 The equivalence relations on E, i.e., reflexive, symmetric and transitive, form a Moore family on $2^{E \times E}$; for every relation $\rho \subseteq E \times E$, the *equivalence generated by* ρ is the least equivalence relation that includes ρ, and is realized as the intersection of all equivalence relations that include ρ.

Exercise 4.4.10 Establish a relationship between the Moore families/closure operators in the previous four Examples.

Example 4.4.5 A groupoid is a pair $(G, \circ)$, where G is a set and $\circ : G \longrightarrow G$ a binary operation on G. A congruence on $(G, \circ)$ is an equivalence relation $\equiv$ on G such that $x \equiv x'$ & $y \equiv y' \Longrightarrow x \circ y \equiv x' \circ y'$ for every $x, y, x', y' \in G$. The congruences of $(G, \circ)$ form a Moore family on $2^{G \times G}$; for every relation $\rho \subseteq G \times G$, the *congruence generated by* ρ is the least congruence that includes ρ and is realized as the intersection of all congruences that include ρ.

Example 4.4.6 The previous example can be generalized to the congruences of other algebraic structures, as semigroups, groups, rings, modules, vector spaces, semilattices, lattices, Boolean algebras, etc. Thus, e.g., a congruence on a lattice $(L, \wedge, \vee)$ is an equivalence $\equiv$ on L such that $x \equiv x'$ & $y \equiv y' \Longrightarrow x \wedge y \equiv x' \wedge y'$ & $x \vee y \equiv x' \vee y'$. As a matter of fact, the construction of the congruence generated by a relation is still valid in universal algebra.

Example 4.4.7 A subgroupoid of a groupoid $(G\circ)$ is a subset $H \subseteq G$ such that $x, y \in H \Longrightarrow x \circ y \in H$. The subgroupoids of $(G, \circ)$ form a Moore family on 2^G; for every subset $X \subseteq G$, the *subgroupoid generated by* X is the least groupoid that includes X and is realized as the intersection of all subgroupoids that include X. As a matter of fact, this is valid in the general case of the subalgebra generated by a set in universal algebra.

Example 4.4.8 Given a topological space T, the mapping $^- : 2^T \longrightarrow 2^T$ which associates with each subset $X \subseteq T$, its closure $\overline{X}$ as defined in topology, is a closure operator in the sense of Definition 4.4.1. This closure operator fulfills the supplementary properties

$$\overline{X \cup Y} = \overline{X} \cup \overline{Y} \,, \tag{4.4.31}$$

$$\overline{\varnothing} = \varnothing \,. \tag{4.4.32}$$

It is important to note that the theory described in this section does not exclude the interest of studying, in each particular case, the internal structure of the closure $\overline{X}$, which is a useful working tool. Thus e.g. the reflexive closure of a relation ρ on E is simply $\rho \cup 1_E$; the transitive closure of a relation can also be described in intrinsic terms and is important in graph theory and its applications; the subspace generated by a non-empty subset X of a vector space is the set of all vectors of th form $\sum_i \lambda_i x_i$, with $x_i \in X$; etc.

Generally speaking, the theory of closure operators turns out to be an important tool in various fields of mathematics. See also the next section.

4.5 Galois Connections

This short section is a kind of prolongation of the previous one. A Galois connection is a certain construction relating two posets and generating a closure operator on each of them; conversely, with each closure operator on a poset is associated a certain Galois connection. A few significant examples will be given in the end of this section and in Chapter 5 §3. See also the theory of formal concepts in the monograph by Ganter and Wille [1999]; see also Chapter 6 §5. The starting point is a Galois connection between "objects" and "attributes", which is a mathematical description of the relationship between the extent and the intent of a concept.

Definition 4.5.1 By a *Galois connection*, also called a *Galois correspondence*, between two posets P and Q is meant a pair $(^+, ^-)$ of mappings $^+ : P \longrightarrow Q$ and $^- : Q \longrightarrow P$ such that for every $x \in P$, $y \in Q$,

$$x \leq y^- \iff y \leq x^+ . \tag{4.5.1}$$

Example 4.5.1 If $\rho \subseteq A \times B$ is a correspondence, where $A \neq \varnothing \neq B$, define $^+ : 2^A \longrightarrow 2^B$ and $^- : 2^B \longrightarrow 2^A$ by

$$X^+ = \{y \in B \mid x \,\rho\, y \;\forall x \in X\} \text{ for } X \subseteq A . \tag{4.5.2}$$

$$Y^- = \{x \in A \mid x \,\rho\, y \;\forall y \in Y\} \text{ for } Y \subseteq B . \tag{4.5.3}$$

Here $P = 2^A, Q = 2^B$ and we prove that $X \subseteq Y^- \iff Y \subseteq X^+$. Indeed, if $X = \varnothing$ then $X^+ = B$ (because $x \in \varnothing \Longrightarrow x \,\rho\, y$), while if $Y = \varnothing$ then $Y^- = A$. Otherwise

$$X \subseteq Y^- \iff \forall x \in X \,\forall y \in Y \; x \,\rho\; y \iff \forall y \in Y \forall x \in X \; x \,\rho\; y \iff Y \subseteq X^+ .$$

The above Galois connection is called a *polarity*, due to the following particular case. Let A be the set of points of a plane, and B the set of lines of that plane. Let further $\sum_{i,j=1}^3 a_{ij} x_i x_j = 0$, where $a_{ij} = a_{ji}$, be the equation of a conic in polar coordinates. For every $(b_1, b_2, b_3) \in A$, the equation $\sum_{i,j=1}^3 a_{ij} b_i x_j = 0$ defines a line called the *polar* of the point (b_1, b_2, b_3), which is known as the *pole* of the line. This defines a correspondence $\rho \subseteq A \times B$ such that for every $x \in A$, $\{x\}^+$ is the polar of x, and for every $y \in B$, $\{y\}^-$ is the pole of y.

The next example is taken from Galois theory and justifies the name of Galois correspondence.

Example 4.5.2 Let $A = K$ be a field and $B = AutK$ the group of automorphisms of K. Define $\rho \subseteq A \times B$ by $x \, \rho \, \alpha \iff \alpha(x) = x$. Let $\mathcal{P}$ be the poset of subfields of K, and $\mathcal{Q}$ the poset of subgroups of $AutK$, both ordered by set inclusion. For every subfield H of K, set $H^+ = \{\alpha \in B \mid \alpha(x) = x \; \forall x \in H\}$, and for every subgroup G of $AutK$, set $G^- = \{x \in A \mid \alpha(x) = x \; \forall \alpha \in G\}$. It is easy to check that H^+ is a subgroup of $AutK$ and G^- is a subfield of K, while $H \subseteq G^- \iff G \subseteq H^+$.

Here are several properties of Galois correspondences.

Theorem 4.5.1 *If* $(^+, ^-)$ *is a Galois connection between the posets* P *and* Q, *then:*

(i) *The mappings* $^+$ *and* $^-$ *are antitone.*

(ii) *The following relations hold:*

$$^+ = {}^{+-+} \text{ and } ^- = {}^{-+-} . \tag{4.5.4}$$

(iii) *The mappings* $^{-+}$ *and* $^{+-}$ *are closure operators on* P *and* Q.

(iv) *The associated closure systems* $C = P^{+-}$ *and* $C' = Q^{-+}$ *are dually isomorphic.*

Proof: Note first that since $x^+ \leq x^+$, we have $x \leq x^{+-}$ by (4.5.1); similarly, $y \leq y^{-+}$.

(i) If $x \leq x'$, then since $x' \leq x'^{+-}$, it follows that $x \leq x'^{+-}$, whence $x'^+ \leq x^+$ follows by (4.5.1). Similar proof for the map $^-$.

(ii) We have $x^+ \leq x^{+-+}$ and $x \leq x^{+-}$ by extensivity, hence $x^{+-+} \leq x^+$ by antitony, therefore $x^+ = x^{+-+}$ and similarly $y^- = y^{-+-}$.

(iii) We have already proved extensivity. Besides, $^{+-}$ and $^{-+}$ are isotone because $^+$ and $^-$ are antitone. Finally, (4.5.4) implies $^{+-} = {}^{+-+-}$ and $^{-+} = {}^{-+-+}$.

(iv) For every $x \in P$, we have $x \in C \iff x = x^{+-}$. If $x \in C$, set $\varphi(x) = x^+$; then $\varphi(x)^{-+} = x^{+-+} = x^+ = \varphi(x)$. So $\varphi : C \longrightarrow C'$ and similarly, setting $\psi(y) = y^-$ for $y \in C'$, we get $\psi : C' \longrightarrow C$. The maps φ and ψ are antitone by (i), and $\psi\varphi(x) = \psi(x^+) = x^{+-} = x$, therefore $\psi\varphi = 1_C$ and similarly we get $\varphi\psi = 1_{C'}$. □

The next two results can be regarded as converses of Theorem 4.5.1 and they provide other situations in which Galois connections occur.

Proposition 4.5.1 *If* P *and* Q *are posets, the mappings* $^+ : P \longrightarrow Q$ *and* $^- : Q \longrightarrow P$ *are antitone and the mappings* $^{-+}$ *and* $^{+-}$ *are extensive, then* $(^+, ^-)$ *is a Galois connection between between* P *and* Q.

Proof: If $x \leq y^-$ then $y^{-x} \leq x^+$ and since $y \leq y^{-+}$, it follows that $y \leq x^+$. The converse implication is proved similarly. □

Theorem 4.5.2 *Let γ be a closure operator on a poset $(P, \leq)$ and $(\mathcal{C}, \leq)$ the associated Moore family. Let further $i : C \longrightarrow P$ be the inclusion mapping $i(x) = x$ and $c : P \longrightarrow \mathcal{C}$ the corestriction $c(x) = \gamma(x)$ of γ to the set $\mathcal{C}$. Then (c, i) is a Galois connection between the poset $(P, \leq)$ and the dual $(\mathcal{C}, \geq)$ of the poset $(\mathcal{C}, \leq)$, such that $c \circ i = 1_C$.*

PROOF: Since $\gamma(x) \in C$ for every $x \in P$, c is well defined. If $x \in P$ and $a \in C$, then $(c \circ i)(a) = \gamma(a) = a$ and $x \leq a \Longleftrightarrow \gamma(x) \leq a$. This can be written $c \circ i = 1_C$ and $x \leq i(a) \Longleftrightarrow a \geq c(x)$. □

The following converse of Example 4.5.1 holds:

Proposition 4.5.2 *If A,B are non-empty sets and $(^+, ^-)$ is a Galois correspondence between 2^A and 2^B, then there is a correspondence $\rho \subseteq A \times B$ such that identities* (4.5.2) *and* (4.5.3) *hold.*

PROOF: For every $x \in A$ and $y \in B$ set

$$x \,\rho\, y \Longleftrightarrow x \in \{y\}^- \Longleftrightarrow y \in \{x\}^+ \,, \tag{4.5.5}$$

which makes sense because $\{x\} \subseteq \{y\}^- \Longleftrightarrow \{y\} \subseteq \{x\}^+$. Therefore it suffices to prove e.g. (4.5.2).

For $X = \varnothing$ (for $Y = \varnothing$) both (4.5.1) and (4.5.2) reduce to $\varnothing^+ = B$ (to $\varnothing^- = A$), as shown in Example 4.5.1. Now suppose $X \neq \varnothing \neq B$. Then

$$y \in X^+ \Longleftrightarrow \{y\} \subseteq X^+ \Longleftrightarrow X \subseteq \{y\}^-$$

$$\Longleftrightarrow \forall x \in X \; x \in \{y\}^- \Longleftrightarrow \forall x \in X \; x \,\rho\, y \,.$$

We have proved that $y \in X^+ \Longleftrightarrow \forall x \in X \; x \,\rho\, y$, which is (4.5.2). □

We have even more:

Proposition 4.5.3 (O. Ore) *The Galois connections between 2^A and 2^B are in bijection with the correspondences $\rho \subseteq A \times B$, via the maps constructed in Example* 4.5.1 *and Proposition* 4.5.2.

PROOF: We have the associations $\rho \overset{(4.5.2,3)}{\mapsto} (^+, ^-) \overset{(4.5.5)}{\mapsto} \rho'$, where

$$x \,\rho'\, y \Longleftrightarrow x \in \{y\}^- \Longleftrightarrow x \,\rho\, y' \; \forall y' \in \{y\} \Longleftrightarrow x \,\rho\, y \,,$$

that is, $\rho' = \rho$. We have also $(^+, ^-) \overset{(4.5.5)}{\mapsto} \rho \overset{(4.5.2,3)}{\mapsto} (^\oplus, ^\ominus)$, where

$$y \in X^\oplus \Longleftrightarrow x \,\rho\, y \; \forall x \in X \Longleftrightarrow x \in \{y\}^- \; \forall x \in X$$

$$\Longleftrightarrow X \subseteq \{y\}^- \Longleftrightarrow y \in X^+ \Longleftrightarrow y \in X^+ \,.$$

that is, $X^\oplus = X^+$, and similarly $Y^\ominus = Y^-$. □

Here are two applications of Proposition 4.5.2.

Example 4.5.3 Let $< >: A \times B \longrightarrow K$ be a bilinear form of the vectors spaces A and B over a field K. Take as $\rho \subseteq A \times B$ the orthogonality correspondence $x \rho y =< x, y >= 0$. Then

$$X^+ = \{y \in B \mid < x, y >= 0 \ \forall x \in X\} = X^\perp$$

is the so-called subspace of B orthogonal to X, and

$$Y^- = \{x \in A \mid < x, y >= 0 \ \forall y \in Y\} = Y^\perp$$

is the subspace of A orthogonal to Y. The closures of X and Y are the double orthogonals $X^{\perp\perp}$ and $Y^{\perp\perp}$, respectively.

For instance, if $A = B =$ the three-dimensional Euclidean space and X is a line passing (not passing) through the origin, then $X^\perp$ is the plane (line) orthogonal to X and passing through the origin, while $X^{\perp\perp} = X$ ($X^{\perp\perp}$ is the plane generated by X and the origin).

Example 4.5.4 Take a poset P as $A = B$ and the order relation $\leq$ as ρ. Then the relation (4.5.4) from Proposition 4.5.3 become

(4.5.6) $$X^+ = \{y \in P \mid x \leq y \ \forall x \in X\}, \ \ Y^- = \{x \in P \mid x \leq y \ \forall y \in Y \ ,$$

for every $X, Y \subseteq P$. In this case the closed sets $X = X^{+-}$ and $Y = Y^{-+}$ are called *normal ideals* and *normal filters*, respectively.

This example will be used in Chapter 5, §3 (the MacNeille completion bv cuts).

Exercise 4.5.1 When is the empty set a normal ideal (filter) ?

Exercise 4.5.2 Let A be a topological space, $B = 2^A$ and the correspondence between A and B defined by $x \rho Y \iff x \in \overline{Y}$, where $\overline{Y}$ is the closure of Y in the space A. Prove that the Galois correspondence provided in Proposition 4.5.3 yields $\{x\}^{+-} = \overline{\{x\}}$. (Hint: $x' \in \{x\}^{+-} \iff \forall Y \ (x \in \overline{Y} \Longrightarrow x' \in \overline{Y})$).

An alteration of Definition 4.5.1 which turns out to be interesting, is the following: change (4.5.1) to

(4.5.1*) $$x \leq y^- \iff x^+ \leq y \ .$$

The pairs ($^+$,$^-$) satisfying (4.5.1*) are called *covariant Galois connections*, while the connections in Definition 4.5.1 are said to be *contravariant Galois connections*.

Exercise 4.5.3 Develop a theory of covariant Galois connections. (Hint: A covariant Galois connection between P and Q is a contravariant Galois connection between P and the dual of Q).

Chapter 5

Representation Theory

The title of this chapter summarizes three types of representation theorems dealt with: representations of the elements of certain lattices as meets/joins of elements from prescribed subsets, isomorphic representation of several types of posets (semilattices, lattices) as posets (semilattices, lattices) of sets with inclusion as partial order, and finally a more sophisticated development of the latter representations in the case of distributive and Boolean lattices: the duality between these categories and certain categories of topological spaces. These types of problems are treated in §§ 2, 3 and 5, respectively. The first section is devoted to ideals and filters both as a preparation to the subsequent sections and in view of the numerous other applications. The topological prerequisites necessary to §5 are collected in §4.

Keywords: Ideal, filter, prime filter, maximal filter, irreducibility, decomposition, set-theoretical embedding, clopen set, Stone space, homomorphism, Priestley space, Priestley duality.

5.1 Ideals and Filters

As it is the case for closure operators, the dual concepts of ideal and filter (the latter are sometimes called dual ideals) constitute one of the lattice-theoretical tools which play an important rôle in other fields of mathematics, as algebraic logic, topology, functional analysis, measure theory via the representation theorem for Boolean algebras, etc. Having these facts in mind, we start here the fundamentals of the dual theories of ideals and filters in posets and semilattices, after which we describe the ideal and filter lattices of a lattice, then the relationship between prime and maximal ideals/filters, with special emphasis on distributive and Boolean lattices.

Definition 5.1.1 An *order filter* or an *upper set* (an *order ideal* or a *lower set*) of a poset P is a non-empty subset $F \subseteq P$ ($I \subseteq P$) such that for every $x, y \in P$,

Sergiu Rudeanu

$$(5.1.1') \qquad x \in F \;\&\; x \leq y \Longrightarrow y \in F$$

(for every $x, y \in P$,

$$(5.1.1'') \qquad x \in I \;\&\; y \leq x \Longrightarrow y \in I \;).$$

Remark 5.1.1 a. An order ideal can be characterized as a non-empty segment; cf. Definition 2.5.1.

b. A subset $X \subseteq P$ is an order filter (order ideal) if and only if $P \setminus X$ is an order ideal (order filter).

Definition 5.1.2 By a *filter* (an *ideal*) of a meet semilattice (join semilattice) S is meant an order filter F (an order ideal I) of S such that for every $x, y \in S$,

$$(5.1.2') \qquad x, y \in F \Longrightarrow x \wedge y \in F$$

(for every $x, y \in S$,

$$(5.1.2'') \qquad x, y \in I \Longrightarrow x \vee y \in I \;).$$

Before giving several examples, we prove a few properties.

Remark 5.1.2 The theory of filters (order filters) in meet semilattices and the theory of ideals (order ideals) in join semilattices are dual to each other. Therefore the *duality principle* holds, to the effect that it suffices to deal with only one of the two theories.

Proposition 5.1.1 *Let S is a meet semilattice and $\varnothing \neq F \subseteq S$. Then F is a filter if and only if*

$$(5.1.3') \qquad x, y \in F \Longleftrightarrow x \wedge y \in F \;;$$

Dual statement for ideals in join semilattices.

PROOF: The implication $\Longrightarrow$ in (5.1.3′) is (5.1.2′), while the converse implication is equivalent to (5.1.1′). □

Remark 5.1.3 In a lattice we can use conditions

$$(5.1.4') \qquad x \in F \Longrightarrow x \vee z \in F \;\forall z \;,$$

$$(5.1.4'') \qquad x \in I \Longrightarrow x \wedge z \in I \;\forall z \;,$$

instead of (5.1.1′) and (5.1.1″), respectively.

Example 5.1.1 A poset (meet semilattice, join semilattice, lattice) is both an order filter and an order ideal (a filter, an ideal, both a filter and an ideal) of itself, known as the *improper* order filter and order ideal (filter, ideal, filter and ideal). The other order filters, order ideals, filters and ideals are said to be *proper*.

Example 5.1.2 For every element a of a meet semilattice (poset) P, the set

(5.1.5′) $$P[a) = \{x \in P \mid a \leq x\}$$

is a filter (an order filter) which we call the *principal (order) filter generated by a*. For every element a of a join semilattice (poset) P, the set

(5.1.5″) $$P(a] = \{x \in P \mid x \leq a\}$$

is an ideal (an order ideal) which we call the *principal (order) ideal generated by a*. Whenever no confusion is possible, the notation $P[a)$ and $P(a]$ is replaced by $[a)$ and $(a]$, respectively.

We have already dealt with principal ideals of well-ordered sets in Chapter 2§4. In Chapter 4, definition (4.4.5), we have associated with each closure system C on a poset P and each element $x \in P$, the set of those elements $c \in C$ which satisfy $x \leq c$. Since the latter condition is equivalent to $\gamma(x) \leq c$, the sets denoted $C[x)$ in Chapter 4 are in fact $P[\gamma(x))$ according to the above notation (5.1.5′). So these sets are principal order filters in posets and principal filters in meet semilattices, including complete lattices and Moore families of sets, which is the framework most frequently encountered in the literature.

As another particular case, note that the definition of limits in a metric space makes use of the principal filters of the chain $\mathbb{N}$ of natural numbers.

Remark 5.1.4 For every element of a poset P and every order filter F (every order ideal I) of P, we have $a \in F \iff [a) \subseteq F$ $(a \in I \iff (a] \subseteq I)$.

Remark 5.1.5 In a non-empty finite meet semilattice (join semilattice), all filters (ideals) are principal: $F = [\bigwedge F)$, resp. $I = (\bigvee I]$.

Example 5.1.3 Let P and P' be posets with 1 and $f : P \longrightarrow P'$ an isotone mapping. Then $f^{-1}(1)$ is an order filter. Dual statement.

Example 5.1.4 Let S and S' be meet semilattices with 1 and $f : S \longrightarrow S'$ a 1-preserving semilattice homomorphism. Then the set $f^{-1}(1)$ is a filter. Dual statement; the set $f^{-1}(0)$ is often called the *kernel* of the morphism f.

Example 5.1.5 The previous example can be generalized as follows. If θ is a congruence on S (cf. Examples 4.4.5, 4.4.6), then the set $\{x \in S \mid x\,\theta\,1\}$ is a filter. Dual statement. (Hint: if $x\,\theta\,1$ and $x \leq y$ then $x = x \wedge y\,\theta\,1 \wedge y = y$, hence $y\,\theta\,1$).

In particular if $f : S \longrightarrow S'$ is a 1-preserving homomorphism and θ is defined by

$$x\,\theta\,y \iff f(x) = f(y)\ ,$$

then one recaptures Example 5.1.4; sometimes it is this congruence which is called the *kernel* of the morphism f.

Example 5.1.6 The segment generated by an element $a \neq 0$ of a poset P (cf. Definition 2.5.1) is an order ideal. However a segment may not be an ideal, even if P is a lattice.

Example 5.1.7 Every non-empty normal filter (normal ideal) of a poset P is an order filter (order ideal); cf. Example 4.5.4. If, moreover, P is a meet (join) semilattice , then every normal filter (normal ideal) of P is a filter (an ideal).

For let $A = A^{-+} \subseteq P$, $a \in A$ and $a \leq x$. Then $A^- \leq a$, hence $A^- \leq x$, therefore $x \in A^{-+} = A$. If, moreover, P is a meet semilattice and $a, b \in A$, then $z \in A^-$ implies $z \leq a, b$, hence $z \leq a \wedge b$, so that $a \wedge b \in A^{-+} + A$; if $A^- = \varnothing$, then $a \wedge b \in P = \varnothing^+ = A^{-+} = A$.

Remark 5.1.6 Every principal order filter (principal order ideal) is a normal filter (normal ideal).

For $[x)^- = \{x\}^- = (x]$ and $(x]^+ = \{x\}^+ = [x)$, therefore $[x)^{-+} = (x]^+ = [x)$ and $(x]^{+-} = [x)^- = (x]$.

Example 5.1.8 If a and b are two elements of a meet semilattice S with 0, the set

$$S(a, b) = \{x \in S \mid a \wedge x = b \wedge x\}$$

is an order ideal. If, moreover, S is a distributive lattice, then $S(a, b)$ is an ideal. Dual statement.

Example 5.1.9 Roughly speaking, algebraic logic associates with each formalized logic system, a lattice known as the *Lindenbaum-Tarski algebra* of the system, in such a way that the properties of the former can be translated as lattice-theoretical properties of the latter. Thus e.g. the set of theorems or derivable formulas of the system correspond to a filter of the Lindenbaum-Tarski lattice; the system is consistent if and only if the filter is proper; etc.

Example 5.1.10 The family of all neighbourhoods $V(\xi)$ of a point ξ of a topological space T is a filter of the lattice $(2^T, \subseteq)$.

Example 5.1.11 The finite subsets of a non-empty (of an infinite) set T form an ideal (a proper ideal) of $(2^T, \subseteq)$.

Many other set-theoretical filters and ideals occur naturally in analysis, measure theory, probabilities and functional analysis; see e.g. Kappos [1969], Sikorski [1964], Vaidyanathaswamy [1960], Vladimirov [1969].

As a matter of fact, other examples of ideals and filters will be provided by some of the theorems which we are going to prove and which construct new ideals (filters) starting from given ones.

Proposition 5.1.2 1^0 *The following conditions are equivalent for an element* 1 *of a poset* P*:*

(i) 1 *is the greatest element of* P *;*

(ii) 1 *belongs to every filter of* P *;*

(iii) $P = (1]$ *;*

(iv) *an order ideal* I *of* P *is proper if and only if* $1 \notin I$.

2^0 *Dual statement.*

PROOF: (i)$\Longleftrightarrow$(iii), (i)$\Longrightarrow$(ii) and (i)$\Longrightarrow$(iv) are obvious.

(ii)$\Longrightarrow$(i): We have $1 \in [x)$ for every $x \in P$, that is, $x \leq 1$.

(iv)$\Longrightarrow$(iii): Because $1 \in (1]$. □

We are now going to show that the (order) filters and (order) ideals from Moore families of sets; in each case we shall give intrinsic characterizations of the associated closure operator $X \mapsto \overline{X}$. As shown in Corollary 4.4.4, a Moore family on $(2^T, \subseteq)$ is a complete lattice, with the join operation given by $\bigvee \mathcal{X} = \overline{\bigcup \mathcal{X}}$, while the meet operation is set-theoretical intersection.

The reason for assuming the existence of 0 or 1 in some of the propositions to be proved below is to ensure, *via* Proposition 5.1.2, that every intersection of filters or of ideals is not empty, which is the first condition for being a filter (an ideal). As a matter of fact, this assumption is not too restrictive, because to any poset, semilattice or lattice without 0 (without 1) can be added a least (greatest) element.

Proposition 5.1.3 *The order filters (order ideals) of a poset with* 1 *(with* 0*) form a Moore family and a complete join subsemilattice of* $(2^P, \subseteq)$. *The order filter (order ideal)* $\overline{X}$ *generated by a subset* $X \subseteq P$ *is described by* $\overline{\varnothing} = \{1\}$ *and*

$$\overline{X} = \bigcup\{[x) \mid x \in X\} \tag{5.1.6'}$$

for $\varnothing \neq X \subseteq P$ *(by* $\overline{\varnothing} = \{0\}$ *and*

$$\overline{X} = \{\bigcup\{(x] \mid x \in X\} \tag{5.1.6''}$$

for $\varnothing \neq X \subseteq P$*).*

Comment In other words, Proposition 5.1.3 states that the order filters (order ideals) of P form a complete lattice with respect to set-theoretical intersections and unions, and this lattice is "almost" a complete sublattice of $(2^P, \subseteq)$, except that its zero is $\{1\}$ (is $\{0\}$) instead of $\varnothing$.

PROOF: P is an order filter by Example 5.1.1. Given a family $(F_i)_{i\in I} \neq \varnothing$ of filters $F_i \subseteq P$, the intersection $\bigcap_{i\in I} F_i$ is not empty because $1 \in F_i$ for all $i \in I$ by Proposition 5.1.2, and it fulfills (5.1.1′) because so does each F_i. Similarly one sees that $\bigcup_{i\in I}$ is also an order filter.

Clearly $\{1\}$ is an order filter included in any other order filter, therefore $\overline{\varnothing} = \{1\}$. If $\varnothing \neq X \subseteq P$, then $\overline{X}$ given by (5.1.6′) is not empty. Moreover, if $y \in \overline{X}$ and $y \leq z$, then $y \in [x)$ for some $x \in X$, hence $z \in [x)$, therefore $x \in \overline{X}$. Thus $\overline{X}$ is an order filter and clearly $X \subseteq \overline{X}$. Finally if $X \subseteq F$ and F is an order filter, then for every $y \in \overline{X}$ we have $x \leq y$ for some $x \in X$, hence $x \in F$, therefore $y \in F$; thus $\overline{X} \subseteq F$.

The proof is achieved by duality. □

Warning The dual of a filter-or-ideal theoretical property which also involves set-theoretical operations (unions, intersections) is obtained by interchanging only the lattice-theoretical operations (joins, meets), but not the set-theoretical ones. Thus the reader may readily check that the following "dual" of (5.1.6′) is false: $\overline{X} = \bigcap\{(x] \mid x \in X\}$.

Definition 5.1.3 Let $\mathcal{F}(S)$ (let $\mathcal{I}(S)$) denote the family of all filters (ideals) of a meet (join) semilattice S.

Proposition 5.1.4 *If S is a meet semilattice with* 1 *(join semilattice with* 0*) then $\mathcal{F}(S)$ (then $\mathcal{I}(S)$) is a Moore family of* $(2^S, \subseteq)$. *The filter (ideal) $\overline{X}$ generated by a subset $X \subseteq S$ is described by $\overline{\varnothing} = \{1\}$ and*

$$(5.1.7') \qquad \overline{X} = \{x \in S \mid x_1 \wedge \ldots \wedge x_n \le x \text{ for some } n \in \mathbb{N} \setminus \{0\} \text{ and some } x_1, \ldots, x_n \in X\}$$

for $\varnothing \ne X \subseteq S$ (is described by $\overline{\varnothing} = \{0\}$ and

$$(5.1.7'') \qquad \overline{X} = \{x \in S \mid x \le x_1 \vee \ldots \vee x_n \text{ for some } n \in \mathbb{N} \setminus \{0\} \text{ and some } x_1, \ldots, x_n \in X\}$$

for $\varnothing \ne X \subseteq S$).

Proof: In view of duality it suffices to prove the first part.

The intersection of every non-empty family of filters is not empty because it contains 1 and clearly it satisfies (5.1.1′) and (5.1.2′), hence this intersection is a filter. Besides, S is the improper filter. Therefore $\mathcal{F}(S)$ is a Moore family by Corollary 4.4.3.

Since $\{1\}$ is the least filter, we have $\overline{\varnothing} = \{1\}$.

Further, let F denote the left side of (5.1.7′). Taking $n := 1$ and $x_1 := x$, we see that $X \subseteq F$, which implies also $F \ne \varnothing$. Obviously F satisfies (5.1.1′) and (5.1.2′). Thus F is a filter. Finally, if G is a filter such that $X \subseteq G$, then in the formula for F we have $x_1, \ldots, x_n \in G$, hence $x \in G$, therefore $F \subseteq G$. □

Comments 1) Formulas (5.1.7) are valid even in the absence of 1 or of 0, which are needed only for $\overline{\varnothing}$. 2) If X is a singleton, (5.1.7) reduce to (5.1.5).

In the next propositions the lattice L is not supposed to have 0 or 1 .

Proposition 5.1.5 1^0 *If L is a lattice then $(\mathcal{F}(L), \subseteq)$ is a lattice and a complete join semilattice $(\mathcal{F}(L), \cap, \bigvee)$, where for every $F_1, \ldots, F_n \in \mathcal{F}(L)$ and every $\mathcal{X} \subseteq \mathcal{F}(L)$,*

$$(5.1.8') \qquad F_1 \cap \cdots \cap F_n = \{x_1 \vee \ldots \vee x_n \mid x_k \in F_k \ (k = 1, \ldots, n)\} ,$$

$$(5.1.9') \qquad \bigvee \mathcal{X} = \{x \in L \mid x_1 \wedge \ldots \wedge x_n \le x \text{ for some } n \in \mathbb{N} \setminus \{0\}, \text{ some pairwise distinct } F_1, \ldots, F_n \in \mathcal{X} \text{ and some } x_k \in F_k \ (k = 1, \ldots, n)\} .$$

2^0 *Dual statement.*

Proof: 1^0 Take n filters $F_1, \ldots, F_n$ and let F denote the right side of (5.1.8′). We have $x_k \le x_1 \vee \ldots \vee x_n$, hence $x_1 \vee \ldots \vee x_n \in F_k$, for each $k = 1, \ldots, n$, therefore $x_1 \vee \ldots \vee x_n \in F_1 \cap \cdots \cap F_n$. Conversely, if $x \in F_1 \cap \cdots \cap F_n$, it follows from $x = x \vee \ldots \vee x$ (n terms) that $x \in F$. Consequently, relation (5.1.8′) holds, hence $F_1 \cap \cdots \cap F_n \ne \varnothing$ and since this intersection obviously satisfies (5.1.1′) and (5.1.2′), it follows that $F_1 \cap \cdots \cap F_n$ is a filter.

It remains to prove that the left side G of (5.1.9′) is the least filter which includes all $F \in \mathcal{X}$. First, $G \neq \varnothing$ and obviously G satisfies (5.1.1′) and (5.1.2′). If $F \in \mathcal{X}$ and $x \in F$, then taking $n := 1$, $F_1 := F$ and $x_1 := x$ in G we see that $F \subseteq G$. Finally, if H is a filter such that $F \subseteq H$ for all $F \in \mathcal{X}$, then in the formula for G we have $F_1, \dots, F_n \subseteq H$, hence $x \in H$; therefore $G \subseteq H$. □

Corollary 5.1.1 *In every lattice L, if the filters $F_1, \dots, F_n$ (the ideals $I_1, \dots, I_n$) are distinct, then*

(5.1.10′)
$$F_1 \vee \dots \vee F_n = \{x \in L \mid x_1 \wedge \dots \wedge x_n \leq x \text{ for some } x_k \in F_k \ (k = 1, \dots, n)\}$$

(then

(5.1.10″)
$$I_1 \vee \dots \vee I_n = \{x \in L \mid x \leq x_1 \vee \dots \vee x_n \text{ for some } x_k \in I_k \ (k = 1, \dots, n)\} \text{).}$$

PROOF: 1^0 For each $m \leq n$ and $i_1, \dots, i_m \in \{1, \dots, n\}$, put

$$S_{i_1 \dots i_m} = \{x \in L \mid x_{i_1} \wedge \dots \wedge x_{i_m} \leq x \text{ for some } x_{i_k} \in F_{i_k} \ (k = 1, \dots, m)\} .$$

Then it follows from Proposition 5.1.5.1^0 that

$$F_1 \vee \dots \vee F_n = \bigcup_{m=1}^{n} \bigcup_{i_1 \dots i_m} S_{i_1 \dots i_m} = S_{1 \dots n} ,$$

because $x_1 \wedge \dots \wedge x_n \leq x_{i_1} \wedge \dots \wedge x_{i_m}$ implies $S_{i_1 \dots i_m} \subseteq S_{1 \dots n}$. □

The above formulas become even simpler in the case of principal filters and principal ideals.

Proposition 5.1.6 *If $a_1, \dots, a_n$ are elements of a lattice L, then the following relations hold in $\mathcal{F}(L)$:*

(5.1.11′)
$$[a_1) \cap \dots \cap [a_n) = [a_1 \vee \dots \vee a_n) ,$$

(5.1.12′)
$$[a_1) \vee \dots \vee [a_n) = [a_1 \wedge \dots \wedge a_n)$$

(hold in $\mathcal{I}(L)$:

(5.1.11″)
$$(a_1] \cap \dots \cap (a_n] = (a_1 \wedge \dots \wedge a_n] ,$$

(5.1.12″)
$$(a_1] \vee \dots \vee (a_n] = (a_1 \vee \dots \vee a_n]) .$$

PROOF: Properties (5.1.11) are mere translations of the definition of sup and inf. Further, $[a_1 \wedge \dots \wedge a_n)$ is a filter, clearly $[a_k) \subseteq [a_1 \wedge \dots \wedge a_n)$ $(k = 1, \dots, n)$ and if a filter F includes all $[a_k)$s then $a_k \in F$ $(k = 1, \dots, n)$, hence $a_1 \wedge \dots \wedge a_n \in F$, therefore $[a_1 \wedge \dots \wedge a_n) \subseteq F$. This proves (5.1.12′) and (5.1.12″) is established similarly. □

Theorem 5.1.1 *Given a lattice L, the principal filters (principal ideals) of L form a sublattice of $\mathcal{F}(L)$ (of $\mathcal{I}(L)$) dually isomorphic (isomorphic) to L via the mapping $x \mapsto [x)$ (the mapping $x \mapsto (x]$).*

Comment In other words, L is *dually embedded* into $\mathcal{F}(L)$ (*embedded* into $\mathcal{I}(L)$).
PROOF: Immediate from Proposition 5.1.6. □

Assuming again the existence of 0 and 1, we are able to sharpen as follows Propositions 5.1.4 and 5.1.5:

Theorem 5.1.2 1^0 *The Moore family $\mathcal{F}(L)$ of all filters of a lattice L with* 1 *is a compactly generated lattice, where the complete operations are set-theoretical intersection and the join given by* (5.1.9′)*, while the compact elements are the principal filters.*
2^0 *Dual statement.*

PROOF: 1^0 We already know that $\mathcal{F}(L)$ is a Moore family of sets by Proposition 5.1.4, hence a complete lattice with intersection as complete meet operation by Corollary 4.4.4, and the complete join operation (5.1.9′) by Proposition 5.1.5.

In view of Lemma 4.4.2, to show that $\mathcal{F}(L)$ is compactly generated, we still have to prove that the union $F = \bigcup \mathcal{D}$ of a directed family of filters $\mathcal{D}$ is again a filter. First note that $F \neq \varnothing$. Further, if $x \in F$ and $x \leq y$, then $x \in D$ for some filter $D \in \mathcal{D}$, hence $y \in D$, therefore $y \in F$. If $x, y \in F$, then $x \in D'$ and $y \in D''$ for some filters $D', D'' \in \mathcal{D}$. Since $\mathcal{D}$ is directed, there is a filter $D \in \mathcal{D}$ such that $D', D'' \subseteq D$, therefore $x, y \in D$, which implies $x \wedge y \in D$ and so $x \wedge y \in F$.

Finally we must determine the compact elements of $\mathcal{F}(L)$.

If F is a compact filter, then from $F \subseteq \bigvee\{[a) \mid a \in F\}$ we deduce the existence of $a_1, \ldots, a_n \in F$ such that $F \subseteq [a_1) \vee \ldots \vee [a_n)$, hence $F \subseteq [a_1 \wedge \ldots \wedge a_n)$ by (5.1.12′). But $a_1 \wedge \ldots \wedge a_n \in F$, hence $[a_1 \wedge \ldots \varepsilon a_n) \subseteq F$, therefore $F = [a_1 \wedge \ldots \wedge a_n)$ is a principal filter. Conversely, the principal filters $[x) = \gamma(\{x\})$ are compact by Lemma 4.4.2. □

Exercise 5.1.1 Prove directly that every principal filter $[x)$ is compact in $\mathcal{F}$.

Now we deal with modularity or distributivity.

Lemma 5.1.1 *If the lattice $\mathcal{F}(L)$ (the lattice $\mathcal{I}(L)$) is modular or distributive, so is the lattice L.*

PROOF: The possible modularity or distributivity of $\mathcal{I}(L)$ is inherited by any sublattice of it (cf. Exercise 4.2.1 or 4.2.5), hence by the sublattice of principal ideals, which is isomorphic to L (cf. Theorem 5.1.1), therefore modularity or distributivity is inherited by L.

A similar argument works for $\mathcal{F}(L)$, if we take into account that in view of the duality principle (cf. Remarks 4.2.1 and 4.2.4), modularity and distributivity are also preserved by dual isomorphisms. □

The following two theorems show that the converse of Lemma 5.1.1 is also valid.

Theorem 5.1.3 *The lattice $\mathcal{F}(L)$ (the lattice $\mathcal{I}(L)$) is modular if and only if the lattice L is modular.*

PROOF: Assume the modularity of L and take $F, G, H \in \mathcal{F}(L)$ such that $F \subseteq H$; we have to prove that $(F \vee G) \cap H \subseteq F \vee (G \cap H)$, as the converse inequality holds in any lattice. Let $z \in (F \vee G) \cap H$, that is, $z \in H$ and $z \in F \vee G$; then (5.1.10′) shows that $x \wedge y \leq z$ for some $x \in F$ and some $y \in G$. But $x \in F$ implies $x \in H$, hence $x \wedge z \in H$, therefore $y \vee (x \wedge z) \in G \cap H$ by (5.1.8′); finally we use the modular law (4.2.3) and obtain

$$z \geq (x \wedge z) \vee (y \wedge x) = ((x \wedge z) \vee y) \wedge x \ ,$$

therefore $z \in (H \wedge G) \vee F$, as desired. □

Theorem 5.1.4 *The lattice $\mathcal{F}(L)$ (the lattice $\mathcal{I}(L)$ is distributive if and only if the lattice L is distributive.*

PROOF: Assume the distributivity of L and take $F, G, H \in \mathcal{F}(L)$; we have to prove that $(F \vee G) \cap (F \vee H) \subseteq F \vee (G \cap H)$, as the converse inequality holds in any lattice. Let $z \in (F \vee G) \cap (F \vee H)$, that is, $z \in F \vee G$ and $z \in F \vee H$. Then $x' \wedge y \leq z$ and $x'' \wedge t \leq z$ for some $x, x' \in F$, some $y \in G$ and some $t \in H$. It follows that $x' \vee (x'' \wedge t) \in F$, $y \vee x'' \in F$ and $y \vee t \in G \cap H$, therefore

$$z \geq (x' \wedge y) \vee (x'' \wedge t) = (x' \vee (x'' \wedge t)) \wedge (y \vee x'') \wedge (y \vee t)$$

implies $z \in F \vee (G \cap H)$, as desired. □

Proposition 5.1.7 1^0 *If L is a distributive lattice, then the arbitrary joins in $\mathcal{F}(L)$ are given by*

$$(5.1.13') \qquad \bigvee \mathcal{X} = \{x_1 \wedge \ldots \wedge x_n \mid x_k \in F_k \in \mathcal{X} \ (k = 1, \ldots, n), \ n \in \mathbb{N} \setminus \{0\}\}$$

for every $\varnothing \neq \mathcal{X} \subseteq \mathcal{F}(L)$, while finite joins are given by

$$(5.1.14') \qquad F_1 \vee \ldots \vee F_n = \{x_1 \wedge \ldots \wedge x_n \mid x_k \in F_k \ (k = (1, \ldots, n)\}$$

for every pairwise distinct $F_1, \ldots, F_n \in \mathcal{F}(L)$ and every $n \in \mathbb{N} \setminus \{0\}$.

2^0 *Dual statement for arbitrary joins in $\mathcal{I}(L)$* (cf. Warning after Proposition 5.1.3).

PROOF: 1^0 Using Proposition 5.1.5 and Corollary 5.1.1, it remains to prove that (5.1.9′) and (5.1.10′) can be written in the simpler forms (5.1.13′) and (5.1.14′), respectively. Indeed, $x = x_1 \wedge \ldots \wedge x_n$ trivially implies $x_1 \wedge \ldots \wedge x_n \leq x$; conversely, if $x_1 \wedge \ldots \wedge x_n \leq x$, with $x_k \in F_k$ $(k = 1, \ldots, n)$, then

$$x = x \vee (x_1 \wedge \ldots \wedge x_n = (x \vee x_1) \wedge \ldots \wedge (x \vee x_) \ ,$$

with $x \vee x_k \in F_k$ $(k = 1, \ldots, n)$. □

Remark 5.1.7 Distributivity is essential in Proposition 5.1.7, as shown by the following example. Let $L = \{0, a, b, c, 1\}$ be one of the two standard non-distributive lattices; cf. Theorems 4.2.1 and 4.2.2, Figures 4.3 and 4.4, respectively. Then $(a] \vee (c] = (a \vee c] = (1] = L$, but (5.1.14″) fails: the element $b \in L$ cannot be written in the form $b = x_1 \vee x_2$ with $x_1 \leq a$ and $x_2 \leq c$.

We now introduce a class of ideals (filters) which is similar to, and in certain cases coincides with a corresponding concept from ring theory, and which plays a fundamental rôle in the representation theory of distributive lattices.

Definition 5.1.4 A proper filter F (proper ideal I) of a lattice L is said to be *prime* if it fulfills

$$x \vee y \in F \Longrightarrow x \in F \text{ or } y \in F \tag{5.1.15'}$$

(if it fulfills

$$x \wedge y \in I \Longrightarrow x \in I \text{ or } y \in I) \ . \tag{5.1.15''}$$

In other words, a proper filter (proper ideal) is prime if it satisfies the converse of the property "$x \in F$ or $y \in F \Longrightarrow x \vee y \in F$" (dual property). The latter property is obviously equivalent to (5.1.4′) (to (5.1.4″)). See also Theorem 5.1.5 below, property (5.1.18).

Example 5.1.12 A proper principal filter $[a)$ is prime if and only if

$$a \leq x \vee y \Longrightarrow a \leq x \text{ or } a \leq y \ . \tag{5.1.16'}$$

This property implies

$$a = x \vee y \Longrightarrow a = x \text{ or } a = y \tag{5.1.17'}$$

because if $a = x \vee y$ then $a \leq x \leq a$ or $a \leq y \leq a$. An element a satisfying (5.1.17′) is said to be *finite-join irreducible.* Property (5.1.16′) is actually stronger than (1.5.17′), as shown by any of the two lattices in Figures 4.3. and 4.4, in which the element b is finite-join irreducible but not prime. However in a distributive lattice the two properties are equivalent. For if a is finitely join irreducible and $a \leq x \vee y$, then $a = a \wedge (x \vee y) = (a \wedge x) \vee (a \wedge y)$, hence $a = a \wedge x$ or $a = a \wedge y$, that is, $a \leq x$ or $a \leq y$.

Similarly, the prime principal ideals $(a]$ are characterized by the dual property of *finite-meet irreducibility.*

Exercise 5.1.2 In a chain all proper principal filters (ideals) are prime. Is the converse true ?

Theorem 5.1.5 1^0 *The following conditions are equivalent for a subset F of a lattice L:*

(i) *F is a prime filter ;*

(ii) *F is a proper filter satisfying, for all $x, y \in L$,*

(5.1.18′) $$x \vee y \in F \Longleftrightarrow x \in F \text{ or } y \in F \;;$$

(iii) *F is a proper filter such that for every two filters G,H,*

(5.1.19′) $$G \cap H \subseteq F \Longrightarrow G \subseteq F \text{ or } H \subseteq F \;;$$

(iv) *$L \setminus F$ is a prime ideal .*

2^0 *Dual statement for prime ideals.*

PROOF: 1^0 (i)$\Longrightarrow$(ii): If $x \in F$ or $y \in F$ then $x \vee y \in F$ by (5.1.1′).

(ii)$\Longrightarrow$ (i): Trivial.

(i)$\Longrightarrow$(iii): We prove that $G \not\subseteq F$ & $H \not\subseteq F \Longrightarrow G \cap H \not\subseteq F$. Indeed, if $x \in G \setminus F$ and $y \in H \setminus F$ then $x \in G$ and $y \in H$, hence $x \vee y \in G \cap H$, but $x \vee y \notin F$ because $x \notin F$ and $y \notin F$.

(iii)$\Longrightarrow$(i): If $x \vee y \in F$ then $[x) \cap [y) \subseteq F$, hence $[x) \subseteq F$ or $[y) \subseteq F$, therefore $x \in F$ or $y \in F$.

(i)$\Longrightarrow$(iv): Let F be a prime filter. Then $\varnothing \neq F \neq L$, hence $\varnothing \neq L \setminus F \neq L$. If $x \in L \setminus F$ and $y \leq x$, then $y \in L \setminus F$ because $y \in F$ would imply $x \in F$. If $x, y \in L \setminus F$ then $x \vee y \in L \setminus F$, otherwise (5.1.15′) would imply $x \in F$ or $y \in F$. If $x \wedge y \in L \setminus F$ then $x \in L \setminus F$ or $y \in L \setminus F$, because $x \in F$ and $y \in F$ would imply $x \wedge y \in F$.

(iv)$\Longrightarrow$(i): From (i)$\Longrightarrow$(iv) by duality. □

Corollary 5.1.2 1^0 *A subset F of a lattice L with 0 and 1 is a prime filter if and only if the characteristic function f of F is a morphism in* **Lat01**.

2^0 *Dual statement* (relating prime ideals to dual morphisms).

PROOF: 1^0 Recall (cf. Exercise 1.2.8) that $f(x) = 1$ for $x \in F$ and $f(x) = 0$ for $x \in L \setminus F$.

In view of Proposition 5.1.1 and Theorem 5.1.5, the desired result follows from the following three properties:

(i) $f(1) = 1$ and $f(0) = 0$ iff $\varnothing \neq F \neq L$.

(ii) f is a meet homomorphism iff (5.1.3′) holds. Indeed, (5.1.3′) can be written $f(x \wedge y) = 1 \Longleftrightarrow f(x) = 1$ and $f(y) = 1 \Longleftrightarrow f(x) \wedge f(y) = 1$, that is, $f(x \wedge y) = f(x) \wedge f(y)$.

(iii) f is a join homomorphism iff (5.1.18′) holds. Indeed, (5.1.18′) can be written $f(x \vee y = 1 \Longleftrightarrow f(x) = 1$ or $f(y) = 1 \Longleftrightarrow f(x) \vee f(y) = 1$, that is, $f(x \vee y) = f(x) \vee f(y)$. □

Remark 5.1.8 The above conditions (5.1.14)-(5.1.19) are obviously equivalent to those obtained from them by replacing two-element meets, joins and intersections, by arbitrary finite meets, joins and intersections, respectively. The same is true for properties (5.1.22) below.

Proposition 5.1.8 (i) *Every prime filter F (prime ideal I) of a lattice L is finite-meet irreducible in the lattice $\mathcal{F}(L)$ (the lattice $\mathcal{I}(L)$), i.e.,*

(5.1.20′) $$G \cap H = F \Longrightarrow G = F \text{ or } H = F \;,$$

$$(5.1.20'') \qquad J \cap K = I \Longrightarrow J = I \text{ or } K = I \,.$$

(ii) *If, moreover, the lattice L is distributive, then prime filters (prime ideals) coincide with finite-meet irreducible proper filters (ideals).*

PROOF: If the lattice L is distributive, then $\mathcal{F}(L)$ and $\mathcal{I}(L)$ are distributive by Theorem 5.1.4. Now everything follows from Example 5.1.12. □

Remark 5.1.9 Distributivity is essential in Proposition 5.1.8(ii), as shown by the following example. Take again one of the two standard non-distributive lattices $L = \{0, a, b, c, 1\}$; cf. Theorem 4.2.2, Figures 4.3 and 4.4. Recall first that all the ideals of L are principal by Remark 5.1.5. The ideal $(a]$ fulfills (5.1.22″) because $(x] \cap (y] = (a]$ means $(x \wedge y] = (a]$ or, equivalently, $x \wedge y = a$, which is possible only if $x = a$ or $y = a$, therefore $(x] = (a]$ or $(y] = (a]$. However $(a]$ is not a prime ideal because $b \wedge c = 0 \in (a]$ although $b \notin (a]$ and $c \notin (a]$.

Finally we introduce

Definition 5.1.5 By a *maximal filter* (*maximal ideal*) of a lattice L is meant a proper filter (proper ideal) of L maximal with respect to set inclusion.

Example 5.1.13 An easy application of the Kuratowski-Zorn lemma shows that in a lattice with 0 (with 1) every proper filter (proper ideal) is included in a maximal filter (maximal ideal).

The chain $\mathbb{Z}$ has no maximal filters and no maximal ideals.

Proposition 5.1.9 *In a distributive lattice every maximal filter (maximal ideal) is prime.*

PROOF: Let F be a maximal filter of a distributive lattice L and $a, b \in L$ two elements such that $a \vee b \in F$ but $a \notin F$; we shall prove that $b \in F$. The filter $G = F \vee [a)$ satisfies $F \subset G$ because $a \in G \setminus F$, hence $G = L$ by the maximality of F. It follows that $b \in G$, therefore Proposition 5.1.7 shows that b is of the form $b = f \wedge c$, where $f \in F$ and $c \geq a$. Therefore

$$b = b \vee (f \wedge c) = (b \vee f) \wedge (b \vee c) \geq f \wedge (b \vee a) \in F \,,$$

hence $b \in F$. □

Remark 5.1.10 Distributivity is essential in Proposition 5.1.9, as shown in Remark 5.1.9, where the ideal $(a]$ is maximal but not prime.

Remark 5.1.11 The converse of Proposition 5.1.9 fails, as shown e.g. in any chain (hence, distributive lattice) L with least element 0 and $| L | \geq 3$: the prime ideal $\{0\}$ is not maximal.

Now we concentrate on the important particular case of Boolean algebras, where several supplementary properties hold, among which the converse of Proposition 5.1.9.

Proposition 5.1.10 *Let* $(B, \wedge, \vee, ', 0, 1)$ *be a Boolean algebra. A subset* $I \subseteq B$ *is an ideal of the lattice* $(B, \wedge, \vee)$ *if and only if it is an ideal of the associated Boolean ring* $(B, +, \cdot, 0, 1)$.

PROOF: Let I be a lattice-theoretical ideal of B. Then $I \neq \varnothing$ and if $x, y \in I$ then $x \wedge y', x' \wedge y \in I$ by (5.1.4″), hence $x - y = x + y = (x \wedge y') \vee (x' \wedge y) \in I$ by (5.1.2″), therefore I is a subgroup of $(B, +)$. If $x \in I$ and $z \in B$, then $xz = x \wedge z \in I$ by (5.1.4″).

Conversely, let I be a ring-theoretical ideal of B. Then $I \neq \varnothing$ because $0 \in I$. If $x, y \in I$ then $xy \in I$, hence $x \vee y = x + y + xy \in I$. If $x \in I$ and $z \leq x$ then $z = z \wedge x = zx \in I$. □

As a consequence of Proposition 5.1.10, the whole of *ideal theory for rings can be applied to the ideals of a Boolean algebra.* Thus, e.g., ideals coincide with kernels of surjective homomorphisms (not only the prime ideals cf. Theorem 5.1.5), etc.

Proposition 5.1.11 *Let* S *be a subset of a Boolean algebra* B *and* $S' = \{x' \mid x \in S\}$. *Then the following conditions are equivalent:*

(i) S *is a proper filter (proper ideal) ;*

(ii) S' *is a proper ideal (proper filter) ;*

(iii) S *is a filter (an ideal) and* $S \cap S' = \varnothing$ *;*

(iv) S *is a filter (an ideal) and for every* $x \in B$, *at most one of the elements* x, x' *belong to* S.

PROOF: Notice first that the uniqueness of the complement of a given element implies the equivalences $x \in S \Longleftrightarrow x' \in S'$ and $x' \in S \Longrightarrow x \in S'$.

(i)$\Longrightarrow$(iv): If $x, x' \in S$ for some x, then $0 = x \wedge x' \in S$; contradiction.

(iv)$\Longrightarrow$(iii): If $x \in S \cap S'$, then $x \in S$ and $x \in S'$, hence $x' \in S$; contradiction.

(iii)$\Longrightarrow$(ii): Since $S \neq \varnothing$, it follows that $S' \neq \varnothing$. Since $S \cap S' = \varnothing$, it follows that $S' \neq B$, otherwise $S = S \cap B = S \cap S' = \varnothing$. If $x, y \in S'$, then $x', y' \in S$, hence $x' \wedge y' \in S$, therefore $x \vee y = (x' \wedge y')' \in S'$. If $x \in S'$ and $z \leq x$, then $z' \geq x' \in S$, hence $z' \in S$, therefore $z \in S'$.

(ii)$\Longrightarrow$(i): Since $S' \neq \varnothing$, it follows that $S \neq \varnothing$, while $S' \neq B$ implies $S \neq B$. If $x, y \in S$, then $x', y' \in S'$, hence $x' \vee y' \in S'$, therefore $x \wedge y = (x' \vee y')' \in S$. If $x \in S$ and $x \leq z$, then $z' \leq x' \in S'$, hence $z' \in S'$, therefore $z \in S$. □

We now prove the announced sharpening of Proposition 5.1.9:

Theorem 5.1.6 *The following conditions are equivalent for a subset* S *of a Boolean algebra* B:

(i) S *is a prime filter (prime ideal) ;*

(ii) S *is a maximal filter (maximal ideal) ;*

(iii) S *is a proper filter (proper ideal) and for every* $x \in B$, *exactly one of the elements* x, x' *belongs to* S.

Comment It can be proved (cf. Monteiro [1947] and Nachbin [1947]) that the coincidence of prime filters (prime ideals) with maximal filters (maximal ideals) characterizes Boolean algebras within the class of distributive lattices with 0 and 1.

PROOF: (i)$\Longrightarrow$(iii): A prime filter S is by definition proper. Now take an element $x \in B$. Then S cannot contain both x and x' by Proposition 5.1.11, but it contains one of these elements, otherwise $x, x' \notin S$ although $x \vee x' = 1 \in S$, which is a contradiction.

(iii)$\Longrightarrow$(ii): Let T be a filter such that $S \subset T$. Take an element $t \in T \setminus S$. Then $t \notin S$, hence $t' \in S$, consequently $t' \in T$, and since $t \in T$ it follows that $0 = t \wedge t' \in T$, which means $T = B$.

(ii)$\Longrightarrow$(i): By Proposition 5.1.9. □

5.2 Meet and Join Decompositions

The problems dealt with in this section are of the following type: represent an arbitrary element of a certain lattice L as a meet(join) of elements from a prescribed subset of L. The main results are the representation of the elements of a compactly generated lattice as meets of meet-irreducible elements and the decomposition of the elements of certain modular lattices as direct joins of indecomposable elements. The approach to the latter result is that of Pierce [1968], but in a more general setting which uses an idea of Schmidt [1970]. We also provide representations as joins of atoms in atomic Boolean algebras and as joins of finite-meet irreducibles in lattices.

Definition 5.2.1 An element x of a poset P is said to be *meet (join) reducible* if it can be represented in the form $x = \bigwedge A$ ($x = \bigvee A$) with $A \neq \varnothing$ an $x \notin A$; otherwise we say that x is *meet (join) irreducible.*

Warning This definition makes indeed sense in an arbitrary poset, because certain (possibly infinite) meets or joins may exist. The weaker properties of finite meet (join) irreducibility introduced in Example 5.1.12 are also meaningful in arbitrary posets. Yet the results to be proved below concern certain types of lattices.

Remark 5.2.1 In every poset, the element 1 (the element 0), if any, is meet (join) irreducible,

Lemma 5.2.1 *An element $m \neq 1$ of a complete lattice L is meet irreducible if and only if*

$$(5.2.1') \qquad m < \bigwedge\{x \in L \mid m < x\}\ ;$$

the dual statement is also valid.

PROOF: Since $m \neq 1$, we have $\{x \in L \mid m < x\} \neq \varnothing$. Since $m \leq \bigwedge\{x \in L \mid m < x\}$, the negation of (5.2.1′) is $m = \bigwedge\{x \in L \mid m < x\}$, which shows that m

is meet reducible. Conversely, if m is meet reducible, then from $m = \bigwedge A$ with $m \notin A \neq \varnothing$ we infer that $x \in A \Longrightarrow m < x$, therefore

$$m \leq \bigwedge\{x \in L \mid m < x\} \leq \bigwedge A = m ,$$

showing that (5.2.1′) fails. □

Theorem 5.2.1 *Every element x of a compactly generated lattice L is the meet of all meet irreducible elements $m \in L$ such that $x \leq m$.*

PROOF: Take $x \in L$ and let M be the set of all irreducible elements $m \geq x$, which is not empty by Remark 5.2.1. Assume $x \neq \bigwedge M$; since $x \leq \bigwedge M$ anyway, it follows that $x < \bigwedge M$ and we are looking for a contradiction.

Note that $\bigwedge M = \bigvee K$ where K is a set of compact elements, and there is an element $k \in K$ such that $k \not\leq x$ (otherwise $\bigwedge M \leq x$).

Let $\mathcal{A}$ be the family of all subsets $A \subseteq L$ such that $x \leq A$ and $k \not\leq A$. If we succeed in proving that

(i) the family $\mathcal{A}$ contains a maximal set B

and we set $m = \bigvee B$, it will follow that $x \leq B \leq m$ and $k \not\leq m$. If we succeed in proving that

(ii) m is meet irreducible,

it will follow that $m \in M$, hence $k \leq \bigwedge M \leq m$, the desired contradiction.

Proof of (i). We show that $\mathcal{A}$ verifies the hypothesis of the Kuratowski-Zorn lemma, variant (e) (cf. Theorem 2.4.1). First, $\mathcal{A} \neq \varnothing$ because $\{x\} \in \mathcal{A}$. Now take a chain $\mathcal{C} \subseteq \mathcal{A}$ and set $A = \bigcup \mathcal{C}$; we have to prove that $A \in \mathcal{A}$. Every $a \in A$ belongs to some $C \in \mathcal{C}$, so that $x \in C$, hence $x \leq a$; therefore $x \leq A$. To prove that $k \not\leq \bigvee A$, we assume the contrary. Since k is compact, there exist $a_1, \ldots, a_n \in A$ such that $k \leq a_1 \vee \ldots \vee a_n$ and there exist $C_1, \ldots, C_n \in \mathcal{C}$ such that $a_i \in C_i$ $(i = 1, \ldots, n)$. But $C_i \subseteq C_j$ for some j and all i, therefore $a_i \in C_j$ $(i = 1, \ldots, n)$, hence $k \leq \bigvee C_j$, in contradiction with $C_j \in \mathcal{A}$.

Proof of (ii). Using Lemma 5.2.1, it suffices to show that $m < \bigwedge\{z \in L \mid m < z\}$. Indeed, take $z > m$; such elements exist, e.g. $m \vee k$. Since $\bigvee B < z$, it follows that $z \notin B$. Setting $D = B \cup \{z\}$, we have $B \subset D$, hence the maximality of B implies $D \notin \mathcal{A}$. But from $x \leq B \leq m < z$ it follows that $x \leq D$, which is the first property defining $\mathcal{A}$. Therefore the second property defining $\mathcal{A}$ fails for D. Thus $m < m \vee k \leq z$ and since this happens for all $z > m$, it follows that $m < m \vee k \leq \bigwedge\{z \in L \mid m < z\}$, thus proving (5.2.1′). □

Theorem 5.2.1 enables us to establish some further properties of filters and ideals in distributive lattices.

Proposition 5.2.1 *Every proper filter (proper ideal) of a distributive lattice with 1 (with 0) is an intersection of prime filters (prime ideals).*

PROOF: Let F be a proper filter. In view of Theorem 5.1.2, $\mathcal{F}(L)$ is a compactly generated lattice, hence it follows from Theorem 5.2.1 that $F = \bigwedge \mathcal{M}$, where $\mathcal{M}$ is a family of meet irreducible filters. Without loss of generality we may suppose that all the filters of $\mathcal{M}$ are proper. They are a fortiori finite-meet irreducible, i.e., they satisfy (5.1.22′). The lattice $\mathcal{F}(L)$ is distributive by Theorem 5.1.4, therefore the elements of $\mathcal{M}$ are prime filters by Proposition 5.1.8(ii). □

Proposition 5.2.2 *If a and b are elements of a distributive lattice with* 1 *(with* 0*) and $a \not\leq b$ ($b \not\leq a$), then there is a prime filter (prime ideal) of L which contains a but not b.*

Comment This *separation property* is important in the representation theory for distributive lattices, as will be shown in Theorem 5.3.2.
PROOF: If $a \not\leq b$, the principal filter $[a)$ is proper, hence it is an intersection of prime filters by Proposition 5.2.1. Each of these filters contains a, whereas at least one of them does not contain b. □

We now turn to join decompositions in a certain class of Boolean algebras. A few prerequisites are needed.

Definition 5.2.2 An *atom* of a poset P with 0 is a minimal element of the subset $P \setminus \{0\}$.

(The dual concept, termed *dual atom*, will not be used in this eBook).

Remark 5.2.2 In other words, an element a of a poset P with 0 is an atom if and only if $a \neq 0$ and there is no element $x \in P$ satisfying $0 < x < a$; or, equivalently, $x \leq a$ implies either $x = 0$ or $x = a$.

Definition 5.2.3 A poset P with 0 is said to be *atomic* , if for every element $x \in P \setminus \{0\}$ there is an atom $a \leq x$.

Example 5.2.1 Every finite poset with 0 is atomic by Proposition 2.2.1′.

Now we can prove:

Theorem 5.2.2 *Every element $x \neq 0$ of an atomic Boolean algebra B is the join of all atoms $a \in B$ such that $a \leq x$.*

PROOF: Let A be the set of all atoms $a \leq x$; then $A \leq x$. To prove that $x = \bigvee A$ we still have to show that $A \leq y$ implies $x \leq y$, i.e., it implies $x \wedge y' = 0$. Indeed, otherwise $x \wedge y' \neq 0$, hence there is an atom $a \leq x \wedge y'$, which implies $a \leq x$ and $a \leq y'$. But the former inequality shows that $a \in A$, hence $a \leq y$, therefore $a \leq y \wedge y' = 0$, a contradiction. □

Warning Although each $x \neq 0$ has a representation of the form $x = \bigvee A$ where the set A may be infinite, the Boolean lattice B need not be complete, not even a complete join semilattice. This is illustrated in:

Example 5.2.2 Let $\mathcal{B}$ be the Boolean algebra of all finite and cofinite subsets of an infinite set T; cf. Example 5.2.12. Then $\mathcal{B}$ is an atomic Boolean algebra, the atoms being the singletons: it is trivial that each $X \subseteq B$ is the union of all the singletons $\{x\} \subseteq X$. To show that $\mathcal{B}$ is not a complete join semilattice, take a sequence $(x_n)_{n \in \mathbb{N}}$ of pairwise distinct elements of T and set $X_n = \{x_0, x_2, x_4, \ldots, x_{2n}\}$ for $n \in \mathbb{N}$. Then all $X_n \in \mathcal{B}$ and there exist upper bounds of $(X_n)_{n\in\mathbb{N}}$ in $\mathcal{B}$, namely all cofinite subsets Y of T such that $(x_{2n})_{n\in\mathbb{N}} \subset Y$. However, since the sets $Y \setminus (x_{2n})_{n\in\mathbb{N}}$ are infinite (because $(x_{2n+1})_{n\in\mathbb{N}}$ cannot be included in the finite set $T \setminus Y$), there is no least such Y.

The rôle of atomicity in the representation theory for Boolean algebras will be seen in Theorem 5.3.3.

Exercise 5.2.1 In a meet semilattice with 0, a principal filter $[a)$ is maximal if and only if a is an atom. Dual statement.

Theorems 5.2.1 and 5.2.2 construct infinite meet or join decompositions in compactly generated lattices and in (not necessarily complete Boolean algebras, respectively. We now wish to establish the existence of a special type of finite join decompositions in modular lattices satisfying a certain type of "finiteness condition". This result, important in universal algebra, needs a more elaborate preparation.

Definition 5.2.4 Let θ be a congruence on a lattice L with 0 (cf. Example 4.4.6). A finite non-empty subset $A \subseteq L$ is said to be *θ-independent* if

(5.2.2) $$a \wedge \bigvee(A \setminus \{a\})\,\theta\,0 \quad (\forall\, a \in A)\,.$$

Remark 5.2.3 Every singleton is θ-independent.

Proposition 5.2.3 *Let A be a θ-independent set in a modular lattice with* 0. *The following properties hold:*

(i) *Every non-empty subset of A is θ-independent.*

(ii) *If $a\,\theta\,p \vee q$ for some $a \in A$ and some θ-independent set $\{p, q\}$, then the set*

(5.2.3) $$A' = (A \setminus \{a\}) \cup \{p, q\}$$

is θ-independent.

Proof: (i) Let $\varnothing \neq B \subseteq A$ and $b \in B$. Then

$$b \wedge \bigvee(B \setminus \{b\}) \leq b \wedge \bigvee(a \setminus \{b\})\,\theta\,0\,,$$

therefore $b \wedge \bigvee(B \setminus \{b\})\,\theta\,0$ on account of the dual of Example 5.1.5.

(ii) For every $b \in A \setminus \{a\}$ we have

$$b \wedge \bigvee(A' \setminus \{b\}) = b \wedge (\bigvee(A \setminus \{a, b\}) \vee p \vee q)$$

$$\theta\, b \wedge (\bigvee(A \setminus \{a, b\}) \vee a) = b \wedge \bigvee(A \setminus \{b\})\,\theta\,0\,,$$

$$p \wedge \bigvee(A' \setminus \{p\}) = p \wedge (p \vee q) \wedge (q \vee \bigvee(A' \setminus \{p, q\}))$$

$$\theta\, p \wedge (q \vee ((p \vee q) \wedge \bigvee(A' \setminus \{p, q\})))$$

$$\theta\, p \wedge (q \vee (a \wedge \bigvee(A \setminus \{a\})))\, \theta\, p \wedge q\, \theta\, 0$$

and similarly $q \wedge \bigvee(A' \setminus \{q\})\, \theta\, 0$. □

At this moment we introduce the join decompositions we are interested in.

Definition 5.2.5 Given a lattice with 0 and a congruence θ on it, the relation

$$a = a_1 \otimes \cdots \otimes a_n \tag{5.2.4}$$

means that $a = a_1 \vee \ldots \vee a_n$ and the set $\{a_1, \ldots, a_n\}$ is θ-independent. We then say that a is the *θ-direct join* of the set $\{a_1, \ldots, a_n\}$, or that (5.2.4) is a *decomposition as a θ-direct join* of the element a. Each a_i is said to be a *θ-direct factor* of a.

Remark 5.2.4 The order of factors in a θ-direct decomposition is irrelevant, that is, the relation (5.2.4) implies

$$a = a_{\pi(1)} \otimes \cdots \otimes a_{\pi(n)} \tag{5.2.5}$$

for every permutation π of $\{a_1, \ldots, a_n\}$.

Remark 5.2.5 The θ-direct decomposition (5.2.4) implies

$$a = a_i \otimes (a_1 \vee \ldots \vee a_{i-1} \vee a_{i+1} \vee \ldots \vee a_n) \quad (i = 1, \ldots, n)\,, \tag{5.2.6}$$

because setting $\bigvee(\{a_1, \ldots, a_n\} \setminus \{a_i\}) = b_i$, we have $a = a_i \vee b_i$ and $a_i \wedge b_i = 0$ by (5.2.2), hence the set $\{a_i, b_i\}$ is θ-independent.

Remark 5.2.6 Every element a satisfies

$$a = a \otimes 0\,. \tag{5.2.7}$$

Remark 5.2.7 If $x\, \theta\, 0 \Longrightarrow x = 0$, then $a = x \otimes y$ and $x > 0$ imply $y < a$ (otherwise $a = x \otimes a$, hence $x = x \wedge a\, \theta\, 0$, therefore $x = 0$).

Definition 5.2.6 Given a lattice with 0 and a congruence θ on it, we say that an element a is *θ-indecomposable* if it has no other θ-direct factors than 0 and itself, otherwise a is said to be *θ-decomposable*.

Our task is to prove that in a modular lattices satisfying a certain finiteness condition, every element is a θ-direct join of θ-indecomposable elements. Moreover, the useless elements can be removed from this decomposition; to express the latter property, we need one more definition.

Definition 5.2.7 A join decomposition (and in particular a θ-join decomposition) $a = \bigvee A$ is said to be *irreducible* if $\bigvee A \neq \bigvee B$ for every proper subset $B \subset A$.

Now everything is ready for establishing the main result.

Theorem 5.2.3 *Let L be a modular lattice with 0 and θ a congruence on L such that $\hat{0} = \{0\}$. Assume further there is no infinite decreasing sequence $x_1 > x_2 > \ldots x_n > \ldots$ of elements of L such that each x_{n+1} be a θ-direct factor of x_n. Then every element of $L \setminus \{0\}$ has an irreducible decomposition as a θ-direct join of θ-indecomposable elements.*

PROOF: We first show that every element $z > 0$ has a decomposition of the form $z = x \otimes y$ for some θ-indecomposable element $x > 0$ and some $y < z$. If z is θ-indecomposable, then $z = z \otimes 0$ is the required decomposition. Otherwise z has a decomposition of the form $z = x_1 \otimes y_1$ with $x_1, y_1 \notin \{z, 0\}$; note that $z > x_1 > 0$. Either x_1 is θ-indecomposable and our assertion is verified, or x_1 has a decomposition of the form $x_1 = x_2 \otimes y_2$ with $x_2, y_2 \notin \{x_1, 0\}$; note that $x_1 > x_2 > 0$. ; etc. In view of the hypothesis, the sequence $z > x_1 > x_2 > \ldots$ must stop after a finite number, say m, of steps. It follows that $x_m > 0$ is θ-indecomposable (otherwise the sequence could have been continued). Now a repeated application of Proposition 5.2.3(ii) yields

$$z = x_1 \otimes y_1 = x_2 \otimes y_2 \otimes y_1 = \cdots = x_m \otimes y_m \otimes \cdots \otimes y_2 \otimes y_1 ,$$

therefore Remark 5.2.5 implies $z = x_m \otimes (y_1 \vee \ldots \vee y_m)$. Finally $y_1 \vee \ldots \vee y_m < z$ by Remark 5.2.7.

To decompose an element $a > 0$ as a θ-direct join of θ-indecomposable elements, we make repeated use of the above result. So $a = b_1 \otimes c_1$, where $b_1 > 0$ is θ-indecomposable and $c_1 < a$. If $c_1 = 0$, then $a = b_1$ is the required decomposition by Remark 5.2.3. If $c_1 > 0$, we get $c_1 = b_2 \otimes c_2$, where $b_2 > 0$ is θ-indecomposable and $c_2 < c_1$. If $c_2 = 0$, then $c_1 = b_2$ is θ-indecomposable and $a = b_1 \otimes c_1$ is the required decomposition. Otherwise $c_1 > c_2 > 0$ and the process continues. We thus obtain a necessarily finite decreasing sequence $c_1 > c_2 \cdots > c_p > c_{p+1} = 0$, where each c_i is decomposed in the form $c_i = b_{i+1} \otimes c_{i+1}$ $(i = 1, \ldots, p)$, where $b_1, \ldots, b_{p+1}$ are positive and θ-indecomposable. We apply again Proposition 5.2.3(ii) and obtain

$$a = b_1 \otimes c_1 = b_1 \otimes b_2 \otimes c_2 = \cdots = b_1 \otimes \cdots \otimes b_p \otimes b_{p+1} .$$

Finally let $\{b_{i_1}, \ldots, b_{i_n}\}$ be a subset of $\{b_1, \ldots, b_{p+1}\}$, minimal with the property $b_{i_1} \vee \ldots \vee b_{i_n} = a > 0$. Then $n \geq 1$, therefore Proposition 5.2.3(i) ensures that the set $\{b_{i_1}, \ldots, b_{i_n}\}$ is θ-independent. Therefore $a = b_{i_1} \otimes \cdots \otimes b_{i_n}$ and this is an irreducible decomposition. □

Remark 5.2.8 The equality $0 = 0$ is a θ-direct decomposition by Remark 5.2.3 and the element 0 is obviously θ-indecomposable, but the above decomposition is not irreducible because $0 = \bigvee \varnothing$.

Theorem 5.2.3 generalizes a classical result from lattice theory, which we recapture below. The particular case is obtained by taking the identity $=$ as congruence θ. The concepts obtained from Definitions 5.2.4-5 2.6 for this choice of θ are called *independent* set, *direct join, direct factor* and *indecomposable* element. We thus obtain:

Corollary 5.2.1 *Let L be a modular lattice with 0 in which there is no infinite decreasing sequence $x_1 > x_2 > \cdots > x_n > \ldots$ such that each x_{n+1} be a direct factor of x_n. Then every element of $L \setminus \{0\}$ has an irreducible decomposition as a direct join of indecomposable elements.*

We conclude this section with a few words about the representation of elements as joins of finite-join irreducible (FJI) elements.

An element which is not FJI is called *finite-join reducible* (FJR).

Lemma 5.2.2 *An element z is FJR if and only if it has a representation of the form $z = x \vee y$ with $x, y \notin \{z, 0\}$ and $x \neq y$.*

PROOF: According to Example 5.1.12, z is FJR iff it has a representation $z = x \vee y$ with $z \neq x$ and $z \neq y$. It follows that $y \neq 0$ and $x \neq 0$. Therefore $x \neq y$, otherwise we get the contradiction $z = x$. □

Proposition 5.2.4 *In a finite lattice every element is a join of finite-join irreducible elements.*

PROOF: The lattice has 0, which is clearly FJI.

Now take $z > 0$. If Z is FJI, the assertion is verified. If not, $z = x_1 \vee x_2$ with $z > x_1 > 0$, $z > x_2 > 0$ and $x_1 \neq x_2$; we have obtained two distinct chains. If x_1 is FJR, we get $x_1 = x_{11} \vee x_{12}$, two distinct chains $z > x_{11} > 0$ and $z > x_{12} > 0$, also distinct from the previous chain $z > x_2 > 0$, and $z = x_{11} \vee x_{12} \vee x_2$. If x_2 is FJR, then $x_2 = x_{21} \vee x_{22}$, the chains $z > x_{21} > 0$ and $z > x_{22} > 0$ are distinct and distinct from the previous ones, and $z = x_{11} \vee x_{12} \vee x_{21} \vee x_{22}$. The process continues as long as some elements are FJR. Since the lattice is finite, every chain of distinct elements between z and 0 is finite; when it stops, its last element > 0 is FJI. Also, the number of distinct chains is finite. Therefore the entire process stops after a finite number of steps and at that moment z is a join of FJI elements. □

Exercise 5.2.2 (Birkhoff [1967], Exercise II.5.4). If all chains in a lattice L are finite, then every $a \in L$ can be represented as a join $a = x_1 \vee \ldots \vee x_n$ of finite-join irreducible elements.

Exercise 5.2.3 Let θ be a congruence on a join semilattice S. Write $x \,\bar{\theta}\, y$ for the negation of $x \,\theta\, y$. Refer to a finite non-empty subset A of S such that $\bigvee A \,\bar{\theta} \bigvee B$ $(\forall B \subset A)$ as a *θ-irreducible* set. Prove that:

(a) A is θ-irreducible if and only if

$$\bigvee A \,\bar{\theta} \bigvee (A \setminus \{a\}) \quad (\forall a \in A)\,.$$

(b) If $A = \{a_1, \ldots, a_n\}$, where $a_1, \ldots, a_n$ are pairwise distinct and $n \geq 2$, then A is θ-irreducible if and only if

$$a_1 \,\bar{\theta}\, a_1 \vee a_2 \,\bar{\theta}\, \ldots \,\bar{\theta}\, a_1 \vee \ldots \vee a_n\,.$$

(c) If S is a lattice, then A is θ-irreducible if and only if

$$\bigvee B \,\bar{\theta}\, (\bigvee B) \wedge \bigvee (A \setminus B) \quad (\forall B \subset A)\,.$$

(Hint. If $\bigvee B \,\theta\, (\bigvee B) \wedge \bigvee(A \setminus B)$ use $\bigvee A = (\bigvee B) \vee \bigvee(A \setminus B)$. If A is not θ-irreducible, then $a \wedge \bigvee(A \setminus \{a\}) \,\theta\, a$.)

It can be proved that the representation in Corolary 5.2 1 is unique up to the order of terms. We omit the rather technical proof of this property (see e.g. Pierce [1968]), but we sketch in Exercises 5.2.4 and 5.2.5 below a simpler proof of the uniqueness of the irreducible representation of an element as a join of finite-join irreducible elements in the case of a distributive lattice.

Exercise 5.2.4 In a distributive lattice, a decomposition $a = a_1 \vee \ldots \vee a_n$, where $a_1, \ldots, a_n$ are finite-join irreducible, is irreducible (cf. Definition 5.2.7) if and only if $a_1, \ldots, a_n$ are pairwise incomparable. (Hint. In view of Exercise 5.2.3, A is irreducible $\Longrightarrow \exists i \ \bigvee A = \bigvee(A \setminus \{a_i\}) \Longrightarrow \exists i \ a_i = \bigvee_{j \neq i}(a_i \wedge a_j) \Longrightarrow \exists i \, \exists j \neq i \ a_i \wedge a_j = a_i$).

Exercise 5.2.5 In a distributive lattice, every irreducible decomposition of an element as a finite join of finite-join irreducible elements, if any, is unique up to the order of terms. (Hint. Same technique as for Exercise 5.2.4. If $a_1 \vee \ldots \vee a_r = b_1 \vee \ldots \vee b_s$ are two such decompositions, then there is a mapping $\varphi : \{1, \ldots, r\} \longrightarrow \{1, \ldots, s\}$ such that $a_i \leq b_{\varphi(i)}$ for all i, and also a mapping $\psi : \{1, \ldots, s\} \longrightarrow \{1, \ldots, r\}$ such that $b_j \leq a_{\psi(j)}$ for all j. The functions φ and ψ are inverse to each other.)

5.3 Set-Theoretical Embeddings

This section exploits two ideas.

One of them is that if every element of a poset P is a join of elements from a fixed subset G of P, then the mapping

$$x \mapsto G(x] = \{g \in G \mid g \leq x\}$$

establishes an isomorphism between P and a poset of subsets of G. This idea can be applied to any poset taking $G := P$; in particular this yields the famous MacNeille completion by cuts of a poset. Other interesting cases correspond to the meet and join decompositions of elements studied in the previous section.

The other idea is to obtain isomorphic representations of distributive and Boolean lattices as lattices of sets by associating with each element the set of all prime filters which contain it. This result is, in fact, a particular aspect of the theory which will be developed in the last section.

Summarizing, our aim is to embed posets and various types of semilattices and lattices into lattices of sets. We first explain this aim in more detail by appropriate definitions.

Definition 5.3.1 We say that a mapping $f : P \longrightarrow P'$ between two psets P and P' *embeds* (*dually embeds*) P into P', or that f is an *embedding* (*dual embedding*) of P into P', if the corestriction $f : P \longrightarrow f(P)$ is an isomorphism (a dual isomorphism).

Remark 5.3.1 The composite of two embeddings (dual embeddings) is an embedding. The composite of an embedding (dual embedding) and a dual embedding (an embedding) is a dual embedding.

Proposition 5.3.1 1^0 *A function $f : P \longrightarrow P'$ between two posets is an embedding if and only if, for every $x, y \in P$,*

(2.1.24) $$x \leq y \Longleftrightarrow f(x) \leq f(y) \, .$$

2^0 *Dual statement.*

PROOF: 1^0 In view of Proposition 2.1.5 with $Q := f(P)$, a map $f : P \longrightarrow P'$ is an isomorphism if and only if it is a bijection and satisfies (2.1.24). But the corestriction is anyway surjective, so that an embedding is the same as an injective map satisfying (2.1.24), which is further equivalent to the single condition (2.1.24), because the latter implies that $f(x) = f(y) \Longrightarrow x = y$, i.e., injectivity. □

In Chapter 4 we have met quite a lot of categories in which isomorphisms coincide with order isomorphisms; cf. Propositions 4.1.5, 4.1.10, 4.2.4, 4.3.7 and 4.3.11. This means that the objects of those categories are algebras endowed with a partial order, and the order isomorphisms preserve the inf's and sup's that exist by the definition of the objects. Similarly, in Proposition 5.3.2 below, property (i) shows that the embeddings preserve all existing inf's and sup's, while property (ii) relaxes Proposition 4.3.6 from complete semilattices to arbitrary posets.

Proposition 5.3.2 1^0 *Let $f : P \longrightarrow P'$ be an embedding of posets. Then:*

(i) *For every $X \subseteq P$ such that $\bigwedge X$ exists, $f(\bigwedge X)$ is the* g.l.b. *of the set $f(X)$ in the poset $f(P)$.*

(ii) *If, moreover, $\bigwedge f(X)$ exists in P', then $f(\bigwedge X) \leq \bigwedge f(X)$.*

2^0 *Dual statement for $\bigvee X$.*

PROOF: By a repeated application of (2.1.24).

(i) If $1 = \bigwedge \varnothing$ exists, the $x \leq 1$ $(\forall x \in X)$, hence $f(x) \leq f(1) \, \forall f(x) \in f(X)$. Further, we suppose $X \neq \varnothing$ and check the properties of g.l.b. for $f(\bigwedge x)$.

We have $\bigwedge X \leq x \, \forall x \in X$, hence $f(\bigwedge X) \leq f(x) \, \forall f(x) \in f(X)$.

Now take $f(z) \in f(P)$ such that $f(z) \leq f(x) \, \forall f(x) \in f(X)$. This implies $z \leq x \, \forall x \in X$, hence $z \leq \bigwedge X$, therefore $f(z) \leq f(\bigwedge x)$.

(ii) For $X = \varnothing$ we have the equality $f(1) = \bigwedge \varnothing$. Now suppose $X \neq \varnothing$. The inequality to be proved amounts to the fact that $f(\bigwedge X)$ is a lower bound of $f(X)$, and this results from $\bigwedge X \leq x \, \forall x \in X$. □

The following types of lattices of sets which will occur in the embedding theorems to be established below are described in:

Definition 5.3.2 By a (*complete*) *ring of sets* on 2^T is meant a (complete) sublattice of the (complete) lattice $(2^T, \subseteq)$. By a *complete*) *field of sets* on 2^T is meant a (complete) Boolean subalgebra of the (complete) Boolean algebra $(2^T, \subseteq)$.

In other words, a (complete) ring of sets is a subset of 2^T closed under finite (under arbitrary) intersections and unions (in particular $\varnothing$ and T belong to every complete ring of sets), while a (complete) field of sets is a (complete) ring of sets closed under complementation; cf. Definitions 4.1.Lat, 4.2.7, 4.3.5, 4.3.6 and Proposition 4.3.9.

Example 5.3.1 1) If the set T is finite, every field of subsets of T is complete.

2) 2^T is a complete field of sets.

3) A subset S of $\mathbb{R}^2$ is called vertical if $(x, y) \in S$ implies $(x, z) \in S$ for every z. The family of vertical sets is a complete field of sets.

Remark 5.3.2 A ring of sets which is also a complete lattice with respect to set inclusion need not be a complete ring of sets on 2^T. Take e.g. the Moore family of all closed sets of a topological space T, which is indeed a complete lattice, yet it is not a complete sublattice of $(2^T, \subseteq)$: in this complete lattice the join of a family of closed sets is the closure of the union of those sets, but the union itself may not be closed, in which case it is strictly included in its closure.

Remark 4.4.5 can be paraphrased to the effect that a Moore family of sets is a complete ring of sets if and only if it is closed under arbitrary unions (including $\varnothing = \bigvee \varnothing$).

Remark 5.3.3 Every (complete) ring of sets is a (complete) distributive lattice. Every (complete) field of sets is a (complete) Boolean algebra. Every ring (field) of sets is embedded into the complete field of sets $(2^T, \subseteq)$.

We will be interested in constructing embeddings and dual embeddings for which property (ii) in Proposition 5.3.2 is verified in the stronger form of equalities. In other words, we are looking for embeddings which preserve meets and/or joins, i.e., $f(\bigwedge X) = \bigwedge f(X)$ and/or $f(\bigvee X) = \bigvee f(X)$ (dual embeddings which interchange meets and/or joins, i.e., $f(\bigwedge X) = \bigvee f(X)$ and/or $f(\bigvee X) = \bigwedge f(X)$) or at least fulfilling the above equalities for all finite X. From this point of view the most remarkable result is perhaps Theorem 5.3.2, stating that every distributive lattice (Boolean algebra) is isomorphic to a ring (field) of sets.

To realize our program, we begin with

Definition 5.3.3 A subset G of a poset P is said to be a *meet (join) generator* of P if every element of P is a meet (join) of elements from P.

Theorems 5.2.1, 5.2.2, 5.2.3 and Proposition 5.2.4 provide examples of generators. Recall the warning after Theorem 5.2.2.

Remark 5.3.4 If $G \subseteq H \subseteq P$ and and G is a meet (join) generator of P, so is H.

The next proposition is the basis for practical work with generators.

Proposition 5.3.3 *A subset G of a poset P is a meet (join) generator of P if and only if $x = \bigwedge G[x)$ ($x = \bigvee G(x]$) for every $x \in P$, where $G[x) = \{g \in G \mid x \leq g\}$ and $G(x] = \{g \in G \mid g \leq x\}$;* cf. Example 5.1.2.

PROOF: Sufficiency is trivial. To prove necessity, let G be a meet generator of P and take $x \in P$. We have $x \leq G[x)$. Now we take $y \leq G[x)$ and we must prove that $y \leq x$. But $x = \bigwedge X$ for some non-empty subset $X \subseteq G$. It follows that $x \leq X$, therefore $X \subseteq G[x)$, hence $y \leq X$, so that $y \leq \bigwedge X = x$. □

We continue our way and prove

Lemma 5.3.1 1^0 *If G is a meet generator of a poset P, then the map $x \mapsto G[x)$ provides a dual isomorphism between P and the poset $(\{G[x) \mid x \in P\}, \subseteq)$ in such a way that for every non-empty subset $X \subseteq P$,*

$$\text{if } \bigvee X \text{ exists, then } G[\bigvee X) = \bigcap \{G[x) \mid x \in X\}\,. \tag{5.3.1$'$}$$

2^0 *Dual statement for a join generator G: the map $x \mapsto G(x]$ is an isomorphism and satisfies*

$$\text{if } \bigwedge X \text{ exists, then } G(\bigwedge X] = \bigcap \{G(x] \mid x \in X\}\,. \tag{5.3.1$''$}$$

PROOF: 1^0 If $x \leq y$, then clearly $G[y) \subseteq G[x)$. Conversely, if $G[y) \subseteq G[x)$, then $x = \bigwedge G[x) \leq \bigwedge G[y) = y$ by Proposition 5.3.3 and Remark 4.3.4. So the map $x \mapsto G[x)$ satisfies condition (2.1.28), therefore it is a dual embedding of P into the poset of all ideals of P by Proposition 5.3.1.2^0; in other words it is the announced dual isomorphism.

To prove (5.3.1$'$), assume $\bigvee X$ exists. For every $g \in G$, the relation $g \in \bigcap\{G[x) \mid x \in X\}$ is equivalent to the implication $x \in X \Longrightarrow x \leq g$, which holds if and only if $g \geq \bigvee X$, or, equivalently, $g \in G[\bigvee X)$. □

Corollary 5.3.1 *For every non-empty subset X of a poset P, the following properties hold:*

$$\text{if } \bigvee X \text{ exists, then } [\bigvee X) = \bigcap \{[x) \mid x \in X\}\,, \tag{5.3.2$'$}$$

$$\text{if } \bigwedge X \text{ exists, then } (\bigwedge X] = \bigcap \{(x] \mid x \in X\}\,. \tag{5.3.2$''$}$$

In the proof of Lemma 5.3.1.1^0 we have used the interpretation of a dual isomorphism as a dual embedding. This point of view can be refined as follows:

Lemma 5.3.2 *Let G be a meet (join) generator of a poset P. The mapping $x \mapsto G[x)$ (the mapping $x \mapsto G(x]$) can also be viewed as a dual embedding (an embedding) of P into a complete ring of sets, namely the lattice consisting of all oder filters (order ideals) of G plus $\varnothing$.*

PROOF: It suffices to show that the empty set and the order filters of a poset G actually form a complete ring of sets. Let $(F_i)_{i \in I}$, where $I \neq \varnothing$, be a family of order filters. Then $\bigcup_{i \in I} F_i$ is obviously an order filter of G, while $\bigcap_{i \in I} F_i$ is either void or an order filter. Finally G is an order filter of itself. □

Lemmas 5.3.1 and 5.3.2 might be referred to as presenting potential or "abstract" (dual) isomorphisms and (dual) embeddings. We are now going to construct several concrete instances of posets P and/or generators G, thus obtaining actual representation theorems.

Theorem 5.3.1 *For every poset P, the mapping $x \mapsto [x)$ (the mapping $x \mapsto (x]$) establishes a dual isomorphism (an isomorphism) between P and the set of all principal order filters (principal order ideals) of P, ordered by set inclusion.*

PROOF: Immediate from Lemma 5.3.1 with $G := P$. □

From Theorem 5.3.1 we obtain several embeddings, according to different targets which we may choose for the mapping.

Corollary 5.3.2 (i) *For every poset P, the mapping $x \mapsto [x)$ dually embeds (the mapping $x \mapsto (x]$ embeds) P into the Moore family of all normal filters (normal ideals) of P;* cf. Example 4.5.4 (an application of Galois connections).

(ii) *Moreover, this dual embedding transforms all existing meets into joins, i.e.,*

$$(5.3.3') \qquad \text{if } \bigwedge X \text{ exists, then } [\bigwedge X) = \bigvee \{[x) \mid x \in X\}$$

(this embedding preserves all existing joins, i.e.,

$$(5.3.3'') \qquad \text{if } \bigvee X \text{ exists, then } (\bigvee X] = \bigvee \{(x] \mid x \in X\}) .$$

PROOF: Property (i) follows from Lemma 5.3.1 applied with $G := P$, taking into account that principal order filters are normal by Remark 5.1.6.

To prove (5.3.3′) we first remark that in the complete lattice of all normal filters, the identity

$$(5.3.4') \qquad \bigvee \{[x) \mid x \in X\} = (\bigvee \{[x) \mid x \in X\})^{-+}$$

holds by (4.4.28) in Corollary 4.4.4 and Example 4.5.4. Now assume that $\bigwedge X = a$ exists; in view of (5.3.4′), our task is to prove $[a) = Y^{-+}$, where we have set $Y = \bigcup \{[x) \mid x \in X\}$.

If $X = \varnothing$, then $a = 1$ and $Y = \varnothing$; since $\varnothing^- = P$ (exercise!), we have $\varnothing^{-+} = P^+ = \{1\} = [1)$.

Now suppose $X \neq \varnothing$. We have $X \subseteq Y$, hence $Y^- \subseteq X^-$; since a is the greatest element of X^-, it follows that $Y^- \leq a$, hence $Y^- \subseteq \{a\}$, therefore $[a) = \{a\}^+ \subseteq Y^{-+}$. On the other hand, if we succeed in proving that $a \in Y^-$, it will follow that $Y^{-+} \subseteq \{a\}^+ = [a)$, thus completing the proof. But for every $y \in Y$ there exists $x \in X$ such that $x \leq y$, and since $a \leq x$, this implies $a \leq y$; thus $a \in Y^-$. □

Exercise 5.3.1 Can properties (5.3.3) be generalized to the sets $G[x)$ and $G(x]$?

Corollary 5.3.3 1^0 (i) *For every poset P, the mapping $x \mapsto [x)$ is a dual embedding of P into a complete ring of sets, namely the lattice of all order filters of P plus $\varnothing$.*

(ii) *Moreover, this dual embedding transforms all existing joins into intersections.*

2^0 *Dual statement: an embedding of P into the complete ring of all order ideals of P plus $\varnothing$, all existing meets being preserved.*

PROOF: From From Lemma 5.3.2 with $G := P$ and Corollary 5.3.1 for $X \neq \varnothing$. For $X = \varnothing$ we use $\bigcap \varnothing = P$, $\bigvee \varnothing = 0$, $[0) = P$ and $\bigwedge \varnothing = 1$, $(1] = P$. □

The merit of Corollary 5.3.3 is that it realizes a dual embedding (an embedding) into a complete ring of sets; we might say that this is a *purely set-theoretical (dual) embedding*. The dual embedding (embedding) in Corollary 5.3.2 is finer, to the effect that if P has a greatest (least) element, then the normal filters and normal ideals are not empty, hence the lattice of all normal filters (normal ideals) is included in the lattice of all order filters (order ideals) by Example 5.1.7; besides, this dual embedding transforms all existing meets into joins and this embedding preserves all existing joins. The price paid for these advantages of Corollary 5.3.2 is that the join operation in the lattice of normal filters (normal ideals) does not coincide with set-theoretical union.

The embedding of a poset P into the lattice of all normal ideals (cf. Corollary 5.3.2) is known as the *MacNeille completion by cuts* of the poset P. The term "cuts" is justified by the fact that in the case of a totally ordered set T for every normal ideal A which is not a principal ideal, the pair (A, A^+) can be thought as a "cut" of T, that is, it satisfies $A \leq A^+$ (i.e., $a \leq b$ for every $a \in A$ and $b \in B$), $A \cup A^+ = T$ and $A \cap A^+ = \varnothing$. Let us prove these three properties. The first one follows from $A = A^{+-} \leq A^+$. Then for every $x \in T$ either $A \leq x$ (which includes the case $A = \varnothing$), in which case $x \in A^+$, or $x < a$ for some $a \in A$, in which case $y \in A^+ \Longrightarrow x < y$, therefore $x \in A^{+-} = A$. Finally, $b \in A \cap A^+$ would imply $b \in A$ and $A \leq b$. But $A \neq \varnothing$, hence A is an ideal, therefore $(b] \subseteq A$ and $A \subseteq (b]$, hence $A = (b]$, in contradiction with the hypothesis.

In the more particular case when T is the set $\mathbb{Q}$ of rational numbers, the set of all normal ideals of $\mathbb{Q}$ is the completed real line $[-\infty, \infty]$; this is the *Dedekind construction of the real numbers*. Thus e.g. if $A = \{x \in \mathbb{Q} \mid x^2 < 2\}$, then $A^+ = \{x \in \mathbb{Q} \mid 2 < x^2\}$ and $A^{+-} = A$; so the real $\sqrt{2}$ can be identified with the cut (A, A^+).

Exercise 5.3.2 (Bourbaki). Prove that every poset P with greatest element 1 is dually isomorphic to a poset of closure operators on P. (Hint. For every $a \in P$, let $\varphi_a(x) = 1$ if $a \leq x, \varphi\ _a(x) = 1$ otherwise. If $\varphi_b(x) \leq \varphi_a(x)$ for all x, in the case $b \neq 1$ take $x := b$.)

Now let us examine what happens if in Theorem 5.3.1 and its corollaries the poset P is assumed to be a (complete) semilattice or a (complete) lattice.

First, by analogy with the concept of a dual lattice homomorphism, we introduce *(complete) dual semilattice homomorphisms* as a common name for the following dual concepts: a *(complete) meet-join homomorphism* is a function from a (complete) meet semilattice to a (complete) join semilattice which transforms all meets (all finite meets) into joins; we refer to the dual concept as a *(complete) join-meet homomorphism*.

Remark 5.3.5 If the poset P is a (complete) meet semilattice, then the dual embedding in Corollary 5.3.2 is a dual (complete) meet-join homomorphism.

Remark 5.3.6 If the poset P is a (complete) join semilattice, then the embedding in Corollary 5.3.2 is a complete join homomorphism, while the dual embedding in Corollary 5.3.3 is a (complete) join-meet homomorphism.

Remark 5.3.7 If the poset P is a lattice, then its dually isomorphic (isomorphic) image in Theorem 5.3.1 is the lattice of principal filters (principal ideals) of P. A similar remark holds for complete lattices.

We continue with other specializations of Lemmas 5.3.1 and 5.3.2.

Proposition 5.3.4 *Let M be the set of all meet-irreducible elements of a compactly generated lattice L. Then:*

(i) *The mapping*

$$x \mapsto M[x) = \{m \in M \mid x \leq m\} \tag{5.3.5}$$

establishes a dual isomorphism between L and the lattice

$$(\{M[x) \mid x \in L\}, \subseteq) \tag{5.3.6}$$

in such a way that $M[\bigvee \varnothing) = M = \bigcap \varnothing$ and for every non-empty set $X \subseteq L$,

$$M[\bigvee X) = \bigcap \{M[x) \mid x \in X\} . \tag{5.3.7}$$

(ii) *The mapping (5.3.5) can be also viewed as a dual embedding of L into a complete ring of sets, namely the lattice of all order filters of M plus $\varnothing$.*

Proof: M is a meet generator by Theorem 5.2.1 and $1 \in M$ by Remark 5.2.1. Now apply Lemmas 5.3.1 and 5.3.1. □

Proposition 5.3.5 *Let A be the set of all atoms of an atomic Boolean algebra B. Then:*

(i) *The mapping*

$$x \mapsto A(x] = \{a \in A \mid a \leq x\} \tag{5.3.8}$$

establishes an isomorphism between B and the Boolean field o subsets of 2^A

$$(\{A(x] \mid x \in B\}, \subseteq) , \tag{5.3.9}$$

in such a way that $A(\bigwedge \varnothing] = P = \bigcap \varnothing$ and for every non-empty subset $X \subseteq B$,

$$\text{if } \bigwedge X \text{ exists, then } A(\bigwedge X] = \bigcap \{A(x] \mid x \in X\} . \tag{5.3.10}$$

(ii) *The mapping (5.3.8) can be also viewed as an embedding of B into a complete ring of sets, namely the lattice of all order ideals of A plus $\varnothing$.*

PROOF: (i) First we prove that (5.3.9) is a field of sets. Clearly $A(0] = \varnothing$, $A(1] = A$ and $A(x] \cap A(y] = A(x \wedge y]$. Also, taking into account that $b \wedge c = 0 \Longleftrightarrow b \leq c'$ and having in view Remark 5.2.2, we see that for every $a \in A$ and $x \in B$,

$$a \leq x' \Longleftrightarrow a \wedge x = 0 \Longleftrightarrow a \wedge x \neq a \Longleftrightarrow a \not\leq x ,$$

showing that $A(x'] = 2^A \setminus A(x]$. It follows by Proposition 4.2.5, (i)$\Longleftrightarrow$ (ii), that $A(x \vee y] = A(x] \cup A(y]$.

We have proved not only that (5.3.9) is a field of sets, but also that the map (5.3.8) is a Boolean homomorphism.

Further, we have

$$x \leq y \Longleftrightarrow A(x) \subseteq A(y] , \tag{5.3.11}$$

because if $A(x] \subseteq A(y]$, Proposition 5.3.3 implies $x = \bigvee A(x] \leq \bigvee A(y] = y$, while the converse implication is clear. Since the map (5.3.8) is also surjective, it follows by Proposition 2.1.5, (i) $\Longleftrightarrow$ (iii), that it is an isomorphism in **Pos**. Therefore, according to Proposition 4.2.4, it is also a Boolean isomorphism.

Property (5.3.10) follows by Lemma 5.3.1.

(ii) In view of Lemma 5.3.2, the map (5.3.8) embeds $B \setminus \{0\}$ into the lattice of all order ideals of A plus $\varnothing$. But $A(0] = \varnothing$, which is not an order ideal. Therefore (5.3.8) realizes the announced embedding. □

So far we have realized various embeddings and dual embeddings into complete rings of sets, using Corollary 5.3.2; this automatically enables us to obtain embeddings and dual embeddings into complete fields of sets; cf. Remarks 5.3.3 and 5.3.1. Yet the (dually) isomorphic image $f(P)$ under such an embedding (a dual embedding) is not itself a ring of sets, even if P is a distributive lattice, because the join operation in the lattice $f(P)$ of principal order ideals (principal order filters) is not the set-theoretical union; cf. Proposition 5.1.6. There is however another idea which enables us to establish that *every distributive lattice (Boolean algebra) is isomorphic to a ring (field) of sets.* More exactly:

Theorem 5.3.2 (i) *Let L be a distributive lattice, $\mathcal{PF}(L)$ the set of all prime filters of L, and for every $x \in L$ set $\varphi(x) = \{P \in \mathcal{PF}(L) \mid x \in P\}$. Then φ establishes an isomorphism between L and the ring $\{\varphi(x) \mid x \in L\}$ of subsets of $\mathcal{PF}(L)$.*

(ii) *If, moreover, L has least element 0 and/or greatest element 1, then $\varphi(0) = \varnothing$ and/or $\varphi(1) = \mathcal{PF}(L)$, respectively.*

(iii) *If, moreover, L is a Boolean algebra, then $\{\varphi(x) \mid x \in L\}$ is a field of sets isomorphic to L.*

PROOF: (i) The surjectivity of φ is trivial, while the injectivity follows from Proposition 5.2.2. Furthermore, for every $F \in \mathcal{PF}(L)$, Proposition 5.1.1 implies

$$F \in \varphi(x \wedge y) \Longleftrightarrow x \wedge y \in F \Longleftrightarrow x \in F \;\&\; y \in F$$

$$\Longleftrightarrow F \in \varphi(x) \;\&\; F \in \varphi(y) \Longleftrightarrow F \in \varphi(x) \cap \varphi(y) ,$$

while Theorem 5.1.5 implies

$$F \in \varphi(x \vee y) \Longleftrightarrow x \vee y \in F \Longleftrightarrow x \in F \text{ or } y \in F$$

$$\Longleftrightarrow F \in \varphi(x) \text{ or } F \in \varphi(y) \Longleftrightarrow F \in \varphi(x) \cup \varphi(y) \ .$$

Therefore $\varphi(x \wedge y) = \varphi(x) \cap \varphi(y)$ and $\varphi(x \vee y) = \varphi(x) \cup \varphi(y)$, which shows that $\{\varphi(x) \mid x \in L\}$ is a ring of sets and φ a bijective homomorphism, hence an isomorphism by Proposition 4.2.4.

(ii) Clearly no proper order filter contains 0, hence in particular no prime filter contains 0, that is, $\varphi(o) = \varnothing$, while 1 belongs to every filter, hence $\varphi(1) = \mathcal{PF}(L)$.

(iii) $\{\varphi(x) \mid x \in L\}$ is a ring of subsets of $\mathcal{PF}(L)$ by (i), containing $\varnothing$ and $\mathcal{PF}(L)$ by (ii). For every $x \in L$, we have $x' \in F \Longleftrightarrow x \notin F$ by Theorem 5.1.6, (i) $\Longleftrightarrow$ (iii), therefore $\varphi(x') = (\varphi(x))'$. Thus $\{\varphi(x) \mid x \in L\}$ is a field of sets and φ is an isomorphism of **Bol** by Remark 4.2.10. □

Theorem 5.3.2 provides us with a mnemonic rule for computing in a Boolean algebra.

Remark 5.3.8 All computation rules with the set-theoretical operations of intersection, union and complementation, are valid in every Boolean algebra.

Exercise 5.3.3 Prove the equivalence of the following conditions for a function $f : L \longrightarrow \{0, 1\}$:

(i) f is a surjective homomorphism in **Lat** ;

(ii) f is a homomorphism in **Lat01** ;

(iii) $F = f^{-1}(1)$ is a prime filter .

(Hint. Use Proposition 5.1.1 and Theorem 5.1.5 or its Corollary 5.1.2).

In §5 we will work with the set $\mathrm{Hom}(L, 2)$ of all surjective lattice homomorphisms $f : L \longrightarrow \{0, 1\}$.

As every Boolean algebra is embedded into a field of sets, it is natural to look for an intrinsic characterization of those Boolean algebras that are isomorphic to a complete field of sets of the form $(2^T, \subseteq)$, i.e., which are isomorphic to the field of *all* subsets of some set. It is interesting that a nice answer to this question is provided by using the idea exploited in Proposition 5.3.5.

Theorem 5.3.3 (i) *A Boolean algebra B is isomorphic to the complete field $(2^T, \subseteq)$ of all subsets of some set T, if and only if it is complete and atomic.*

(ii) *When this is the case, B is isomorphic to $(2^A, \subseteq)$, where A is the set of atoms of B.*

Proof: Every field of the form $(2^T, \subseteq)$ is obviously complete and atomic, the atoms being the singletons of T; moreover, completeness and atomicity are preserved by isomorphisms. It remains to prove that if an atomic Boolean algebra B is complete and A stands for the set of all atoms of B, then B is isomorphic to $(2^A, \subseteq)$.

Let $\varphi : B \longrightarrow 2^A$ be the map which works as (5.3.8) in Proposition 5.3.5, that is, $\varphi(x) = A(x]$. We already know that (5.3.8) satisfies (5.3.11), which now reads $x \leq y \iff \varphi(x) \subseteq \varphi(y)$. If we only prove that φ is surjective, it will follow by Proposition 2.1.5, (i) $\iff$ (iii), that φ is an isomorphism in **Pos**, hence an isomorphism in **CBol** by Proposition 4.3.11.

So, for every $Y \in 2^A$, let us prove that $Y = A(\bigvee Y]$. If $Y = \varnothing$, note that $A(\bigvee \varnothing] = A(0] = \varnothing$. Now suppose $Y \neq \varnothing$. Clearly $Y \subseteq A(\bigvee Y]$. The converse implication is trivial if $Y = A$. Otherwise take an element $z \in A \setminus Y$ and note that $z \wedge y = 0$ for all $y \in Y$ by Remark 5.2.2. Using also Proposition 4.3.10, we get $z \wedge \bigvee Y = \bigvee(z \wedge Y) = 0$, therefore $z \not\leq \bigvee Y$, hence $z \notin A(\bigvee Y]$. From $A \setminus Y \subseteq A \setminus A(\bigvee Y]$ we get $A(\bigvee Y] \subseteq Y$. □

Corollary 5.3.4 *Every finite Boolean algebra B is isomorphic to the field $(2^A, \subseteq)$ of all subsets of the set A of atoms of B; the number of elements of B is $2^{|A|}$.*

PROOF: B is complete by an obvious extension of Example 4.3.1, and atomic by Example 5.2.1. □

5.4 Stone Spaces

The aim of this section is to establish the topological background necessary to the next section. Whereas in the previous sections we have illustrated certain concepts by topological examples, in this section we begin with the very definition of a topological space. Then we introduce the more advanced prerequisites to the next section, by making use of some results established so far in this eBook. As a matter of fact, we will study certain closure and interior operators (cf. Exercise 4.4.8) on the complete field of sets $(\mathcal{P}(T), \subseteq)$.

The reader desirous of learning more about topology may consult e.g. the conventional textbook by Kelley [1957], or the book by Vaidyanathaswamy [1960] with emphasis on lattices, or the modern treatise by Preuß [1972], strongly based on categories.

There are several equivalent definitions of a topological space; in each of them T is a non-empty set (the *subjacent set*). The elements of T are called *points*. A topological space may be introduced as a pair $(T, \mathcal{C})$, where the family $\mathcal{C} \subseteq 2^T$ of *closed sets* is closed under arbitrary intersections and finite unions, and $\varnothing, T \in \mathcal{C}$. Alternatively, a topological space is a pair $(T, ^-)$, where $^- : 2^T \longrightarrow 2^T$ is a *topological closure*, i.e., it satisfies the properties $X \subseteq \overline{X}$, $\overline{\overline{X}} = \overline{X}$, $\overline{X \cup Y} = \overline{X} \cup \overline{Y}$, and $\overline{\varnothing} = \varnothing$. Dually, a topological space is a pair $(T, \mathcal{O})$, where the family $\mathcal{O}$ of *open sets* is closed under finite intersections and arbitrary joins, and $\varnothing, T \in \mathcal{O}$; or, equivalently, a pair $(T, ^\circ)$, where $^\circ : 2^T \longrightarrow 2^T$ is a *topological interior*, i.e., it satisfies the properties $X^\circ \subseteq X$, $X^{\circ\circ} = X^\circ$, $(X \cap Y)^\circ = X^\circ \cap Y^\circ$, and $T^\circ = T$.

No matter which alternative has been chosen as definition, the above concepts are related as follows: the closed sets are the sets of the form $\overline{X}$, the open sets are the sets of the form X°, and the complementary of a closed set (an open set) is an open set (a closed set).

The more general background of the above concepts and interrelations will be revealed in Ch.6§1.

A *neighbourhood* of a point $x \in T$ is a set V including an open set which contains x. An *open neighbourhood* of x is an open set G containing x; then G is a neighbourhood of every point of it.

Let $\mathcal{V}(x)$ be the family of all neighbourhoods of x (cf. French: *voisinages*). Then $\mathcal{V}(x)$ is a filter on $(2^T, \subseteq)$ and the above remarks can be formalized as follows:

$$(5.4.1) V \in \mathcal{V}(x) \Longrightarrow x \in V \;\&\; \exists G \,(G \in \mathcal{V}(x) \;\&\; G \subseteq V \;\&\; (y \in G \Longrightarrow G \in \mathcal{V}(y))).$$

This yields one more definition of topological spaces, *via* neighbourhoods: a topological space can be defined as a pair $(T, (\mathcal{V}(x))_{x\in T})$, where each $\mathcal{V}(x)$ is a filter on $(2^T, \subseteq)$ satisfying (5.4.1). We have just seen how filters can be defined in terms of open sets. Conversely, if we start with the definition in terms of $\mathcal{V}(x)$ satisfying (5.4.1), then the family $\mathcal{O} \subseteq 2^T$ of all subsets $G \subseteq T$ that satisfy the property

$$(5.4.2) \qquad x \in G \Longrightarrow G \in \mathcal{V}(x)$$

fulfills the axioms of open sets in a topological space.

Definition 5.4.1 By a *base of open sets* in a topological space is meant a family $\mathcal{B}$ of open sets such that every non-empty open set is a union of sets from $\mathcal{B}$.

Remark 5.4.1 If $\mathcal{B}$ is a base of open sets, then every neighbourhood of a point includes a neighbourhood from $\mathcal{B}$ (because it includes an open neighbourhood).

Proposition 5.4.1 (i) *Every base $\mathcal{B}$ of open sets in a topological space T satisfies*

$$(5.4.3) \qquad (\forall x \in T)\ (\exists A \in \mathcal{B})\ x \in A\ ,$$

$$(5.4.4) \qquad A, B \in \mathcal{B} \;\&\; x \in A \cap B \Longrightarrow (\exists C \in \mathcal{B})\ x \in C \subseteq A \cap B\ .$$

(ii) *Conversely, if a family of subsets $\mathcal{B}$ fulfills* (5.4.3) *and* (5.4.4), *then the family $\mathcal{O}$ of all unions of sets from $\mathcal{B}$ plus $\varnothing$ defines a topology $(T, \mathcal{O})$ on T having $\mathcal{O}$ as family of open sets.*

PROOF: (i) Remark 5.4.1 implies (5.4.3), while (5.4.4) holds because $A \cap B$ is an open neighbourhood of x.

(ii) It follows from (5.4.3) that for each $x \in T$ there is a set $A_x \in \mathcal{B}$ such that $x \in A_x$; this implies $T = \bigcup\{A_x \mid x \in T\} \in \mathcal{O}$. Clearly, if $\mathcal{A} \subseteq \mathcal{O}$ then $\bigcup \mathcal{A} \in \mathcal{O}$. Finally, let $A, B \in \mathcal{O}$; then $A = \bigcup \mathcal{X}$ and $B = \bigcup \mathcal{Y}$ for some $\mathcal{X}, \mathcal{Y} \in \mathcal{B}$, therefore $A \cap B$ is the union of all sets of the form $X \cap Y$ with $X \in \mathcal{X}$ and $Y \in \mathcal{Y}$. It follows from (5.4.4) that for every $t \in A \cap B$ there is a set $C_t \in \mathcal{B}$ such that $t \in C_t \subseteq X \cap Y \subseteq A \cap B$ for some $X \in \mathcal{X}$ and some $Y \in \mathcal{Y}$. This implies that $A \cap B = \bigcup\{C_t \mid t \in A \cap B\} \in \mathcal{O}$. □

Note that, unlike what happens in the previous properties, there may be several bases defining the same topology; in particular the family of all open sate is a base of open sets.

Example 5.4.1 For every set $T \neq \varnothing$ we can take 2^T as family of open sets, thus obtaining the so-called *discrete topology* on T. The discrete topology is also characterized by the fact that all sets are closed.

Example 5.4.2 For every set $T \neq \varnothing$, there is a topology on T defined by taking $\{\varnothing, Y\}$ as family of open sets; this is sometimes called the *rubbish topology.* The sets $\varnothing$ and T are also the only closed sets in the rubbish topology.

Example 5.4.3 Let T be a topological space and $\varnothing \neq S \subseteq T$. Then we can define a topology on S by taking as open sets all the sets of the form $S \cap G$, where G is an open set of T. This is called the topology *induced* on S by the topology of T; alternatively, S endowed with the induced topology is said to be a *subspace* of T. The closed sets of the subspace S are the sets of the form $S \cap F$, where F is a closed set of T.

Remark 5.4.2 The neighbourhoods of a point $x \in S$ in the induced topology coincide with the intersections between S and the neighbourhoods of x in T. If $\mathcal{B}$ is a base of open sets in T, then the family $\{B \cap S \mid B \in \mathcal{B}\}$ is a base of open sets in S.

Example 5.4.4 Let $T_i)_{i\in I}$ be a non-empty family of topological spaces and $T = \prod_{i\in I} T_i$. Let $\mathcal{B}$ be the family of all sets $B = \prod_{i\in I} B_i$ such that each *i-component* B_i of B is a non-empty open set of T_i and $B_i \neq T_i$ holds for at most a finite number of indices $i \in I$. Then it is immediately seen that $\mathcal{B}$ fulfills conditions (5.4.3) and (5.4.4) in the stronger forms : $T \in \mathcal{B}$, and if $A, B \in \mathcal{B}$ then $A \cap B \in \mathcal{B}$. The topology on T having $\mathcal{B}$ as a base of open sets is known as the *product topology* of T.

Remark 5.4.3 Let $T = \prod_{i\in I} T_i$ be a product of topological spaces and $(p_i)_{i\in I}$ the canonical projections. If G is an open set of T, then for each $i \in I$ the set $p_i(G) = \{p_i(x) \mid x \in G\}$ is an open set of T_i.

Indeed, for the sets $B \in \mathcal{B}$ this follows from $p_i(B) = B_i$, while an arbitrary open set G of T is of the form $G = \bigcup_{k\in K} B^k$ with all $B^k \in \mathcal{B}$, therefore $p_i(G) = \bigcup_{k\in K} p_i(B^k)$ is an open set of T_i.

Further examples of topological spaces will be given in the following, though limited to what we need in the next section.

Let us now give a characterization of the closure of a set in terms of neighbourhoods.

Proposition 5.4.2 *For every point x and every subset A of a topological space T, $x \in \overline{A}$ if and only if $V \cap A \neq \varnothing$ for every $V \in \mathcal{V}(x)$.*

PROOF: First let G be an open set of T. Then $T \setminus G$ is closed, hence

$$G \cap A = \varnothing \iff A \subseteq T \setminus G \iff \overline{A} \subseteq T \setminus G \iff G \subseteq T \setminus \overline{A}\,.$$

Further let $\mathcal{B}$ be a base of open sets. Then for every $x \in T$,

$$\exists V \in \mathcal{V}(x)\ \ V \cap A = \varnothing \Longleftrightarrow \exists G \in \mathcal{V}(x) \cap \mathcal{B}\ \ G \cap A = \varnothing$$

$$\Longleftrightarrow \exists G \in \mathcal{V}(x) \cap \mathcal{B}\ \ G \subseteq T \setminus \overline{A} \Longleftrightarrow x \in T \setminus \overline{A}\,.$$

□

Corollary 5.4.1 *Let $\mathcal{B}$ be a base of open sets. Then for every $x \in T$ and every subset $A \subseteq T$, we have*

$$x \in \overline{A} \Longleftrightarrow G \cap A \neq \varnothing \quad (\forall G \in \mathcal{V}(x) \cap \mathcal{B})$$

In the sequel we concentrate on certain particular types of topological spaces.

Definition 5.4.2 A topological space T is said to be *separated* or a *Hausdorff space* if for every two distinct points x, y of T there is a neighbourhood V of x and a neighbourhood W of y such that $V \cap W = \varnothing$.

Remark 5.4.4 In the above definition the (arbitrary) neighbourhoods V and W may be replace by open neighbourhoods V and W, or by neighbourhoods V and W from a given base $\mathcal{B}$ of open sets.

Remark 5.4.5 The discrete topology is separated, while the rubbish topology is not separated if $|\ T\ | \geq 2$.

Remark 5.4.6 It follows easily from Remark 5.4.2 that every subspace of a separated space is separated.

Remark 5.4.7 The product topology of a family of separated spaces is separated.

Definition 5.4.3 A subset X of a topological space T is said to be *compact* if every open covering of X includes a finite covering, that is, for every family $\mathcal{C}$ of open sets such that $X \subseteq \bigcup \mathcal{C}$ there is a finite subfamily $\mathcal{C}_0 \subseteq \mathcal{C}$ such that $X \subseteq \bigcup \mathcal{C}_0$. We say that a topological space T is *compact*, or that its topology is *compact*, if T is a compact subset of T, that is, if $T = \bigcup \mathcal{C}$ implies $T = \bigcup \mathcal{C}_0$ for some finite subfamily $\mathcal{C}_o \subseteq \mathcal{C}$.

There is no ambiguity in this definition, as shown in:

Remark 5.4.8 A subset $X \neq \varnothing$ of a topological space T is compact if and only if the subspace X is compact.

Example 5.4.5 Every finite topological space is compact.

Example 5.4.6 The rubbish topology is compact.

Proposition 5.4.3 *The following properties are equivalent for a topological space T:*

(i) *T is compact ;*

(ii) *there exists a base $\mathcal{B}$ of open sets such that if $\mathcal{C} \subseteq \mathcal{B}$ satisfies $T = \bigcup \mathcal{C}$, then $T = \bigcup \mathcal{C}_0$ for some finite $\mathcal{C}_0 \subseteq \mathcal{C}$;*

(iii) *for every family $\mathcal{K}$ of closed sets of T : if $\bigcap \mathcal{K} = \varnothing$, then $\bigcap \mathcal{K}_0$ for some finite $\mathcal{K}_0 \subseteq \mathcal{K}$;*

(iv) *for every family $\mathcal{K}$ of closed sets of T : if $\bigcap \mathcal{K}_0 \neq \varnothing$ for every finite $\mathcal{K}_0 \subseteq \mathcal{K}$, then $\bigcap \mathcal{K} \neq \varnothing$.*

Comment The property "$\bigcap \mathcal{K}_0 \neq \varnothing$ for every finite $\mathcal{K}_0 \subseteq \mathcal{K}$" is sometimes called the *finite intersection property* for $\mathcal{K}$.

PROOF: (i)$\Longrightarrow$(ii) : Trivial.

(ii)$\Longrightarrow$(i) : Let $T = \bigcup_{D\in\mathcal{D}} D$, where $\mathcal{D}$ is a family of open sets. Then for each $D \in \mathcal{D}$ we have $D = \bigcup_{B\in\mathcal{B}_D} B$, where $\mathcal{B}_D \subseteq \mathcal{B}$. So $T = \bigcup_{C\in\mathcal{C}} C$, where $\mathcal{C} = \{B \subseteq T \mid (\exists\, D \in \mathcal{D})\ \ B \in \mathcal{B}_D\} \subseteq \mathcal{B}$. Then there exists a finite covering $T = C_1 \cup \cdots \cup C_n$, where for each $i \in \{1, \ldots, n\}$ there exists $D_i \in \mathcal{D}$ such that $C_i \in \mathcal{B}_{D_i}$, hence $C_i \subseteq D_i$. Finally $T \subseteq D_1 \cup \cdots \cup D_n \subseteq T$.

(i)$\Longrightarrow$(iii) : If $\bigcap \mathcal{K} = \varnothing$, where $\mathcal{K}$ is a family of closed sets, then $T = T \setminus \bigcap \mathcal{K} = \bigcup \{T \setminus F \mid F \in \mathcal{K}\}$ by the De Morgan laws. The sets $T \setminus F$ being open, it follows that $T = (T \setminus F_1) \cup \cdots \cup (T \setminus F_m) = T \setminus (F_1 \cap \cdots \cap F_m)$ for some $F_1, \ldots, F_m \in \mathcal{K}$, therefore $F_1 \cap \cdots \cap F_m = \varnothing$.

(iii)$\Longrightarrow$(i) : If $T = \bigcup_{D\in\mathcal{D}} D$, where $\mathcal{D}$ is a family of open sets, then $\varnothing = T \setminus \bigcup_{D\in\mathcal{D}} D = \bigcap_{D\in\mathcal{D}} (T \setminus D)$, where the sets $T \setminus D$ are closed. Therefore there exist $n > 0$ and $D_1, \ldots, D_n \in \mathcal{D}$ such that

$$\varnothing = (T \setminus D_1) \cap \cdots \cap (T \setminus D_n) = T \setminus \bigcup_{i=1}^{n} D_i \,,$$

hence $T = \bigcup_{i=1}^{n} D_i$.

(iii)$\Longleftrightarrow$(iv) : Trivial. □

To establish Proposition 5.4.4 below, we first prove:

Lemma 5.4.1 (i) *If $\mathcal{M} \subseteq 2^T$ is a maximal family with the finite intersection property, then*

$$\bigcap \mathcal{M}_0 \in \mathcal{M} \text{ for every finite } \mathcal{M}_0 \subseteq \mathcal{M} \,, \tag{5.4.5}$$

$$S \in \mathcal{M} \Longleftrightarrow S \cap M \neq \varnothing \ \ (\forall\, M \in \mathcal{M}) \,. \tag{5.4.6}$$

(ii) *Conversely, if $\mathcal{M} \subseteq 2^T$ is a family with the finite intersection property and*

$$S \cap M \neq \varnothing \ \ (\forall\, M \in \mathcal{M}) \Longrightarrow S \in \mathcal{M} \,, \tag{5.4.7}$$

then $\mathcal{M}$ is maximal.

PROOF: (i) If $\bigcap \mathcal{M}_0 \notin \mathcal{M}$ for some finite $\mathcal{M}_0 \subseteq \mathcal{M}$, then $\mathcal{M} \subset \mathcal{M} \cup \{\bigcap \mathcal{M}_0\}$ and every finite intersection of the latter family is a finite intersection of $\mathcal{M}$, hence it is not empty; this contradicts the maximality of $\mathcal{M}$. The property (5.4.5) being established, $S \cap M \neq \varnothing$ $(\forall M \in \mathcal{M})$ implies $S \in \mathcal{M}$, otherwise $\mathcal{M} \cup \{S\}$ would contradict the maximality of $\mathcal{M}$. The converse is trivial.

(ii) If $\mathcal{M} \cup \{S\}$ has the finite intersection property, then $S \in \mathcal{M}$ by (5.4.10). □

Proposition 5.4.4 (Tikhonov) *The product topology of every family of compact spaces is compact.*

PROOF: Assume that a family $\mathcal{K}$ of closed sets of $T = \prod_{i \in I} T_i$ has the finite intersection property (f.i.p.); the task is to prove that $\bigcap \mathcal{K} \neq \varnothing$.

It is easy to see that the set of all families of subsets of T together with f.i.p. verifies the hypothesis of the Tuckey lemma (condition (g^*) in Theorem 2.4.1), therefore there is a maximal family $\mathcal{L} \subseteq 2^T$ with f.i.p. and such that $\mathcal{K} \subseteq \mathcal{L}$. Clearly $\bigcap \mathcal{L} \neq \varnothing \Longrightarrow \bigcap \mathcal{K} \neq \varnothing$, therefore we will suppose, without toss of generality, that $\mathcal{K}$ is maximal with f.i.p.

Now for every $i \in I$, every $X \subseteq T$ and every $\mathcal{X} \subseteq 2^T$, we put $p_i(X) = \{p_i(x) \mid x \in X\}$ and $p_i(\mathcal{X}) = \{p_i(X) \mid X \in \mathcal{X}\}$, where p_i are the canonical projections; let us prove that $p_i(\mathcal{K}) \subseteq 2^{T_i}$ is a maximal family satisfying f.i.p. The fact that for every $\mathcal{K}_0 \subseteq \mathcal{K}$ we have $p_i(\bigcap \mathcal{K}_0) \subseteq \bigcap p_i(\mathcal{K}_0)$ shows that $p_i(\mathcal{K})$ has f.i.p. To prove maximality we use Lemma 5.4.1 with $\mathcal{M} := p_i(\mathcal{K})$ and $S := S_i \in 2^{T_i}$. Assume that $S_i \cap p_i(K) \neq \varnothing$ for every $K \in \mathcal{K}$; we have to prove that $S_i \in p_i(\mathcal{K})$. Set $S = S_i \times \prod_{j \neq i} T_j$. For every $K \in \mathcal{K}$ there is $x \in K$ such that $p_i(x) \in S_i$, hence $x \in S \cap K$. We have proved that $S \cap K \neq \varnothing$ for every $K \in \mathcal{K}$, hence $S \in \mathcal{K}$ again by Lemma 5.4.1, therefore $S_i \in p_i(\mathcal{K})$.

Since $\{p_i(K) \mid K \in \mathcal{K}\}$ has f.i.p., the family of closed sets $\{\overline{p_i(K)} \mid K \in \mathcal{K}\}$ has also f.i.p. Hence, for each $i \in I$ we can choose a point $a_i \in \bigcap\{\overline{p_i(K)} \mid K \in \mathcal{K}\}$, because the latter set is not empty by Proposition 5.4.3,(i)$\Longleftrightarrow$(iv). Set $a = (a_i)_{i \in I}$. If we succeed in proving that for every $K \in \mathcal{K}$ and every neighbourhood of a of the form $V = \prod_{i \in I} V_i \in \mathcal{B}$ (cf. Example 5.4.4) we have $V \cap K \neq \varnothing$, Corollary 5.4.1 will imply that for every $K \in \mathcal{K}$ we have $a \in \overline{K} = K$, therefore $\bigcap \mathcal{K} \neq \varnothing$, as desired.

For every $i \in I$ and every $K \in \mathcal{K}$ we have $a_i \in \overline{p_i(K)}$, therefore $V_i \cap p_i(K) \neq \varnothing$ by Proposition 5.4.2; this implies $V_i \in p_i(\mathcal{K})$ by Lemma 5.4.1. Let I_0 be a finite subset of I such that $V_i = T_i$ for $i \in I \setminus I_0$. For every $i \in I_0$ choose $K^i \in \mathcal{K}$ such that $V_i = p_i(K^i)$. The set $K^* = \bigcap\{K^i \mid i \in I_0\} \subseteq \mathcal{K}$ is not empty and $K^* \in \mathcal{K}$ by (5.4.5). If $x \in K^*$, then for $i \in I_0$ we have $p_i(x) \in p_i(K^i) = V_i$, while for $i \in I \setminus I_0$ we have $p_i(x) \in T_i = V_i$. This proves that $K^* \subseteq \prod_{i \in I} V_i = V$. Taking also into account that $\mathcal{K}$ has f.i.p., we get $\varnothing \neq K \cap K^* \subseteq K \cap V$, therefore $K \cap V \neq \varnothing$. □

Proposition 5.4.5 *Every closed subset of a compact space is compact.*

PROOF: Let F be a closed subset of a compact space T and $F \subseteq \bigcup \mathcal{G}$ an open covering of F. Then $\mathcal{G} \cup \{T \setminus F\}$ is an open covering of T, hence there is a finite $\mathcal{G}_0 \subseteq \mathcal{G}$ such that $T = (T \setminus F) \cup \bigcup \mathcal{G}_0$. This implies $F = F \cap T = F \cap \bigcup \mathcal{G}_0$, that is, $F \subseteq \bigcup \mathcal{G}_0$. □

Proposition 5.4.6 *A compact subset of a separated space is closed.*

PROOF: Let K be a compact subset of a separated space T. If $K = T$ the conclusion is established; otherwise take an arbitrary $y \in T \setminus K$. For every $x \in K$ there are two disjoint open neighbourhoods U_x, V_x of x and y, respectively; cf. Remark 5.4.4. From the open covering $K \subseteq \bigcup \{U_x \mid x \in K\}$ one gets $K \subseteq \bigcup \{U_x \mid x \in K_0\}$ for some finite $K_0 \subseteq K$. Then $W_y = \bigcap \{V_x \mid x \in K_0\}$ is an open neighbourhood of y which fulfills

$$W_y \cap \bigcup \{U_x \mid x \in K_0\} = \bigcup \{W_y \cap U_x \mid x \in K_o\} \subseteq \bigcup \{V_x \cap U_x \mid x \in K_0\} = \varnothing .$$

We have thus proved that every $y \in T \setminus K$ has an open neighbourhood W_y such that $W_y \cap K = \varnothing$. It follows that $y \in W_y \subseteq T \setminus K$ for every $y \in T \setminus K$, therefore $T \setminus K = \bigcup \{W_y \mid y \in T \setminus K\}$ is an open set, consequently K is closed. □

Definition 5.4.4 A mapping $f : T \longrightarrow T'$ between two topological spaces is said to be *continuous* provided for every open set G' of T', the set $f^{-1}(G') = \{x \in T \mid f(x) \in G'\}$ is open in T. A *homeomorphism* between two topological spaces T and T' is a continuous bijection $f : T \longrightarrow T'$ such that the inverse mapping $f^{-1} : T' \longrightarrow T$ is also continuous. The continuous functions and the homeomorphisms are the morphisms and the isomorphisms of the category **Top**, respectively (cf. Example 1.3.13).

Proposition 5.4.7 *The following conditions are equivalent for a function $f : T \longrightarrow T'$ between two topological spaces:*

(i) *f is continuous ;*

(ii) *for every closed set F' of T', the set $f^{-1}(F')$ is closed in T ;*

(iii) *for every $x \in T$ and every neighbourhood V' of $f(x)$, there is a neighbourhood V of x such that $f(V) \subseteq V'$.*

PROOF: (i)$\Longleftrightarrow$(ii): Obvious from the identity $f^{-1}(T' \setminus X') = T \setminus f^{-1}(X')$.

(i)$\Longrightarrow$(iii): Take an open neighbourhood G' of $f(x)$ such that $G' \subseteq V'$. Then $V = f^{-1}(G')$ is an open neighbourhood of x and $f(V) \subseteq G' \subseteq V'$.

(iii)$\Longrightarrow$(i): Let G' be an open set of T' and $G = f^{-1}(G')$. For every $x \in G$, we have $f(x) \in G'$, therefore G' is a neighbourhood of $f(x)$, hence there is a neighbourhood V_x of x such that $f(V_x) \subseteq G'$; if W_x is an open neighbourhood of x included in V_x, then a fortiori $f(W_x) \subseteq G'$. It follows that $x \in W_x \subseteq f^{-1}(G') = G$ for every $x \in G$, hence $G = \bigcup \{W_x \mid x \in G\}$ is an open set. □

Remark 5.4.9 Whenever one verifies that a mapping $f : T \longrightarrow T'$ is continuous, it suffices to prove that $f^{-1}(G')$ is open for every set G' belonging to a fixed base of open sets in T'. If continuity is to be verified with the aid of property (iii) in Proposition 5.4.7, it suffices to take an open neighbourhood V' or even a neighbourhood V' from a fixed base of open sets in T; cf. Remark 5.4.4.

Proposition 5.4.8 (i) *The image $f(K)$ of a compact set K of a space T by a continuous mapping $f : T \longrightarrow T'$ is compact in T'.*
(ii) *If, moreover, T' is separated, then $f(K)$ is closed.*

PROOF: (i) Left to the reader.
(ii) From (i) and Proposition 5.4.6. □

Corollary 5.4.2 *If T is a compact space, T' a separated space and $f : T \longrightarrow T'$ a continuous bijection, then f is a homeomorphism.*

Warning A continuous bijection is not always a homeomorphism. In other words, in the category **Top** there exist bijective morphisms that are not isomorphisms; cf. Example 1.3.26.
PROOF: To prove that $f^{-1} : T' \longrightarrow T$ is continuous, we apply Proposition 5.4.7. Let K be a closed set of T; we have to prove that $(f^{-1})^{-1}(K)$ is closed in T'. Indeed, $(f^{-1})^{-1}(K) = f(K)$ and K is compact by Proposition 5.4.5, hence $f(K)$ is compact by Proposition 5.4.8, therefore $f(K)$ is closed by Proposition 5.4.6. □

Definition 5.4.5 By a *clopen* set of a topological space is meant a set which is both closed and open. A space in which the clopen sets form a base of open sets is termed a *totally disconnected* space. A compact totally disconnected space is called a *Stone space*.

Remark 5.4.10 In every topological space T, the sets $\varnothing$ and T are clopen. Moreover, the clopen sets form a field of sets.

Remark 5.4.11 Any totally disconnected space is separated: every x has a clopen neighbourhood K by Remark 5.4.1; then $T \setminus K$ is a neighbourhood of every $y \neq x$. In particular Stone spaces are separated.

Example 5.4.7 A closed subset of a Stone space is a Stone space in the induced topology (cf. Example 5.4.3, Remark 5.4.6 and Proposition 5.4.5).

Example 5.4.8 The finite discrete spaces are Stone spaces (cf. Examples 5.4.1, 5.4.5 and Remark 5.4.5; total disconnectedness is trivial).

Example 5.4.9 Every product of a family of finite discrete spaces is a is a Stone space (cf. Example 5.4.8, Remark 5.4.7 and Proposition 5.4.4).

Example 5.4.10 In particular if all $T_i = 2 = \{0,1\}$ are endowed with the discrete topology, one gets the Stone space 2^I known as the *Cantor discontinuum*. In this case there is a nice description of the base $\mathcal{B}$ in Example 5.4.4: since $\{0\}$ and $\{1\}$ are the only sets V_i with $\varnothing \neq V_i \neq 2$, the neighbourhoods in $\mathcal{B}$ can be described with the aid of several parameters: distinct $i_1, \ldots, i_n \in I, \varepsilon_1, \ldots, \varepsilon_n \in \{0,1\}$, and $n \geq 1$; they are the sets of the form

$$V^{i_1,\ldots,i_n}_{\varepsilon_1,\ldots,\varepsilon_n} = \prod_{i \in I} V_i\,,\ V_i = \{\varepsilon_i\}\ (i = i_1, \ldots, i_n)\,,\ \text{else } V_j = 2\,. \tag{5.4.8}$$

If we use the representation $2^I = \{f \mid f : L \longrightarrow 2\}$, then

(5.4.9) $$V^{i_1,\dots,i_n}_{\varepsilon_1,\dots,\varepsilon_n} = V^{i_1}_{\varepsilon_1} \cap \dots \cap V^{i_n}_{\varepsilon_n} ,$$

(5.4.10) $$V^i_\varepsilon = \{f \in 2^I \mid f(i) = \varepsilon\} = p_i^{-1}(\varepsilon) .$$

Example 5.4.11 Let T be an infinite set and $t_o \in T$. Take as family $\mathcal{O}$ of open sets all the sets which do not contain t_0 plus the cofinite sets that contain t_0. This space is compact: if $T = \bigcup \mathcal{C}$ is an open covering, then there is $G \in \mathcal{C}$ such that $t_o \in G$, hence G is cofinite, say $T \setminus G = \{c_1, \dots, c_n\}$; for each $i \in \{1, \dots, n\}$ there is $G_i \in \mathcal{C}$ such that $c_i \in G_i$, hence $T = G \cup G_1 \cup \dots \cup G_n$. The closed sets are all the sets which contain t_0 plus the finite sets that do not contain t_0. Finally the clopen sets are all the finite sets which do not contain t_0 plus the cofinite sets that contain t_0; it is clear that they form a base of open sets. Therefore the space defined in this way is a Stone space.

Remark 5.4.12 Every closed set of a Stone space is a finite union of clopen sets (cf. Proposition 5.4.5).

Definition 5.4.6 A field of sets $\mathcal{SF}$ on $(2^T, \subseteq)$ is said to be *separating*, if it *separates* points, that is, for every $x, y \in T$ with $x \neq y$ there is a set $S \in \mathcal{SF}$ such that $x \in S$ and $y \in T \setminus S$.

Proposition 5.4.9 (i) *The clopen sets of a Stone space form a separating field.*

(ii) *Conversely, if T is a compact space and $\mathcal{SF}$ a separating field of clopen sets from T, then T is a Stone space and $\mathcal{SF}$ coincides with the field of all clopen sets of T.*

PROOF: (i) Obvious.

(ii) Since the sets in $\mathcal{SF}$ are clopen, it follows from Definition 5.4.6 that T is separated.

Further, we show that $\mathcal{SF}$ separates points from closed sets, i.e., for every closed set F of T and every $y \in T \setminus F$ there is a set $S \in \mathcal{SF}$ such that $F \subseteq S$ and $y \notin S$. Indeed, we can associate with each $x \in F$ a set $S_x \in \mathcal{SF}$ such that $x \in S_x$ and $y \notin S_x$, whence $F \subseteq \bigcup\{S_x \mid x \in F\}$, and since F is compact by Proposition 5.4.5, it follows that $F \subseteq \bigcup\{S_x \mid x \in F_0\} = S$ for some finite subset $F_0 \subseteq F$, and this implies $y \notin S \in \mathcal{SF}$.

Now we prove that $\mathcal{SF}$ is a base of open sets. Let G be an open set. Then for every $x \in G$ we can find a set $R_x \in \mathcal{SF}$ which separates x from the closed set $T \setminus G$, that is, $x \in R_x$ and $R_x \cap (T \setminus G) = \varnothing$. Consequently $x \in R_x \subseteq G$ for every $x \in G$, therefore $G = \bigcup\{R_x \mid x \in G\}$.

Thus T is a totally disconnected space, hence a Stone space. In particular any clopen set K is a union of sets from $\mathcal{SF}$, hence K is a finite union of sets from $\mathcal{SF}$, therefore $K \in \mathcal{SF}$. □

Corollary 5.4.3 *If $\mathcal{SF}$ is both a field of clopen sets and a base of open sets of a separated compact space T, then T is a Stone space and $\mathcal{SF}$ is the field of all clopen sets of T.*

PROOF: Clearly $\mathcal{SF}$ is a separating field. □

5.5 Topological Duality for Distributive Lattices

We have seen in Theorem 5.3.2 that every distributive lattice L is isomorphic to a ring of sets, namely the ring $\mathcal{PF}(L)$ of prime filters of L; in particular every Boolean algebra is isomorphic to a field of sets. In this section we follow Priestley [1970] beyond the above result by endowing the set $\mathcal{PF}(L)$ with a structure of ordered topological space. This yields a dual equivalence between the category of distributive lattices with 0 and 1 and the category of the so-called totally order-disconnected Stone spaces, and in particular a dual equivalence between the category of Boolean algebras and the category of Stone spaces. These basic theorems are related to other categorical (dual) equivalences dealt with e.g. in functional analysis. We might say that the lattice-theoretical tool is thus inserted, at a higher level, into the set-theoretical framework it comes from.

Definition 5.5.1 A lattice is said to be *bounded* if it has 0 and 1. Let **Dis01** be the category having as objects all bounded distributive lattices and as morphisms all 0-1-preserving lattice homomorphisms between them. For every $L, L' \in \mid \mathbf{Dis01} \mid$ we denote by $\mathrm{Hom}(L, L')$ the set of all morphisms $f : L \longrightarrow L'$.

The fact that we only deal with bounded lattice is not really too restrictive; as a matter of fact, it is possible to drop it, but this would result in certain complications which are not worthwhile.

The starting point is Theorem 5.3.2, which is expressed in terms of prime filters. But Exercise 5.3.3 establishes a bijection between $\mathcal{PF}(L)$ and $\mathrm{Hom}(L, 2)$, namely $P \longleftrightarrow f$ holds iff for all $x \in L$ we have $x \in P \iff f(x) = 1$. Having this in mind, the embedding of L into $2^{\mathcal{PF}(L)}$ can be reformulated as an embedding of L into $2^{\mathrm{Hom}(L,2)}$:

Proposition 5.5.1 (i) *Let L be a bounded distributive lattice. The mapping φ_L which associates with every $x \in L$ the set*

$$\varphi_L(x) = \{f \in \mathrm{Hom}(L, 2) \mid f(x) = 1\} \qquad (5.5.1)$$

establishes an isomorphism between L and the ring $\varphi_L(L) = \{\varphi_L(x) \mid x \in L\}$ of subsets of $\mathrm{Hom}(L, 2)$.

(ii) *If L is a Boolean algebra, then $\varphi_L(L)$ is a field of sets.*

The next step is to endow $\mathrm{Hom}(L, 2)$ with a topology. Since $\mathrm{Hom}(L, 2) \subset 2^L$ and the latter is already a topological space, namely the Cantor discontinuum (cf. Example 5.4.10, the particular case of Example 5.4.4), it is natural to equip $\mathrm{Hom}(L, 2)$ with the induced topology.

Proposition 5.5.2 *For every distributive lattice L with 0 and 1, the subspace* $\mathrm{Hom}(L, 2)$ *of the Cantor discontinuum 2^L is a Stone space*

PROOF: In view of Example 5.4.7, it suffices to show that $\mathrm{Hom}(L, 2)$ is a closed subset of 2^L. But $\mathrm{Hom}(L, 2)$ is the intersection of the following four sets :

$$H_\circ = \{f \in 2^L \mid f(x \circ y) = f(x) \circ f(y) \ (\forall x, y \in L)\} \quad (\circ \in \{\wedge\, , \vee\})\ ,$$

$$H_\varepsilon = \{f \in 2^L \mid f(\varepsilon) = \varepsilon\} \quad (\varepsilon \in \{0,1\}) ,$$

therefore it remains to prove that the sets $H_\wedge, H_\vee, H_0$ and H_1 are closed.

For every $x, y \in L$, define $\alpha_x, \alpha_x \circ \alpha_y : 2^L \longrightarrow 2$ $(\circ \in\{\wedge , \vee\})$ by

$$\alpha_x(f) = f(x) \quad (\forall\, f \in 2^L) ,$$

$$(\alpha_x \circ \alpha_y)(f) = \alpha_x(f) \circ \alpha_f(y) \quad (\forall\, f \in 2^L) .$$

Then we have

$$H_\varepsilon = (\alpha_\varepsilon)^{-1}(\varepsilon) \quad (\varepsilon \in \{0,1\}) ,$$

while for $\circ \in\{\wedge , \vee\}$,

$$H_\circ = \{f \in 2^L \mid \alpha_{x\circ y}(f) = (\alpha_x \circ \alpha_y)(f) \ (\forall\, x, y \in L)\}$$

$$= \bigcap_{(x,y)\in L\times L} \bigcup_{\varepsilon\in\{0,1\}} \{f \in 2^L \mid \alpha_{x\circ y}(f) = (\alpha_x \circ \alpha_y)(f) = \varepsilon\}$$

$$= \bigcap_{(x,y)\in L\times L} \bigcup_{\varepsilon\in\{0,1\}} \{(\alpha_{x\circ y})^{-1}(\varepsilon) \cap (\alpha_x \circ \alpha_y)^{-1}(\varepsilon)\} .$$

Since the singleton $\{\varepsilon\}$ is a closed set in the discrete topology of 2, in view of Proposition 5.4.7 it suffices to prove the continuity of the mappings $\alpha_x, \alpha_x \wedge \alpha_y$ and $a_x \vee \alpha_y$ for every $x, y \in L$. The same Proposition shows that $\alpha : 2^L \longrightarrow L$ is continuous iff for every $f \in 2^L$ and every neighbourhood V' of $\alpha(f)$ there is a neighbourhood V of f such that $\alpha(g) \in V'$ for every $g \in V$.

For every $f \in 2^L$, the only neighbourhoods of the point $\alpha_x(f) \in 2$ are 2 itself and $\{f(x)\}$. For every g in the neighbourhood $V^x_{f(x)}$ of f (cf. (5.4.10) in Example 5.4.10) we have $\alpha_x(g) = g(x) = f(x) \in \{f(x)\}$. Also, for $\circ \in\{\wedge , \vee\}$ the only neighbourhoods of the point $(\alpha_x \circ \alpha_y)(f)$ are 2 and $\{f(x) \circ f(y)\}$. For every g in the neighbourhood $V^{x,y}_{f(x),f(y)}$ of f (cf. (5.4.9) in Example 5.4.10) we have $(\alpha_x \circ \alpha_y)(g) = g(x) \circ g(y) = f(x) \circ f(y) \in \{f(x) \circ\ f(y)\}$. □

Summarizing, every bounded distributive lattice (every Boolean algebra) L is is isomorphic to a ring (field) of subsets of $\mathrm{Hom}(L, 2)$. In other terms, this is an embedding of L in the Stone space $\mathrm{Hom}(L, 2)$, which, in its turn is included in 2^T and is a subspace of the Cantor discontinuum $T = 2^I$. Note that several non-isomorphic bounded distributive lattices L may be embedded into the same Stone space $\mathrm{Hom}(L, 2)$. Thus, for instance, – using the prime-filter language – both the three-element chain $\{0, \frac{1}{2}, 1\}$ and the four-element Boolean algebra $\{0, a, a', 1\}$ have two prime filters: $\{\frac{1}{2}, 1\}, \{1\}$ and $\{a, 1\}, \{a', 1\}$, respectively. Therefore they are associated with the two-element Stone space 2.

Exercise 5.5.1 Wite down explicitly the basis $V^1_0, V^1_1, \ldots, V^{1,2,3}_{1,1,1}$ of the Cantor discontinuum $2^{\{1,2,3\}}$ (denote the elements of $2^{\{1,2,3\}}$ just by concatenation: $000, 001, \ldots, 111$).

Exercise 5.5.2 Prove that a Cantor discontinuum is a discrete space if and only if it is finite.

Priestley [1970] has enriched the structure of a Stone space by introducing an order relation on it; a bijection between bounded distributive lattices and certain ordered analogs of Stone spaces is thus obtained up to an isomorphism, to the effect that an equivalence (cf. the comment between Proposition 1.3.11 and Definition 1.3.14) of the corresponding two categories is established; cf. Theorems 5.5.1 and 5.5.2 below.

Definition 5.5.2 By an *ordered topological space* is meant a topological space T equipped with a partial order $\leq$. We denote by $\mathcal{U}$ (by $\mathcal{L}$) the family of all clopen order filters (clopen order ideals) of T plus the empty set $\varnothing$. [1] We refer to the elements of $\mathcal{U}$ and $\mathcal{L}$ as $\mathcal{U}$*-sets* and $\mathcal{L}$*-sets*, respectively.

Remark 5.5.1 Let T be an ordered topological space. Then $\mathcal{U}$ and $\mathcal{L}$ are rings of sets and $\varnothing, T \in \mathcal{U} \cap \mathcal{L}$. Besides, a subset $X \subseteq T$ is an $\mathcal{U}$-set (an $\mathcal{L}$-set) if and only if $T \setminus X$ is an $\mathcal{L}$-set (an $\mathcal{U}$-set) (cf. Remark 5.1.1.b).

Proposition 5.5.3 *The following conditions are equivalent for an ordered topological space T:*

i) *If $x, y \in T$ $x \not\leq y$, there exist two disjoint sets $U \in \mathcal{U}$ and $V \in \mathcal{L}$ such that $x \in U$ and $y \in V$.*

(ii) *If $x, y \in T$ and $x \not\leq y$, there is an $\mathcal{U}$-set U such that $x \in U$ and $y \notin U$.*

(iii) *If $x, y \in T$ and $x \not\leq y$ there is an $\mathcal{L}$-set V such that $y \in V$ and $x \notin V$.*

PROOF: (ii)$\Longrightarrow$(i) and (iii)$\Longrightarrow$(i) follow by Remark 5.5.1. The converse implications are trivial. □

Definition 5.5.3 An ordered topological space is said to be *totally order-discon-nected* if it satisfies the equivalent conditions in Proposition 5.5.3. A *Priestley space*, also called an *ordered Stone space*, is a compact totally order-disconnected space.

The next Proposition clarifies the relationship between Stone spaces and Priestley spaces, which justifies the alternative denomination "ordered Stone spaces" for the latter spaces.

Proposition 5.5.4 (i) *Every Priestley space is a Stone space in which $\mathcal{U} \cup \mathcal{L}$ is the field of all clopen sets.*

(ii) *Conversely, Stone spaces coincide with Priestley spaces with respect to equality taken as partial order.*

PROOF: (i) It follows from Remark 5.5.1 that $\mathcal{U} \cup \mathcal{L}$ is a field of clopen sets, while Definition 5.5.3 shows that this field is separating (cf. Definition 5.4.6), whence the conclusion follows from Proposition 5.4.9(ii).

(ii) Follows from (i) and the easy remark that every Stone space is a Priestley space with $\mathcal{U} = \mathcal{L} =$ the field of all clopen sets. □

The theory to be developed below will furnish non-trivial examples of totally order-disconnected spaces (i.e., with a partial order different from the equality).

[1] The notation is reminiscent of "upper set" and "lower set" in Definition 5.1.1. Davey and Priestley [1990] use the converse notation.

Lemma 5.5.1 *Every clopen set K of a Priestley space T can be written in the form $K = \bigcup_{i=1}^{n}(U_i \cap V_i)$, for some $n \geq 1$ and some $U_i \in \mathcal{U}$, $V_i \in \mathcal{L}$ $(i = 1, \ldots, n)$.*

PROOF: The property is trivial for $\varnothing$ and T, by Remark 5.5.1. Now let K be a clopen set with $\varnothing \neq K \neq T$. If for every $x \in K$ we succeed in finding $U_x \in \mathcal{U}$ and $V_x \in \mathcal{L}$ such that $x \in U_x \cap V_x \subseteq K$, it will follow that $K = \bigcup_{x \in K}(U_x \cap V_x)$, whence the desired conclusion will follow because K is compact by Proposition 5.4.5.

Take $x \in K$ and $y \in T \setminus K$. Since $\mathcal{U} \cup \mathcal{L}$ is a separating field by Proposition 5.5.4, there is a set $S \in \mathcal{U} \cup \mathcal{L}$ such that $x \in S$ and $y \notin S$. If $S \subseteq K$, take $U_x = S, V_x = T$ or $U_x = T, V_x = S$ as $S \in \mathcal{U}$ or $S \in \mathcal{L}$. If $S \not\subseteq K$, apply again the separation property to x and every $y \in S \setminus K$, thus finding new sets $T(y) \in \mathcal{U} \cup \mathcal{V}$ such that $y \in T(y)$ and $x \notin T(y)$. Since K and the sets $T(y)$ for $y \in S \setminus K$, are open sets covering the compact set S, there exists a finite covering $S \subseteq K \cup T_1 \cup \cdots \cup T_m$, hence $x \in S \cap (T \setminus T_1) \cap \cdots \cap (T \setminus T_m) \subseteq K$. Now take as U_x (as V_x) the intersection of those sets from $\{S, T \setminus T_1, \ldots, T \setminus T_m\}$ that belong to $\mathcal{U}$ (to $\mathcal{L}$). □

Proposition 5.5.5 *Every open set of a Priestley space T is a union of sets of the form $U \cap V$ with $U \in \mathcal{U}$ and $V \in \mathcal{V}$.*

PROOF: From Lemma 5.5.1, since every open set of the Stone space T is a union of clopen sets. □

Definition 5.5.4 We denote by **Pr** the category of all Priestley spaces and all continuous isotone mappings between them.

The next step is to construct a contravariant functor from **Dis01** to **Pr**.

Definition 5.5.5 The *dual* of a bounded distributive lattice L is the Stone space $\mathrm{Hom}(L, 2)$ (cf. Proposition 5.5.2) endowed with the pointwise order

(5.5.2) $$f \leq g \iff f(x) \leq g(x) \quad (\forall\, x \in L)\ .$$

The *dual* of a morphism $\alpha : L \longrightarrow L'$ in **Dis01** is the mapping $\beta : \mathrm{Hom}(L', 2) \longrightarrow \mathrm{Hom}(L, 2)$ determined by

(5.5.3) $$\beta(f') = f' \circ \alpha \quad (\forall\, f' \in \mathrm{Hom}(L', 2))\ .$$

Remark 5.5.2 The sets

(5.5.4) $$W^{x_1,\ldots,x_n}_{\varepsilon_1,\ldots,\varepsilon_n} = V^{x_1,\ldots,x_n}_{\varepsilon_1,\ldots,\varepsilon_n} \cap \mathrm{Hom}(L, 2)\ ,$$

defined for every pairwise distinct $x_1, \ldots, x_n \in L$, every $\varepsilon_1, \ldots, \varepsilon_n \in 2$ and every $n \geq 1$, are clopen and form a base of open sets in $\mathrm{Hom}(L, 2)$; cf. Example 5.4.10 and Remark 5.4.2. Moreover,

(5.5.5) $$W^{x_1,\ldots,x_n}_{1,\ldots,1} \in \mathcal{U} \text{ and } W^{x_1,\ldots,x_n}_{0,\ldots,0} \in \mathcal{L}$$

for every pairwise distinct $x_1, \ldots, x_n$ and every $n \geq 1$.

Proposition 5.5.6 (i) *The dual of a bounded distributive lattice is a Priestley space.*

(ii) *The dual of a morphism in* **Dis01** *is a continuous isotone mapping.*

PROOF: (i) We have to complete Proposition 5.5.2 by the property of total order-disconnectedness. If $f \not\leq g$ for $f, g \in \mathrm{Hom}(L, 2)$, then $f(x) = 1$ and $g(x) = 0$ for some $x \in L$. It follows from Remark 5.5.2 that $f \in W_1^x \in \mathcal{U}$ and $g \in W_0^x \in \mathcal{L}$, where $W_1^x \cap W_0^x = \varnothing$.

(ii) Let $\alpha : L \longrightarrow L'$ be a morphism of **Dis01** and β the mapping given by (5.5.3). Clearly if $f', g' \in \mathrm{Hom}(L', 2)$ and $f' \leq g'$, then $f' \circ \alpha \leq g' \circ \alpha$; therefore β is isotone. To prove continuity, take a point of the form $\beta(f')$ in $\mathrm{Hom}(L, 2)$ and a neighbourhood $W_{\varepsilon_1,\ldots,\varepsilon_n}^{x_1,\ldots,x_n}$ of $\beta(f')$; cf. Remarks 5.4.9 and 5.5.2. Then

$$(f' \circ \alpha)(x_i) = \beta(f')(x_i) = \varepsilon_i \ \ (i = 1, \ldots, n) \ ,$$

therefore $(W')_{\varepsilon_1,\ldots,\varepsilon_n}^{\alpha(x_1),\ldots,\alpha(x_n)}$, where W' indicates a (5.5.2)-like set in $\mathrm{Hom}(L', 2)$, is a neighbourhood of f' in $\mathrm{Hom}(L', 2)$ and if g' is another point in this neighbourhood, then

$$\beta(g')(x_i) = (g' \circ \alpha)(x_i) = g'(\alpha(x_i)) = \varepsilon_i \ \ (i = 1, \ldots, n) \ ,$$

therefore $\beta(g') \in W_{\varepsilon_1,\ldots,\varepsilon_n}^{x_1\ldots,x_n}$. □

Corollary 5.5.1 *A contravariant functor* Φ : **Dis01** $\longrightarrow$ **Pr** *is obtained by setting* $\Phi L = \mathrm{Hom}(L, 2)$ *for every* $L \in |$ **Dis01** $|$ *and* $(\Phi\alpha)(f') = f' \circ \alpha$ $(\forall f' \in \mathrm{Hom}(L', 2)$ *for every* $\alpha \in$ **Dis01**(L, L').

PROOF: From Proposition 5.5.6 plus routine verification of commutativity of diagrams. □

The next task is the construction of a contravariant functor from **Pr** to **Dis01**.

Definition 5.5.6 The *dual* of a Priestley space T is the family $\mathcal{U}$ of all clopen order filters of T plus $\varnothing$, endowed with the set-theoretical operations of union and intersection; cf. Remark 5.5.1. The *dual* of a morphism $\delta : T \longrightarrow T'$ in **Pr** is the mapping $\varepsilon : \mathcal{U}' \longrightarrow \mathcal{U}$ between the duals $\mathcal{U}'$ and $\mathcal{U}$ of T and T', respectively, determined by

$$\varepsilon(K') = \delta^{-1}(K') \ \ (\forall K' \in \mathcal{U}') \ . \tag{5.5.6}$$

Proposition 5.5.7 (i) *The dual of a Priestly space is a bounded distributive lattice.*

(ii) *The dual of a continuous isotone mapping between two Priestley spaces is a morphism of* **Dis01**.

PROOF: (i) This is a restatement of the first part of Remark 5.5.1.

(ii) First we must prove that $\varepsilon(K') \in \mathcal{U}$ for every $K' \in \mathcal{U}'$. But $\varepsilon(K') = \delta^{-1}(K')$ is clopen by the continuity of δ. Besides, if $x \in \varepsilon(K')$ and $x \leq y$, then $\delta(x) \in K'$ and $\delta(x) \leq \delta(y)$, therefore $\delta(y) \in K'$, hence $y \in \delta^{-1}(K') = \varepsilon(K')$. Moreover, δ^{-1} is a morphism of **Dis01** by Exercise 1.2.5 (i), (iii), (vi), and $\delta^{-1}(T') = T$, because δ is coinjective and cosurjective. □

Corollary 5.5.2 *A contravariant functor* $\Psi : \mathbf{Pr} \longrightarrow \mathbf{Dis01}$ *is obtained by setting* $\Psi T = (\mathcal{U}, \cap, \cup, \varnothing, T)$ *for every* $T \in \mid \mathbf{Pr} \mid$ *and* $(\Psi\delta)(K') = \delta^{-1}(K')$ $(\forall K' \in \mathcal{U}')$ *for every* $\delta \in \mathbf{Pr}(T, T')$.

PROOF: From Proposition 5.5.7 plus routine verification of commutativity of diagrams. □

Like the previous Propositions 5.5.6 and 5.5.7, the subsequent Propositions 5.5.8 and 5.5.9 have obviously their own interest, although they constitute steps towards the proof of the main result: Theorem 5.5.1.

Proposition 5.5.8 *Every bounded distributive lattice* L *is isomorphic to the dual* $\Psi\Phi L$ *of its dual space* ΦL.

PROOF: Taking into account Proposition 5.5.1 and using the same notation, it suffices to show that $\varphi_L(L) = \Psi\Phi L$, in other words that $\varphi_L(L)$ coincides with the set $\mathcal{U}$ of all clopen order filters ofΦ $L = \mathrm{Hom}(L, 2)$. But for every $x \in L, \varphi \quad_L x) = W_1^x \in \mathcal{U}$ by (5.5.1), (5.4.10), (5.5.4) and (5.5.5); therefore $\varphi_L(L) \subseteq \mathcal{U}$. It remains to prove the converse inclusion, i.e., that every $K \in \mathcal{U}$ is of the form $\varphi_L(x)$.

As $\varnothing = \varphi_L(0)$ and $\mathrm{Hom}(L, 2) = \varphi_L(L)$, we may assume $\varnothing \neq K \neq \mathrm{Hom}(L, 2)$. Take an element $f \in K$. Since K is an order filter, it follows that for every $g \in \mathrm{Hom}(L, 2) \setminus K$ we have $f \not\leq g$, hence $f \in W_1^x$ and $g \in W_0^x$ for some $x \in L$, as shown in the proof of Proposition 5.5.6(i). Now the compact setΦ $L \setminus K$ is covered by a family of open sets of the form W_0^x, hence there is a finite covering

$$\Phi L \setminus K \subseteq \bigcup_{i=1}^{m} W_0^{x_i} = \bigcup_{i=1}^{m} \{g \in \mathrm{Hom}(L, 2) \mid g(x_i) = 0\}$$

$$= \bigcup_{i=1}^{m} \{g \in \mathrm{Hom}(L, 2) \mid \neg(g(x_1) = \cdots = g(x_m) = 1)\}$$

$$= \{g \in \mathrm{Hom}(L, 2) \mid g(x_1) \wedge \ldots \wedge g(x_m) = 0\} = W_0^y \, ,$$

where we have set $y = x_1 \wedge \ldots \wedge x_m$. So $h \in \Phi L \setminus K$ implies $h \in W_0^y$, hence $h \notin W_1^y$; on the other hand, $f \in W_1^{x_i}$ $(i = 1, \ldots, m)$ implies $f(y) = f(x_1) \wedge \ldots \wedge f(x_m) = 1$. We have thus proved that for every element $f \in K$ there is an element $y \in L$ such that $f \in W_1^y$. Therefore the compact set K is covered by a family of open sets $W_1^y \subseteq K$, hence there is a finite covering

$$K = \bigcup_{j=1}^{n} W_1^{y_j} = \bigcup_{j=1}^{n} \{f \in \mathrm{Hom}(L, 2) \mid f(y_j) = 1\}$$

$$= \{f \in \mathrm{Hom}(L, 2) \mid f(y_1) \vee \ldots \vee f(y_n) = 1\}$$

$$= \{f \in \mathrm{Hom}(L, 2) \mid f(y_1 \vee \ldots \vee f(y_n) = 1\} = \varphi_L(y_1 \vee \ldots \vee y_n) \, .$$

□

Proposition 5.5.9 *Every Priestley space T is homeomorphic and order isomorphic to the dual $\Phi\Psi T$ of its dual ΨT.*

PROOF: For every $x \in T$ and $K \in \Psi T$, define

$$\psi_T(x)(K) = \begin{cases} 1 & \text{if } x \in K\,, \\ 0 & \text{if } x \notin K\,. \end{cases} \tag{5.5.6}$$

Then it is immediately seen that $\psi_T(x) : \Psi T \longrightarrow 2$ is a morphism in **Dis01**, namely $\psi_T(x) \in \mathrm{Hom}(\Psi T, 2) = \Phi\Psi T$. Therefore $\psi_T : T \longrightarrow \Phi\Psi T$.

To prove injectivity,take $x, y \in T$ with $x \neq y$, say $x \not\leq y$. Then Proposition 5.5.3 shows that $x \in K$ and $y \notin K$ for some $K \in \Psi T$, hence $\psi_T(x)(K) \neq \psi_T(y)(K)$ for that K, therefore $\psi_T(x) \neq \psi_T(y)$.

To prove isotony, take $x, y \in T$ with $x \leq y$. Then for every $K \in \Psi T$, the equality $\psi_T(x)(K) = 1$ implies $x \in K$, hence $y \in K$, therefore $\psi_T(y)(K) = 1$. It follows that $\psi_T(x) \leq \psi_T(y)$.

To prove surjectivity, take $f \in \Phi\Psi T$. Then

$$K_0 = (\bigcap\{K \in \Psi T \mid f(K) = 1\}) \cap \bigcap\{T \setminus K \mid K \in \Psi T \;\&\; f(K) = 0\} \tag{5.5.7}$$

is an intersection of closed sets of the compact space T. If $K_0 = \varnothing$, then Proposition 5.4.3 implies the existence of a finite subfamily having a non-empty intersection, say

$$K_1 \cap \cdots \cap K_m \cap (T \setminus K_{m+1}) \cap \cdots \cap (T \setminus K_n) = \varnothing\,.$$

If $1 \leq m < n$ then $K_1 \cap \cdots \cap K_m \subseteq K_{m+1} \cap \cdots \cap K_n$, thus implying

$$1 = f(K_1) \wedge \ldots \wedge f(K_m) = f(K_1 \cap \cdots \cap K_m)$$

$$\subseteq f(K_{m+1} \cup \cdots \cup K_n) = f(K_{m+1}) \vee \ldots \vee f(K_n) = 0\,,$$

which is a contradiction. If $1 \leq m = n$ we get the contradiction $1 = f(K_1 \cap \cdots \cap K_m) = f(\varnothing) = 0$, and if $m = 0 < n$ the contradiction is $f(T) = f(K_1 \cup \cdots \cup K_n) = 0$. Therefore $K_0 \neq \varnothing$, which means the existence of a point $x \in T$ such that, for every $K \in \Psi T$, if $f(K) = 1$ then $x \in K$, while if $f(K) = 0$ then $x \notin K$. Comparing to (5.5.6), we see that $f(K) = \psi_T(x)(K)$ for every $K \in \Psi T$, that is, $f = \psi_T(x)$.

To prove the isotony of ψ_T^{-1}, take $f, g \in \Phi\Psi T$ with $f \leq g$. Then $x = \psi_T^{-1}(f)$ is the unique element of the set K_0 constructed in (5.5.7) and similarly for $y = \psi_T^{-1}(g)$. To prove $x \leq y$, assume the contrary. Then in view of Proposition 5.5.3, there exists an $\mathcal{U}$-set K such that $x \in K$ and $y \notin K$. But $K \in \Psi T$ and $\psi_T(x)(K) = 1$, while $\psi_T(y)(K) = 0$. This means that $f(K) = 1$ and $g(K) = 0$, which is a contradiction.

In view of Corollary 5.4.2, it only remains to prove that ψ_T is continuous. As indicated by Remarks 5.4.9 and 5.5.2, we take a point $\psi_T(x) \in \Phi\Psi T$ and a neighbourghood $W_{1,\ldots,1,0,\ldots,0}^{U_1,\ldots,U_p,U_{p+1},\ldots,U_r}$ of it, and we must find a neighbourhood of x such that for every y belonging to that neighbourhood, $\psi_T(y)$ should belong to the above neighbourhood of $\psi_T(x)$. Indeed, from $\psi_T(x)(U_i) = 1$ $(i = 1, \ldots, p)$ and $\psi_T(x)(U_j) = 0$ $(j = p + 1. \ldots, r)$, we infer that

$$K = U_1 \cap \cdots \cap U_p \cap (T \setminus U_{p+1}) \cap \cdots \cap (T \setminus U_r)$$

is a neighbourhood of x. For every $y \in K$ we have $\psi_T(y)(U_i) = 1$ $(i = 1, \ldots, p)$ and $\psi_T(y)(U_j) = 0$ $(j = p + 1, \ldots, r)$, which shows that $\psi_T(y)$ belongs to the given neighbourhood of $\psi_T(x)$. □

We now come to the main result.

Theorem 5.5.1 *The functors constructed in Proposition* 5.5.6 *and* 5.5.7 *establish a dual equivalence between the category of bounded distributive lattices and the category of Priestley spaces.*

PROOF: We have to construct a couple of natural isomorphisms $\varphi : I_{\mathbf{Dis01}} \longrightarrow \Psi\Phi$ and $\psi : I_{\mathbf{Pr}} \longrightarrow \Phi\Psi$; cf. Definitions 1.3.14 and 1.3.12. We define φ_L by (5.5.1) for every $L \in |\, \mathbf{Dis01}\, |$, and we define ψ_T by (5.5.6) for every $T \in |\, \mathbf{Pr}\, |$. Then $\varphi_L : L \longrightarrow \Psi\Phi L$ and $\psi_T : T \longrightarrow \Phi\Psi T$ are isomorphisms in **Dis01** and **Pr**, respectively; cf. Propositions 5.5.8 and 5.5.9. It remains to prove the commutativity of two diagrams of the form described in Remark 1.3.9 :

$$\begin{array}{ccc} L & \xrightarrow{\varphi_L} & \Psi\Phi L \\ \downarrow{\scriptstyle\alpha} & & \downarrow{\scriptstyle\Psi\Phi\alpha} \\ L' & \xrightarrow{\varphi_{L'}} & \Psi\Phi L' \end{array} \qquad\qquad \begin{array}{ccc} T & \xrightarrow{\psi_T} & \Phi\Psi T \\ \downarrow{\scriptstyle\delta} & & \downarrow{\scriptstyle\Phi\Psi\delta} \\ T' & \xrightarrow{\psi_{T'}} & \Phi\Psi T' \end{array}$$

Fig. 5.1

Let $\alpha : L \longrightarrow L'$ be a morphism of **Dis01**. Then for every $x \in L$,

$$((\Psi\Phi\alpha) \circ \varphi_L)(x) = (\Psi\Phi\alpha)(\varphi_L(x)) = (\Phi\alpha)^{-1}(\{f \in \mathrm{Hom}(L, 2) \mid f(x) = 1\})$$

$$= \{f' \in \mathrm{Hom}(L', 2) \mid (\Phi\alpha)(f')(x) = 1\} = \{f' \in \mathrm{Hom}(L', 2) \mid (f' \circ \alpha)(x) = 1\}$$

$$= \varphi_{L'}(\alpha(x)) = (\varphi_{L'} \circ \alpha)(x) \ ,$$

thus proving $(\Psi\Phi\alpha) \circ \varphi_L = \varphi_{L'} \circ \alpha$.

Let $\delta : T \longrightarrow T'$ be a morphism of **Pr**. Then for every $x \in T$ and every $K' \in \Psi T'$ we have $(\Phi\Psi\delta) \circ \psi_T : T \longrightarrow \Phi\Psi T'$, therefore $(\Phi\Psi\delta \circ \psi_T)(x) : \Psi T' \longrightarrow 2$, and

$$((\Phi\Psi\delta) \circ \psi_T)(x)(K') = (\Phi\Psi\delta)(\psi_T(x))(K')$$

$$= ((\psi_T(x)) \circ \Psi\delta)(K') = (\psi_T(x))(\delta^{-1}(K')) = F(x),$$

where $F : T \longrightarrow 2$ satisfies $F(x) = 1 \iff x \in \delta^{-1}(K') \iff \delta(x) \in K'$, therefore $F(x) = \psi_{T'}(\delta(x))(K')$. Since K' and x are arbitrary, it follows that $(\Phi\Psi\delta) \circ \psi_T)(x) = (\psi_{T'} \circ \delta)(x)$ and finally $(\Phi\Psi\delta) \circ \psi_T = \psi_{T'} \circ \delta$. □

Exercise 5.5.3 Prove that in Fig. 5.1 we have :

(i) α is injective $\Longleftrightarrow$ $\Psi\Phi\alpha$ is injective ,

(ii) α is surjective $\Longleftrightarrow$ $\Psi\Phi\alpha$ is surjective,

and similarly for δ and $\Phi\Psi\delta$. (Hint. For (ii),$\Longrightarrow$, take $z \in \Psi\Phi L'$, set $b = \varphi_{L'}^{-1}(z)$ and take $a \in L$ such that $\alpha(a) = b$. For (ii),$\Longleftarrow$, take $b \in L'$, $z = \varphi_{L'}(b)$, then y such that $(\Psi\Phi\alpha)(y) = z$, and $a = \varphi_L^{-1}(y)$; prove that $z = \varphi_{L'}(\alpha(a))$).

To obtain a correspondent of Theorem 5.5.1 for Boolean algebras, we need:

Definition 5.5.7 Let **St** be the category of all Stone spaces and all continuous mappings between them.

We have seen in Proposition 5.5.4 that every Stone space is a Priestley space equipped with equality as partial order. Since every function is isotone with respect to the partial order =, it follows from Definitions 5.5.4 and 5.5.7 that **St** is a full subcategory of **Pr**.

Theorem 5.5.2 *The restrictions of the functors* Φ *and* Ψ *in Theorem* 5.5.1 *to the category* **Bol** *of Boolean algebras and the category* **St** *of Stone spaces, respectively, establish a dual equivalence between* **Bol** *and* **St**.

Comment It was Stone who established the bijection between Boolean algebras and totally disconnected compact spaces; cf. Stone [1937].

PROOF: Let B be a Boolean algebra; we have to prove that $\Phi B \in |$ **St** $|$, i.e., that if $f, g \in \mathrm{Hom}(L, 2)$ and $f \le g$, then $f = g$. Take an arbitrary $x \in B$; then $f(x) \le g(x)$ and

$$g(x) \wedge (f(x))' = g(x) \wedge f(x') \le g(x) \wedge g(x') = g(x \wedge x') = 0 ,$$

hence $g(x) \le f(x)$, consequently $f(x) = g(x)$.

Let T be a Stone space; we have to prove that $\Psi T \in |$ **Bol** $|$. But ΨT is the ring of all clopen filters of T plus $\varnothing$. Since every non-empty set is trivially an order filter with respect to =, it follows that ΨT is the ring of all clopen sets, which is in fact a field of sets.

If α is a morphism in the full subcategory **Bol** of **Dis01**, then $\Phi\alpha$ is a morphism in **Pr**, hence it is a morphism in the full subcategory **St** of **Pr**. One proves similarly that if δ is a morphism in **St**, then $\Psi\delta$ is a morphism in **Bol**. □

Theorems 5.5.1 and 5.5.2 enable a transfer of properties between the involved categories. Here is an elementary sample.

Proposition 5.5.10 *The following properties hold for every morphism* α *in* **Dis01** *and every morphism* δ *in* **Pr**:

(i) α $(\Psi\delta)$ *is injective* $\Longleftrightarrow$ $\Phi\alpha$ (δ) *is surjective.*

(ii) α $(\Psi\delta)$ *is surjective* $\Longleftrightarrow$ $\Phi\alpha$ (δ) *is an order isomorphism between its domain and range.*

PROOF: (a) We prove that α injective $\Longrightarrow$ $\Phi\alpha$ surjective.

Assume that α is injective and $\Phi\alpha$ is not surjective. Then $\Phi L \setminus (\Phi\alpha)(\Phi L')$ is a non-empty set by Proposition 4.8.(ii), hence Proposition 5.5.5 implies that

$\varnothing \neq U \cap V \subseteq \Phi(L) \setminus (\Phi\alpha)(\Phi L')$ for some $\mathcal{U}$-set U and some $\mathcal{L}$-set V. The latter inclusion means $(\Phi\alpha)^{-1}(U \cap V) = \varnothing$. Using computation rules with functions f and with f^{-1}, we obtain in turn $(\Phi\alpha)^{-1}(U) \cap (\Phi\alpha)^{-1}(V) = \varnothing$, so that $(\Phi\alpha)^{-1}(U)$ is a subset ofΦ $L' \setminus (\Phi\alpha)^{-1}(V) = (\Phi\alpha)^{-1}(\Phi L \setminus V)$. The latter inclusion can be rewritten in the form $(\Psi\Phi\alpha)(U) \subseteq (\Psi\Phi\alpha)(\Phi L \setminus V$. Using also the injectivity of $\Psi\Phi\alpha$ (cf. Exercise 5.5.3,), we get

$$(\Psi\Phi\alpha)(U \cap (\Phi L \setminus V)) = (\Psi\Phi\alpha)(U) \cap (\Psi\Phi\alpha)(\Phi L \setminus V) = (\Psi\Phi\alpha)(U) ,$$

then $U \cap (\Phi L \setminus V) = U$, which contradicts $U \cap V \neq \varnothing$.

(b) We prove thatΨ δ injective $\Longrightarrow \delta : T \longrightarrow \delta(T)$ is an order isomorphism.

We already know that $x \leq y \Longrightarrow \delta(x) \leq \delta(y)$; now we take $x, y \in T$ with $x \not\leq y$ and must prove that $\delta(x) \not\leq \delta(y)$. Take $U \in \mathcal{U}$ such that $x \in U$ and $y \notin U$. But $U = (\Psi\delta)(U') = \delta^{-1}(U')$ for some $U' \in \Psi T'$. So $\delta(x) \in U'$ and $\delta(y) \notin U'$, hence $\delta(x) \not\leq \delta(y)$ because U' is an order filter.

(c) We prove that δ surjective $\Longrightarrow \Psi\delta$ injective.

Suppose $(\Psi\delta)(H') = (\Psi\delta)(K')$. The reader is urged to prove, passing through $(\Psi\delta)(H')$, that $\delta^{-1}(H' \cap (T' \setminus K')) = \varnothing$. Since δ is surjective, this implies $H' \cap (T' \setminus K') = \varnothing$, that is, $H' \subseteq K'$. Similarly $K' \subseteq H'$.

(d) We prove that δ order isomorphism $\Longrightarrow \Psi\delta$ surjective

Given $U \in \Psi T$, we must find $U' \in \Psi T'$ such that $U = (\Psi\delta)(U') = \delta^{-1}(U')$. Without loss of generality we may assume $\varnothing \neq U \neq T$. Take $x \in U$. For every $y \in T \setminus U$ we have $x \not\leq y$, hence $\delta(x) \not\leq \delta(y)$, therefore $\delta(x) \in U'_{x,y}$ and $\delta(y) \in T \setminus U'_{x,y}$ for some $U'_{x,y} \in \Psi T'$. Consequently $x \in \delta^{-1}(U'_{x,y})$ and $y \in T \setminus \delta^{-1}(U'_{x,y})$. This yields an open covering of $T \setminus U$, then a finite covering

$$T \setminus U \subseteq \textstyle\bigcup_{i=1}^{m}(T \setminus \delta^{-1}(U'_{x,y_i})) = \delta^{-1}(V'_x) , \tag{5.5.8}$$

where $V'_x = \bigcup_{i=1}^{m}(T' \setminus U'_{x,y_i}) \in \mathcal{L}$. On the other hand, $x \in \bigcap_{i=1}^{m} \delta^{-1}(U'_{x,y_i}) = \delta^{-1}(T' \setminus V'_x)$. This yields an open covering of U, hence a finite covering $U \subseteq \bigcup_{j=1}^{n} \delta^{-1}(T' \setminus V'_{x_j}) = \delta^{-1}(U')$, where $U' = \bigcup_{j=1}^{n}(T' \setminus V'_{x_j}) \in \Psi T'$. But $\delta^{-1}(U') = \bigcup_{j=1}^{n}(T \setminus \delta^{-1}(V'_{x_j})) \subseteq U$ by (5.5.8), hence $U = \delta^{-1}(U')$.

The remaining properties from (i) and (ii) are obtained from (a)-(d) *via* Exercise 5.5.3:

$\Psi\delta$ injective $\Longrightarrow$ $\Phi\Psi\delta$ surjective $\Longrightarrow$ δ surjective,
α surjective $\Longrightarrow$ $\Psi\Phi\alpha$ surjective $\Longrightarrow$ $\Phi\alpha$ order isomorphism,
$\Phi\alpha$ surjective $\Longrightarrow$ $\Psi\Phi\alpha$ injective $\Longrightarrow$ α injective,
$\Phi\alpha$ order isomorphism $\Longrightarrow$ $\Psi\Phi\alpha$ surjective $\Longrightarrow$ α surjective. □

Several deep examples showing transfers of properties between the categories **Bol** and **St** can be found in Halmos [1963].

Chapter 6

Applications

We have selected in five chapters what we believe to be the core of sets and order in mathematics. Of course,we are aware that this was an ambitious plan and that our selection could not be infallible. Yet we did our best and we conclude the eBook with several examples which illustrate the applications of lattices to topology (§§1-3), to universal algebra (§4), to a new field of applied mathematics called formal concept analysis (§5) and to logic (§6). Besides the references to the literature given in the sections of this chapter, we recommend a nice paper by Birkhoff [1970], with the suggestive title "What can lattices do for you?".

Keywords: Closure algebra, topological closure, almost discrete space, Heyting algebra, algebraic lattice, equivalence lattice, congruence lattice, formal context, formal concept, Lindenbaum-Tarski algebra, completeness theorem.

6.1 Closure Algebras

In Ch.4,§4 we have studied closure operators at several levels, from posets to complete lattices. We have already mentioned that the latter level generalizes the definition of topological spaces in terms of a family of closed sets or in terms of a closure operator. As a matter of fact, there are two other equivalent definitions of topological spaces, in terms of a family of open sets (much used!) or in terms of an interior operator. In this section we generalize the equivalence of these four definitions of topological spaces to arbitrary Boolean algebras (not necessarily of the form 2^T). The equivalence established in Theorem 6.1.1 mirrors the conventional case of topological spaces and is a sample of the spirit of closure algebras, to which we refer in the end of the section.

We briefly recall that a *closure operator* on a poset P is a unary operation $^-$ which satisfies the properties $x \leq \overline{x}$, $\overline{\overline{x}} = \overline{x}$ and $x \leq y \Longrightarrow \overline{x} \leq \overline{y}$. A *closure system* or a *Moore family* is a subset $C \subseteq P$ such that for every $x \in P$, the set

Sergiu Rudeanu

$C[x) = \{c \in C \mid x \leq c\}$ has least element. According to Theorem 4.4.1, closure operators are in bijection with closure systems via the mappings

$$(6.1.1) \qquad {}^{-} \mapsto C = \{x \in P \mid x = \overline{x}\} ,$$

$$(6.1.2) \qquad C \mapsto x \rightsquigarrow \overline{x} = \text{least element of } C[x) .$$

The dual concepts are: *interior operator*, a unary operation $^\circ$ satisfying $x^\circ \leq x$, $x^{\circ\circ} = x^\circ$, and $x \leq y \Longrightarrow x^\circ \leq y^\circ$; and *interior system*, a subset $D \subseteq P$ such that for every $x \in P$, the set $D(x] = \{d \in D \mid d \leq x\}$ has greatest element. According to the dual of Theorem 4.4.1, interior operators are in bijection with interior systems via the mappings

$$(6.1.3) \qquad {}^\circ \mapsto D = \{x \in P \mid x = x^\circ\} ,$$

$$(6.1.4) \qquad D \mapsto x \rightsquigarrow x^\circ = \text{greatest element of } D(x] .$$

The elements $x = \overline{x}$ are said to be *closed* and the elements $x = x^\circ$ are called *open*.

Now we specialize the discussion to bounded lattices.

We have already noted that $\overline{1} = 1$ (Remark 4.4.4) and dually, $0^\circ = 0$. But the constant closure operator $\overline{x} = 1$ is an example showing that 0 need not be closed, and similarly, 1 need not be open.

To generalize topology, we need concepts stronger than than the above closure and interior.

Definition 6.1.1′ A topological closure operator, or a *Kuratowski closure operator* on a bounded lattice L is a function $^- : L \longrightarrow L$ satisfying $x \leq \overline{x}$, $\overline{\overline{x}} = \overline{x}$, $\overline{x \vee y} = \overline{x} \vee \overline{y}$ and $\overline{0} = 0$.

A Kuratowski closure operator is indeed a closure operator, because $x \leq y \Longrightarrow \overline{x} \vee \overline{y} = \overline{x \vee y} = \overline{y} \Longrightarrow \overline{x} \leq \overline{y}$.

Definition 6.1.2′ A *topological closure system* or a *topological Moore family* on a bounded lattice L is a closure system $C \subseteq L$ closed with respect to finite joins, that is, $0 \in C$ and $x, y \in C \Longrightarrow x \vee y \in C$.

Definition 6.1.1″ A *topological interior operator* on a bounded lattice L is a map $^\circ : L \longrightarrow L$ satisfying $x^\circ \leq x$, $x^{\circ\circ} = x^\circ$, $(x \wedge y)^\circ = x^\circ \wedge y^\circ$ and $1^\circ = 1$.

Definition 6.1.2″ A *topological interior system* on a bounded lattice L is an interior system $D \subseteq L$ closed with respect to finite meets, that is, $1 \in D$ and $x, y \in D \Longrightarrow x \wedge y \in D$.

So, 0 and 1 belong to any topological closure system and any topological interior operator.

Proposition 6.1.1 *1^0 The bijection* (6.1.1), (6.1.2) *between closure operators and closure systems applied to a bounded lattice, restricts to a bijection between topological closure operators and topological closure systems.*

2^0 Dual statement.

PROOF: If $^-$ is a topological closure operator, then the associated closure system C is topological. Indeed, if $\overline{x}, \overline{y} \in C$, then $\overline{x} \vee \overline{y} = \overline{x \vee y} \in C$: besides, $0 \in C$ because $\overline{0} = 0$.

If C is a topological closure system, then the associated closure operator $^-$ is topological. Indeed, from $x, y \leq x \vee y$ we infer that anyway $\overline{x} \vee \overline{y} \leq \overline{x \vee y}$. The converse inequality follows from $x \vee y \leq \overline{x} \vee \overline{y}$ and $\overline{x} \vee \overline{y} \in C$, which imply $\overline{x \vee y} \leq \overline{\overline{x} \vee \overline{y}} = \overline{x} \vee \overline{y}$. Finally $\overline{0} = 0$ because $0 \in C$.

Statement 2^0 and its proof are left to the reader. □

Summarizing, Theorem 4.4.1, Proposition 6.1.1.1^0 and the dual of Theorem 4.4.1, Proposition 6.1.1.2^0 provide four bijections which can be summarized in compact form:

(topological) closure operators $\longleftrightarrow$ (topological) closure systems ,

(topological) interior) operators $\longleftrightarrow$ (topological) interior systems .

Propositions 6.1.2 and 6.1.3 below provide the four missing "vertical" bijections in the above table, but only for Boolean algebras.

Proposition 6.1.2 *In a Boolean algebra* $(B, \wedge, \vee, ', 0, 1)$ *there is a bijection between (topological) closure operators and (topological) closure systems, via the mappings*

$$^- \mapsto \, '^{-}{}' \, , \tag{6.1.5}$$

$$^\circ \mapsto \, '^{\circ}{}' \, . \tag{6.1.6}$$

PROOF: Let us write x^- instead of $\overline{x}$.

If $^-$ is a closure operator, then from $x' \leq x'^-$ we infer $x'^{-\,\prime} \leq x'' = x$. Further, $x'^{-\,\prime\prime-\,\prime} = x'^{--\,\prime} = x'^{-\,\prime}$ and $x \leq y \Longrightarrow y' \leq x' \Longrightarrow y'^- \leq x'^- \Longrightarrow x'^{-\,\prime} \leq y'^{-\,\prime}$. Therefore $'^{-}{}'$ is an interior operator and one proves similarly that if $^\circ$ is an interior operator, then $'^{\circ}{}'$ is a closure operator.

To prove that (6.1.5) and (6.1.6) are inverse to each other, we compute

$$x''^{-\,\prime\prime} = x''^{-\prime\prime} = x^- \text{ and } x''^{\circ\,\prime\prime} = x''^{\circ\prime\prime} = x^\circ \, .$$

To prove that (6.1.5) and (6.1.6) are inverse to each other, we compute

$$((x')'^{-\,\prime})' = x''^{-\prime\prime} = x^- \text{ and } ((x')'^{\circ\,\prime})' = x''^{\circ\prime\prime} = x^\circ \, .$$

If $^-$ is a topological closure operator then the interior operator $'^{\circ}{}'$ satisfies $1'^{\circ\,\prime} = 0^{-\prime} = 0' = 1$ and $(x \wedge y)'^{-\,\prime} = (x' \vee y')^{-\prime} = (x'^- \vee y'^-)' = x'^{-\,\prime} \wedge y'^{-\,\prime}$, therefore $'^{-}{}'$ is a topological interior operator. One proves similarly that if $'^{\circ}{}'$ is a topological interior operator, then $'^{-}{}'$ is a topological closure operator. □

To go further, for every subset X of a Boolean algebra B, set

$$X' = \{x' \mid x \in X\} \, . \tag{6.1.7}$$

Proposition 6.1.3 *In a Boolean algebra* $(B, \wedge, \vee, ', 0, 1)$ *there is a bijection between (topological) closure systems and (topological) interior systems, via the mappings*

(6.1.8) $$C \mapsto C' ,$$

(6.1.9) $$D \mapsto D' .$$

PROOF: For the bijection between closure systems and interior systems, suppose first that C is a closure system. Then

$$C' = \{x' \in B \mid x \in C\} = \{x' \in B \mid x = x^-\}$$

$$= \{x \in B \mid x' = x'^-\} = \{x \in B \mid x = x'^{-\prime}\} ,$$

which is the interior system corresponding to $'^{-}{}'$. Similarly, if D is an interior system, then D' is the closure system corresponding to $'^{\circ}{}'$. Therefore the mappings (6.1.8) and (6.1.9) are correctly defined, and since $X'' = X$, they are inverse to each other.

In view of Proposition 6.1.2, the above bijection can be restricted to a bijection between topological closure systems and topological interior systems.

□

We have thus obtained:

Theorem 6.1.1 *In a Boolean algebra there is a bijection between the structures* $(B,^-)$, $(B,^- \circ)$, (B, C) *and* (B, D), *as described in Propositions* 6.1.1, 6.1.2 *and* 6.1.3.

We have thus generalized the equivalence of four (out of five) well-known definition of topological spaces to the level of arbitrary Boolean algebras.

As a matter of fact, Boolean algebras equipped with a Kuratowski closure operator, introduced by McKinsey and Tarski [1944] and called *closure algebras*, are studied by many authors. Some of them prefer the dual structure of a *topological Boolean algebra*, meaning a Boolean algebra endowed with a topological interior operator. Rasiowa and Sikorski [1963] have a chapter on topological Boolean algebras, and Sikorski [1964] provides references to the literature of closure algebras, among which Nöbeling [1954].

In §6.6 we will mention the application of closure algebras in mathematical logic.

6.2 Posets and Almost Discrete T_0 Spaces

The connection between posets and topological spaces T_0 goes back at least to the 1961 edition of Birkhoff's book [1967]. Vaidyanathaswamy [1960], Exercise V.18 [1] noted that the spaces occurring in this connection satisfy the property that arbitrary unions of closed sets are closed. Gierz et al. [1980] lifted the connection to a functor from the category of posets to the category of T_0 spaces.

[1] The first 1948 edition of Birkhoff's book and the first 1947 edition of Vaidyanathaswamy's book are not available to us.

In this section we combine the above results and obtain an isomorphism between the category of posets and a full subcategory of the category of T_0 spaces.

A few prerequisites are necessary.

Exercise 6.2.1 If T is a topological space, then for every $x, y \in T$ the following conditions are equivalent (where we have noted $\overline{x}$ instead of $\overline{\{x\}}$):

(i) $\overline{x} \subseteq \overline{y}$;

(ii) $x \in \overline{y}$;

(iii) $x \in U \Longrightarrow y \in U$ for every open set U.

(Hint. For (iii) use Proposition 5.4.2 and (5.4.1)).

We recall that axiom T_0 (which is weaker than T_2) says that for every two distinct points, one of them has a neighbourhood which does not contain the other point.

Exercise 6.2.2 Axiom T_0 is equivalent to the following condition: for every x, y, $\overline{x} = \overline{y} \Longrightarrow x = y$. (Hint. Use Exercise 6.2.1. If $\overline{x} = \overline{y}$ for some x, y with $x \neq y$, then T_0 fails. If $x \neq y$ in a T_0 space then $\overline{x} \neq \overline{y}$.)

Definition 6.2.1 By an *almost discrete space* we mean a topological space in which the closed sets form a complete ring of sets, that is, all intersections and unions of closed sets are closed (cf. Definition 5.3.2 and subsequent explanations).

This is precisely the property mentioned by Vaidyanathaswamy[2]. We have noted in Example 5.3.1 that even the stronger property of being a complete field of sets is weaker than being the field 2^T.

Proposition 6.2.1 (i) *A* T_0 *space* $(T, ^-)$ *is made into a poset* $(T, \leq)$ *by setting*

$$x \leq y \Longleftrightarrow x \in \overline{y} \quad (\forall x, y \in T) . \tag{6.2.1}$$

(ii) *A poset* $(P, \leq)$ *is made into an almost discrete* T_0 *space* $(P, ^-)$ *by setting*

$$\overline{X} = \{t \in P \mid t \leq x \text{ for some } x \in X\} \quad (\forall X \subseteq P) . \tag{6.2.2}$$

PROOF: (i) The reflexivity of (6.2.1) is clear. In the following we repeatedly use Proposition 5.4.2. If $x \leq y$ and $y \leq z$ then $z \in \overline{x} \Longleftrightarrow z \in \overline{y}$ by Proposition 5.4.2 and Exercise 6.2.1(iii), hence $x = y$ by Exercise 6.2.2. Finally suppose $x \leq y$ and $y \leq z$; to prove $x \leq z$ we take an arbitrary neighbourhood V of x and must show that $z \in V$. According to (5.4.1), there is an open neighbourhood G of x which is included in V. It follows that $y \in G$ and G is a neighbourhood of y as well, therefore $z \in G$, hence $z \in V$.

---

[2] Vaidyanathaswamy [1970], 13.11 quite improperly refers to such topologies as discrete topologies.

(ii) Clearly $\overline{\varnothing} = \varnothing$, $X \subseteq \overline{X}$; in particular $\overline{X} \subseteq \overline{\overline{X}}$. If $t \in \overline{\overline{X}}$ then $t \leq y$ for some $y \in \overline{X}$. But $y \leq x$ for some $x \in X$, hence $t \leq x$, therefore $t \in \overline{X}$. Thus $\overline{\overline{X}} \subseteq \overline{X}$, therefore $\overline{X} = \overline{\overline{X}}$. Furthermore,

$$t \in \overline{\bigcup_{i \in I} X_i} \Longleftrightarrow \exists x \in \bigcup_{i \in I} X_i \;\; t \leq x \Longleftrightarrow \exists i \in I \; \exists x \in X_i \;\; t \leq x \Longleftrightarrow t \in \bigcup_{i \in I} \overline{X_i} \, .$$

So $\overline{\bigcup X_i} = \bigcup \overline{X_i}$; in particular $\overline{X \cup Y} = \overline{X} \cup \overline{Y}$. We have thus proved that $(P, ^-)$ is a topological space in which arbitrary joins of closed sets are closed; since arbitrary intersections of closed sets are anyway closed, it follows that $(P, ^-)$ is an almost discrete topological space. Finally (6.2.2) implies that $\overline{x} = (x]$, hence $\overline{x} = \overline{y} \Longleftrightarrow (x] = (y] \Longleftrightarrow x = y$, therefore the space P is T_0 by Exercise 6.2.2.

□

Let $\mathbf{Top}_0$ denote the full subcategory of **Top** whose objects are the T_0 spaces. Let further $\mathbf{ADT}_0$ denote the full subcategory of $\mathbf{Top}_0$ whose objects are the almost discrete T_0 spaces.

Proposition 6.2.2 (i) *A functor* $\Phi : \mathbf{Top}_0 \longrightarrow \mathbf{Pos}$ *is constructed by the prescription:* $\Phi(T, ^-) = (T, \leq)$ *defined by* (6.2.1), *and* $\Phi f = f$.

(ii) *A functor* $\Psi : \mathbf{Pos} \longrightarrow \mathbf{Top}_0$ *is constructed by the prescription:* $\Psi(P, \leq) = (P, ^-)$ *defined by* (6.2.2), *and* $\Psi g = g$.

(iii) $\Phi\Psi = I_{\mathbf{Pos}}$ *and if* $(T, ^-) \in \mid \mathbf{ADT}_0 \mid$ *then* $\Psi\Phi(T, ^-) = (T, ^-)$.

PROOF: (i) and (ii). Having in view Proposition 6.2.1, it remains to prove that Φ and Ψ are correctly defined on morphisms. The verification of commutativity of diagrams is left to the reader, as well as the situations in which we take an element in a certain set, when the proof must be completed with the case when that set is empty, unless that set cannot be empty.

(i) Let $f : (T, ^-) \longrightarrow (T', \widehat{\ })$ be a continuous function. We must prove that $f : (T, \leq) \longrightarrow (T, \leq')$ is isotone. Recall first that according to Proposition 5.4.7, for every $x \in T$ and every neighbourhood V' of $f(x)$, there is a neighboourhood V of x such that $f(V) \subseteq V'$. Now we use (6.2.1) in $(T, \leq)$ and $(T', \leq')$. If $x \leq y$, that is, $x \in \overline{y}$, we have $y \in V$ by Proposition 5.4.2, hence $f(y) \in V'$, but since V' was arbitrary, the same proposition implies $f(x) \in \widehat{f(y)}$, that is, $f(x) \leq' f(y)$.

(ii) Let $g : (P, \leq) \longrightarrow (P', \leq')$ be an isotone function. We must prove that $g : (P, ^-) \longrightarrow (P', \widehat{\ })$ is continuous. According to Satz 1.4.3 in Preuß[1972], this happens iff $g(\overline{X}) \leq' \widehat{g(X)}$. Indeed, take $y' \in g(\overline{X})$. Then $y' = g(y)$ for some $y \in \overline{X}$, hence by (6.2.2) $y \leq x$ for some $x \in X$. Therefore $y' = g(y) \leq' g(x) \in g(X)$, so that (6.2.2) applied to $(P', \widehat{\ })$ shows that $y' \in \widehat{g(X)}$.

(iii) It remains to prove the desired equalities on objects.

Indeed, $\Psi(P, \leq) = (P, ^-)$, where (6.2.2) implies that $t \in \overline{x} \Longleftrightarrow t \leq x$. Further, $\Phi(P, ^-) = (P, \leq')$ where $t \leq' x \Longleftrightarrow t \in \overline{x}$ by (6.2.1). Therefore $\Phi\Psi(P, \leq) = (P, \leq')$, where $t \leq' x \Longleftrightarrow t \in \overline{x} \Longleftrightarrow t \leq x$, that is, $\leq'$ coincides with $\leq$, so that $\Phi\Psi(P, \leq) = (P, \leq)$.

If $(T,^{-})$ is an almost discrete T_0 space, then $\Psi\Phi(T,^{-}) = \Psi(T,\leq) = (T,\widehat{\ })$, where

$$t \in \widehat{X} \Longleftrightarrow \exists x \in X\ t \leq x \Longleftrightarrow \exists x\ x \in X \,\&\, t \in \overline{x} \Longleftrightarrow t \in \bigcup_{x\in X} \overline{x} = \overline{\bigcup_{x\in X} \{x\}} = \overline{X}\ .$$

□

Theorem 6.2.1 *The restriction* $\Phi : \mathbf{ADT_0} \longrightarrow \mathbf{Pos}$ *and the corestriction* $\Psi : \mathbf{Pos} \longrightarrow \mathbf{ADT_0}$ *of the functors constructed in Proposition* 6.2.2 *establish an isomorphism between the categories* $\mathbf{ADT_0}$ *and* $\mathbf{Pos}$.

PROOF: In view of Proposition 6.2.1, the corestriction of Ψ to $\mathbf{ADT_0}$ exists. Now part (iii) of Proposition 6.2.2 reads $\Phi\Psi = I_{\mathbf{Pos}}$ and $\Psi\Phi = I_{\mathbf{ADT_0}}$. □

6.3 Heyting Algebras in Topology

The lattices known as Heyting algebras have been invented in the context of non-classical logics, then it turned out that they were equally important to topology. In this section we try to sketch the topological interpretation of Heyting algebras, while in §6.6 there will be just a few words about their rôle in logic.

A lattice $(L,\wedge,\vee)$ is said to be *relatively pseudocomplemented* if for every $a,b \in L$, the set $\{x \in L \mid a \wedge x \leq b\}$ has greatest element, usually denoted by $a \to b$ and called the *relative pseudocomplement* of a with respect to b. In other words,

$$a \wedge x \leq b \Longleftrightarrow x \leq a \to b\ . \tag{6.3.1}$$

This implies that L has greatest element $1 = a \to a$. A relatively pseudocomplemented lattice with least element 0 is said to be a *Heyting algebra*. Rasiowa and Sikorski prefer the term *pseudo-Boolean algebra*, because every Boolean algebra satisfies (6.3.1) with $a \to b = a' \vee b$. The dual of a Heyting algebra is sometimes called a *Brouwer algebra*. Birkhoff uses the terms *Brouwerian lattice* and *dually Brouwerian lattice* for Heyting algebras and Brouwer algebras, respectively.

Proposition 6.3.1 *Every relatively pseudocomplemented lattice is distributive*

PROOF: Given a,b,c, set $x = (a \wedge b) \vee (a \wedge c)$. Then $a \wedge b \leq x$ and $a \wedge c \leq x$, hence $b \leq a \to x$ and $c \leq a \to x$, therefore $a \wedge (b \vee c) \leq a \wedge (a \to x) \leq x$ (for the last inequality take $b := x$ and $x := a \to x$ in (6.3.1)). Since $x \leq a \wedge (b \vee c)$ in any lattice, we have $a \wedge (b \vee c) = x$. □

Proposition 6.3.2 *Every complete relatively pseudocomplemented lattice satisfies the infinite distributivity law*

$$a \wedge \bigvee_{i\in I} x_i = \bigvee_{i\in I}(a \wedge x_i) \tag{6.3.2}$$

and conversely, every complete lattice which satisfies (6.3.2) *is relatively pseudocomplemented.*

Comment The dual of identity (6.3.2) need not hold.

PROOF: For the first statement the proof is similar to that of Proposition 6.3.1. Conversely, assume L is a complete lattice satisfying (6.3.2) and take $a, b \in L$. Setting $X = \{x \in L \mid a \wedge x \leq b\}$ and $\bigvee X = c$, let us prove that $c = a \rightarrow b$. Firstly, $X \neq \varnothing$ because $b \in X$, and for every $x \in X$ we have $x \leq c$. Besides, $a \wedge c = \bigvee_{x \in X}(a \wedge x) \leq b$, showing that $c \in X$. Therefore c is the greatest element of X. □

Corollary 6.3.1 *Every complete Heyting algebra satisfies* (6.3.2).

Suppose L is a lattice with 0. If $a \in L$ satisfies the property that the nonempty set $X = \{x \in L \mid a \wedge x = 0\}$ has greatest element a^*, then a^* is called the *pseudocomplement* of a. If every element of L has a (necessarily unique!) pseudocomplement, then L is said to be *pseudocomplemented.*

Proposition 6.3.3 *Every Heyting algebra is pseudocomplemented, with*

$$a^* = a \rightarrow 0\,. \tag{6.3.3}$$

The map * *is antitone and* ** *is a closure operator.*

PROOF: The first claim is immediate. If $x \leq y$, then from $y \wedge y^* = 0$ we infer $x \wedge y^* = 0$, hence $y^* \leq x^*$. Besides, $x \leq y^* \Longrightarrow x \wedge y = 0 \Longrightarrow y \leq x^*$, therefore $y \leq x^* \Longrightarrow x \leq y^*$. We have thus proved that $(^*, ^*)$ is a Galois connection, which implies that ** is a closure operator by Theorem 4.5.1. □

Theorem 6.3.1 *The family* $\mathcal{C}\ell(T)$ *of closed sets of a topological space* T *is a complete Heyting algebra, with the relative pseudocomplement* $A \rightarrow B = (\mathbf{C}A \cup B)^0$ *and the pseudocomplement* $A^* = (\mathbf{C}A)^0$.

PROOF: We already know that $\mathcal{C}\ell(T)$ is a Moore family, hence a complete lattice by Corollary 4.4.4. Now let $A, B, X \in \mathcal{C}\ell(T)$. It is well known that in the Boolean algebra $(2^T, \cap, \cup, \mathbf{C}, \varnothing, T)$ we have $A \cap X \subseteq B \Longleftrightarrow X \subseteq \mathbf{C}A \cup B$. But for an arbitrary subset $Y \subseteq T$, the largest set $X \in \mathcal{C}\ell(T)$ satisfying $X \subseteq Y$ is the interior Y^0 of Y. Therefore in $\mathcal{C}\ell(T)$ we have

$$A \cap X \subseteq B \Longleftrightarrow X \subseteq (\mathbf{C}A \cup B)^0\,.$$

In particular $A^* = A \rightarrow \varnothing = (\mathbf{C}A)^0$. □

We can say a bit more if the topological space satisfies the separation axiom T_1.

Axiom T_1, stronger than T_0 but still weaker than T_2, says that if x and y are distinct points, then each of them has a neighbourhood which does not contain the other point.

This axiom is equivalent to the property that every singleton is a closed set. Indeed, if x is an arbitrary point of a T_1 space T, then for every $y \neq x$ there is a neighbourhood of y included in $T \setminus \{x\}$, therefore $T \setminus \{x\}$ is an open set. Conversely, if every singleton of T is closed, then if x and y are distinct points of T, it follows that $T \setminus \{y\}$ and $T \setminus \{x\}$ are neighbourhoods as required in axiom T_1.

In T_1 spaces Theorem 6.3.1. can be strengthened.

We need the following definition. A complete lattice is called *algebraic* if every non-zero element is a join of compact elements (cf. Definition 4.3.9). The motivation for this name will appear in the next Section.

Theorem 6.3.2 *In every topological space satisfying* T_1, *the lattice* $\mathcal{C}\ell(T)$ *of closed sets is an atomic algebraic Heyting lattice. The topological space is determined by* $\mathcal{C}\ell(T)$ *up to a homeomorphism, via formula*

$$\overline{X} = \bigvee_{x \in X} \{x\} \quad (\forall X \subseteq T)\,. \tag{6.3.4}$$

PROOF: We already know that $\mathcal{C}\ell(T)$ is a complete Heyting lattice by Theorem 6.3.1 and we have noted that every singleton $\{x\}$ is in $\mathcal{C}\ell(T)$. Then the following facts are obvious: the singletons coincide with the atoms (cf. Definition 5.2.2) of $\mathcal{C}\ell(T)$, the lattice $\mathcal{C}\ell(T)$ is atomic (cf. Definition 5.2.3), and the atoms are compact (cf. Definition 4.3.9). It remains to prove (6.3.4), which implies both the fact that $\mathcal{C}\ell(T)$ is atomic and the uniqueness of the corresponding topological space T.

Indeed, $\bigvee_{x \in X} \in \mathcal{C}\ell(T)$. Further,

$$x \in X \Longrightarrow \{x\} \subseteq \bigvee_{x \in X} \{x\} \Longrightarrow x \in \bigvee_{x \in X} \{x\}$$

therefore $X \subseteq \bigvee_{x \in X}\{x\}$. Finally, if $F \in \mathcal{C}\ell(T)$ and $X \subseteq F$, then $x \in X \Longrightarrow x \in F \Longrightarrow \{x\} \subseteq F$, therefore $\bigvee_{x \in X} \subseteq F$. □

As to the family $\mathcal{O}(T)$ of open sets of a topological space T, it is a bounded distributive lattice closed with respect to arbitrary unions. Note that every open set U satisfies $(\overline{U})^0 \subseteq U$. An open set U is called *regular* if $U = (\overline{U})^0$. A well-known theorem says that *the family of regular open sets of a topological space is a complete Boolean algebra.*

We conclude this section by observing that topology derives from geometry and we refer the reader e.g. to Birkhoff [1967], Ch. IV, "Geometric lattices", for a survey of lattice-theoretical methods applied to geometry.

6.4 Algebraic Lattices

In the introduction to Chapter 4 we have already referred to universal algebra, which owes much to lattice theory. This idea is developed in a lecture given by Grätzer [1970], in which he emphasizes that in fact the connection between the two fields is reciprocal: there are many results in lattice theory inspired by universal algebra.

In this Section we follow the first line indicated above and we concentrate on algebraic lattices, already defined in the previous Section. Due to their basic rôle in universal algebra, the name "algebraic" will appear to be perfectly justified.

Now we are going to introduce the closely related concept of an algebraic closure operator, which however needs one more auxiliary concept, which is a particular case of Definition 4.3.10:

A family $\mathcal{F}$ of subsets of a set A is said to be *directed*, if for every $A, B \in \mathcal{F}$ there is $C \in \mathcal{F}$ such that $A, B \subseteq C$.

Lemma 6.4.1 *If $(X_i)_{i\in I}$ is a directed family of subsets of a set A, and a finite set Y satisfies $Y \subseteq \bigcup_{i\in I} X_i$, then $Y \subseteq X_i$ for some $i \in I$.*

PROOF: If $Y = \{y_1, \ldots, y_n\}$ and $y_k \in X_{i_k}$ $(k = 1, \ldots, n)$, then there is $i \in I$ such that $X_{i_k} \subseteq X_i$ $(k = 1, \ldots, n)$, hence $Y \subseteq Y_i$. □

Theorem 6.4.1 *Let $^-$ be a closure operator on a set 2^A and $\mathcal{C}$ the associated closure system. Then the following conditions are equivalent:*

(i) $\overline{X} = \bigcup\{\overline{Y} \mid Y \subseteq X,\ Y \text{ finite}\}$ $(\forall X \subseteq A)$;

(ii) *for every directed family $(X_i)_{i\in I}$ of subsets of A, $\overline{\bigcup_{i\in I}} = \bigcup_{i\in I} \overline{X_i}$;*

(iii) *if $(F_i)_{i\in I} \subseteq \mathcal{C}$ is a directed family, then $\bigcup_{i\in I} F_i \in \mathcal{C}$.*

PROOF: (Davey and Priestley [1990]) (i)$\Longrightarrow$(ii): Using (4.4.23) and Lemma 6.4.1, we compute

$$\bigcup_{i\in I} \overline{X_i} \subseteq \overline{\bigcup_{i\in I} X_i} = \bigcup\{\overline{Y} \mid Y \subseteq \bigcup_{i\in I} X_i,\ Y \text{ finite}\}$$

$$= \bigcup\{\overline{Y} \mid Y \text{ finite and } \exists\, i \in I\ Y \subseteq X_i\} = \bigcup_{i\in I}\{\overline{Y} \mid Y \text{ finite and } Y \subseteq X_i\} \subseteq \bigcup_{i\in I} \overline{X_i}\ .$$

(ii)$\Longrightarrow$(iii) : Taking $X_i := F_i$ we get $\overline{\bigcup_{i\in I} F_i} = \bigcup_{i\in I} F_i$.

(iii)$\Longrightarrow$(i) : Given $X \subseteq A$, set $\mathcal{X} = \{\overline{Y} \mid Y \subseteq X,\ X \text{ finite}\}$. Take two members of $\mathcal{X}$, say $\overline{Y_1}, \overline{Y_2}$, where Y_1 and Y_2 are finite subsets of X. Then $\overline{Y_1 \cup Y_2} \in \mathcal{X}$ and $\overline{Y_1}, \overline{Y_2} \subseteq \overline{Y_1 \cup Y_2}$, which proves that $\mathcal{X}$ is a directed family of closed sets, therefore $\bigcup \mathcal{X} \in \mathcal{C}$. For each $x \in X$ we have $\overline{\{x\}} \in \mathcal{X}$, hence $\overline{\{x\}} \subseteq \bigcup \mathcal{X}$, therefore $X \subseteq \bigcup_{x\in X} \overline{\{x\}} \subseteq \bigcup \mathcal{X}$. It follows that

$$\overline{X} \subseteq \overline{\bigcup \mathcal{X}} = \bigcup \mathcal{X} = \bigcup\{\overline{Y} \mid Y \subseteq X,\ Y \text{ finite}\} \subseteq \overline{X}\ .$$

□

Definition 6.4.1 A closure operator on a set 2^A is called an *algebraic closure operator* if it satisfies the equivalent conditions in Theorem 6.4.1.

Recall that an *algebraic lattice* is a complete lattice in which every non-zero element is a join of compact elements (cf. §6.3).

Theorem 6.4.2 *If $\mathcal{C}$ is an algebraic closure operator, then $(\mathcal{C}, \subseteq)$ is an algebraic lattice in which the compact elements are the closures of the finite sets.*

PROOF: It follows from Proposition 4.4.3 applied to $(2^A, \subseteq)$, that $(\mathcal{C}, \subseteq)$ is a complete lattice in which the complete meet operation is set-theoretical intersection, while Proposition 4.4.4 implies that $\bigvee_{\mathcal{C}} X = \overline{\bigcup X}$. Having also in view condition (i) from Theorem 6.4.1, it remains to prove the equivalence : closure of a finite set $\Longleftrightarrow$ compact in $\mathcal{C}$.

The empty set $\varnothing$ is finite by definition and $\overline{\varnothing}$ is a compact element of $\mathcal{C}$ because $\overline{\varnothing} \subseteq F$ for every F. Now take $X = \{x_1, \ldots, x_n\}$ and let $\overline{X} \subseteq \bigvee_{i \in I} F_i$ be a covering of $\overline{X}$, hence of X, in $\mathcal{C}$. Then

$$X \subseteq \overline{\bigcup_{i \in I} F_i} = \bigcup \{\overline{Y} \mid Y \subseteq \bigcup_{i \in I} F_i,\ Y \text{ finite}\} .$$

It follows that for each $k = 1, \ldots, n$ there is a finite subset $Y_k \subseteq \bigcup_{i \in I} F_i$ such that $x_k \in \overline{Y_k}$. Further, for each $k = 1, \ldots, n$ there is a finite subset $I_k \subseteq I$ such that $Y_k \subseteq \bigcup_{i \in I_k} F_i$. Set $J = I_1 \cup \cdots \cup I_n$ and $Y = Y_1 \cup \cdots \cup Y_n$. Then the finite subsets $J \subseteq I$ and $Y \subseteq \bigcup_{j \in J} F_j$ satisfy

$$X \subseteq \overline{Y_1} \cup \cdots \cup \overline{Y_n} \subseteq \overline{Y} \subseteq \overline{\bigcup_{j \in J} F_j} = \bigvee_{j \in J} F_j ,$$

therefore $\overline{X} \subseteq \bigvee_{j \in J} F_j$, which is a finite covering.

Conversely, let F be a compact element of $\mathcal{C}$. Then, since

$$F = \overline{F} = \bigcup \{\overline{Y} \mid Y \subseteq F,\ Y \text{ finite}\} \subseteq \bigvee \{\overline{Y} \mid Y \subseteq F,\ Y \text{ finite}\} ,$$

there exists a finite family $\{Y_1, \ldots, Y_p\}$ of finite subsets $Y_h \subseteq F$ $(h = 1, \ldots, p)$ such that $F \subseteq \overline{Y_1} \vee \ldots \vee \overline{Y_p}$. Set $Y = \bigcup_{h=1}^{p} Y_h$. Since Y is a subset of F, it follows that $\overline{Y} \subseteq F$. On the other hand, from $Y_h \subseteq Y$ $(h = 1, \ldots, p)$, we infer that $\overline{Y_h} \subseteq \overline{Y}$ $(h = 1, \ldots, p)$, therefore $F \subseteq \overline{Y}$. So $F = \overline{Y}$, where Y is a finite set. □

As a first application, let us prove:

Proposition 6.4.1 *The family* Eq(A) *of equivalence relations of a set A is an algebraic lattice*

PROOF: Clearly Eq(A) is a Moore family. In view of Theorem 6.4.2, it remains to prove that the associated closure operator is algebraic. We check condition (iii) in Theorem 6.4.1. Let $(\theta_i)_{i \in I}$ be a directed family of equivalence relations and $\theta = \bigcup_{i \in I} \theta_i$. The reflexivity and symmetry of θ are immediate. To prove transitivity, suppose $x\theta y$ and $y\theta z$. Then $x\theta_i y$ and $y\theta_j z$ for some $i, j \in I$; but $\theta_i, \theta_j \subseteq \theta_k$ for some $k \in I$, therefore $x\theta_k y$ and $y\theta_k z$, which imply $x\theta_k z$ and finally $x\theta z$. □

Now it is time to briefly and roughly recall a few basic concepts of universal algebra.

We begin with a particular case. An *algebra* of *type* (2,0) is a couple (A, F), where A is a non-empty set and $F = (\circ, e)$, where $\circ$ is a binary operation $\circ : A^2 \longrightarrow A$ and e is a nullary operation $e \in A$. Another algebra of the same type is written (A', F'), where $F' = (\circ', e')$ with $\circ' : A'^2 \longrightarrow A'$ and $e' \in A'$. Roughly speaking, the *signature* of these algebras is the "abstract" sequence $(\circ, e)$, regardless of a specific algebra. If $n_1, \ldots, n_p$ are natural numbers, an *algebra of type* $(n_1, \ldots, n_p)$ is a pair (A, F), where $F = (f_1, \ldots, f_p)$, with $f_i : A^{n_i} \longrightarrow A$ $(i = 1, \ldots, p)$; possibly some $n_i = 0$, i.e., $f_i \in A$. When no confusion is possible an algebra $\mathcal{A} = (A, F)$ is denoted simply by its *carrier* or *support* A.

A *closed subset* of an algebra (A, F), usually and rather improperly called a *subalgebra* of A, is a subset $S \subseteq A$ such that for every $i \in \{1, \ldots, p\}$, if $x_1, \ldots, x_{n_i} \in S$, then $f_i(x_1, \ldots, x_{n_i}) \in S$. The subalgebra $< X >$ *generated* by a subset $X \subseteq A$ is the smallest subalgebra which includes X. According to the theory of closure operators, $< X >$ is the set-theoretical intersection of all subalgebras that include X; yet in many situations we may be interested in finding an intrinsic characterization of $< X >$.

A *congruence* of the algebra (A, F) is an equivalence relation $\sim$ on A such that for every $i \in \{1, \ldots, p\}$,

$$x_k \sim y_k \ (k = 1, \ldots, n_i) \Longrightarrow f_i(x_1, \ldots, x_{n_i}) \sim f_i(y_1, \ldots, y_{n_i}) \ .$$

If $\sim$ is a congruence, the *quotient algebra* $(A/\sim, [F])$ is the algebra defined on the quotient set $A/\sim$ by the functions

$$[f_i]([x_1], \ldots, [x_{n_i}]) = [f_i(x_1, \ldots, x_{n_i})] \quad (i = 1, \ldots, p) \ ,$$

where $[x]$ denotes the class of x modulo $\sim$; however one may write simply f_i instead of $[f_i]$.

A *homomorphism* between two *similar* algebras (i.e., algebras of the same type) (A, F) and (A', F') is a function $h : A \longrightarrow A'$ such that the following identities hold:

$$h(f_i(x_1, \ldots, x_{n_i})) = f'_i(h(x_1), \ldots, h(x_{n_i})) \quad (i = 1, \ldots, p) \ .$$

Recall that an *isomorphism* in a category is a morphism $f : A \longrightarrow B$ such that there exists a morphism $g : B \longrightarrow A$ satisfying $g \circ f = 1_A$ and $f \circ g = 1_B$ (cf. Definition 1.3.4). One proves easily, as in Example 1.3.27, the more general property that *every bijective homomorphism between two algebras is an isomorphism*.

The *direct product* of a family $(A_i, F_i)_{i \in I}$ of similar algebras is the direct product $\prod_{i \in I} A_i$ of their support sets endowed with an algebraic structure which we illustrate for a binary operation $\circ \in F$: if the corresponding operation of A_i is denoted by $\circ_i$, then the operation $\circ$ corresponding to the direct product is defined by $(x_i)_{i \in I} \circ (y_i)_{i \in I} = (x_i \circ y_i)_{i \in I}$.

The following analogue of Proposition 6.4.1 is important and has a quite similar proof.

Proposition 6.4.2 *The family* Sub(A) *of subalgebras of an algebra* A *is an algebraic lattice.*

PROOF: Clearly Sub(A) is a Moore family. We check that the associated closure operator $<\ >$ verifies condition (iii) in Theorem 6.4.1. We must prove that if $(S_i)_{i\in I}$ is a directed family of subalgebras, then $S = \bigcup_{i\in I} S_i$ is a subalgebra. Indeed, let $f : A^n \longrightarrow A$ be an operation in the signature of A. If $x_1, \ldots, x_n \in S$, then then for each $h \in \{1, \ldots, n\}$ there exists $i_h \in I$ such that $x_h \in S_{i_h}$. On the other hand there exists $j \in I$ such that $S_{i_h} \subseteq S_j$ $(h = 1, \ldots, n)$. It follows that $x_h \in S_j$ $(h = 1, \ldots, n)$, therefore $f(x_1, \ldots, x_n) \in S_j \subseteq S$. □

For lattices $\mathcal{L} \subseteq 2^A$, the converse of Proposition 6.4.2 also holds.

Theorem 6.4.3 (Birkhoff-Frink) *A lattice* $\mathcal{L} \subseteq 2^A$ *is algebraic if and only if it is isomorphic to* Sub(A) *for a certain algebra of support* A.

PROOF: Let $^-$ be the algebraic closure operator associated with $\mathcal{L}$. For every $n \geq 1$, every $\alpha = (a_1, \ldots, a_n) \in A^n$ and every $b \in \overline{\{a_1, \ldots, a_n\}}$, define

$$f_b^\alpha(x_1, \ldots, x_n) = \begin{cases} b\,, & \text{if } (x_1, \ldots, x_n) = \alpha\,, \\ x_1\,, & \text{if } (x_1, \ldots, x_n) \neq \alpha\,. \end{cases}$$

and $f_b = b$ if $b \in \overline{\varnothing}$. We will prove that $< X >= \overline{X}$ for every $X \subseteq A$.

To prove $< X > \subseteq \overline{X}$, we apply a theorem from universal algebra which in this case says that $\overline{X} = \bigcup_{p\geq 0} X_p$, where $X_0 = X$ and $X_{p+1} = X_p \cup Y_p$, where Y_p is the set of all f_b and all $f_b^\alpha(x_1, \ldots, x_n)$ with $x_1, \ldots, x_n \in X_p$. We prove by induction that $X_p \subseteq \overline{X}$ for all p.

We have $X_0 = X \subseteq \overline{X}$. Now suppose $X_p \subseteq \overline{X}$. Note first that $f_b = b \in \overline{\varnothing} \subseteq \overline{X}$. For f_b^α and $x_1, \ldots, x_n \in X_p$, two cases may occur. If $(x_1, \ldots, x_n) = \alpha$, then

$$f_b^\alpha(x_1, \ldots, x_n) = b \in \overline{\{a_1, \ldots, a_n\}} = \overline{\{x_1, \ldots, x_n\}} \subseteq \overline{\overline{X}} = \overline{X}\ ,$$

while if $(x_1, \ldots, x_n) \neq \alpha$, then $f_b^\alpha(x_1, \ldots, x_n) = x_1 \in \overline{X}$.

To prove $\overline{X} \subseteq< X >$, consider first the case $X = \varnothing$. If $\overline{\varnothing} = \varnothing$, then $\overline{\varnothing} \subseteq< \varnothing >$. If $\overline{\varnothing} \neq \varnothing$, then the elements $b \in \overline{\varnothing}$, being the constants $b = f_b$ of the algebra, belong to any subalgebra; in particular $b \in< \varnothing >$, proving that $\overline{\varnothing} \subseteq< \varnothing >$.

The next case which we consider is when X is a finite non-empty set, say $X = \{x_1, \ldots, x_n\}$. Then for every $b \in \overline{X}$ we have

$$b = f_n^{(x_1,\ldots,x_n)}(x_1, \ldots, x_n) \in< X >\ ,$$

therefore $\overline{X} \subseteq< X >$.

We have thus proved that $\overline{Y} \subseteq< X >$ for every finite subset Y of X. Since $^-$ is an algebraic closure operator, condition (i) in Theorem 6.4.1 implies that $\overline{X} \subseteq< X >$ for every $X \subseteq A$. □

Proposition 6.4.3 *The family* Con(A) *of congruences of an algebra* A *is an algebraic lattice.*

PROOF 1 : Having in view Theorem 6.4.2, we work with algebraic closure operators. Define on $A \times A$ an algebra of the same type as A by setting

$$f'((x_1, y_1), \ldots, (x_n, y_n)) = (f(x_1, \ldots, x_n), f(y_1, \ldots, y_n))$$

for every operation f in the signature of A. If $\theta \in \mathrm{Con}(A)$, then

$$(x_1, y_1), \ldots, (x_n, y_n) \in \theta \Longrightarrow$$

$$\Longrightarrow f'((x_1, y_1), \ldots, (x_n, y_n)) = (f(x_1, \ldots, x_n), f(y_1, \ldots, y_n)) \in \theta\ ,$$

therefore $\mathrm{Con}(A) = \mathrm{Eq}(A) \cap \mathrm{Sub}(A \times A)$. But $\mathrm{Eq}(A)$ and $\mathrm{Sub}(A \times A)$ are algebraic closure operators on $2^{A\times A}$, which implies easily that their intersection is also an algebraic closure operator on $2^{A\times A}$.
PROOF 2 : One establishes a formula for the join operation in $\mathrm{Con}(A)$, which shows that $\mathrm{Con}(A)$ is a complete sublattice of $\mathrm{Eq}(A)$, therefore condition (iii) in Theorem 6.4.1 shows that $\mathrm{Con}(A)$ is an algebraic lattice. □

A result similar to Theorem 6.4.3 holds for congruences : a lattice of subsets of a set A *is algebraic if and only if it i isomorphic to* $\mathrm{Con}(A)$ *for a certain algebra of support* A.

If an algebra A is *with commuting congruences*, i.e., if $\theta_1 \circ \theta_2 = \theta_2 \circ \theta_1$ for every two congruencies θ_1 and θ_2, then *the lattice* $\mathrm{Con}(A)$ *is modular.* This fact enables the transfer from certain possible properties of the lattice $\mathrm{Con}(A)$ to properties of the algebra A.

For instance, consider the following two possible properties for a lattice. The *restricted chain condition* says that there is no infinite decreasing sequence $x_0 > x_1 > \cdots > x_n > \ldots$ with the property that for every $n \geq 1$ there is y_n such that $x_{n-1} = x_n \vee y_n$ with $x_n \wedge y_n = 0$. The *descending* (*ascending*) chain condition says that there is no infinite decreasing (increasing) sequence. If $\mathrm{Con}(A)$ satisfies the restricted chain condition, then A *can be decomposed as a finite direct product of indecomposable algebras.* If $\mathrm{Con}(A)$ satisfies the ascending and descending chain conditions, then *this decomposition is unique up to an isomorphism.* This means that if $A = B_1 \times \cdots \times B_r = C_1 \times \cdots \times C_s$, then $r = s$ and there is a bijection $B_i \longleftrightarrow C_j$ such that the corresponding factors satisfy $B_i \cong C_j$. Also, A has a decomposition as a direct sum of simple algebras (i.e., with only two congruences) if and only if $\mathrm{Con}(A)$ is complemented.

The book by Pierce [1968] reports several transports of properties from $\mathrm{Con}(A)$ to A. See also Schmidt [1969].

While the congruence lattice plays an important rôle for algebras in general, it is the subgroup lattice which turns out to be very useful in the study of groups. The monograph by Suzuki [1956] has the title "Structure of a Group and the Structure of its Lattice of Subgroups".

Several significant result can be found in Ch. VIII of Birkhoff [1967]. For instance, a preliminary result is that if G is an Abelian group then $\mathrm{Con}(G) = \mathrm{Sub}(G)$. It is easy to see that $\mathrm{Sub}(G)$ is finite if and only if the group G is finite. Birkhoff appreciates that "it is of interest to classify groups according to their subgroup lattices". Thus, Theorem VII.14 says that Sub(G) is distributive if and only if G is cyclic, while Theorem VII.19 establishes that if G is finite and Sub(G) is modular, then G is solvable.

Last but not least, let us mention the works of the Romanian mathematician Benado in this line. He studied the theory of regular products of operator groups introduced by Golovin and the Schreier-Zassenhaus and Jordan-Hölder theorems for groups, by devising lattice-theoretical models beyond the previously used framework of modular lattices. Instead, Benado worked with arbitrary lattices and even with posets or with his multilattices. He obtained very general results which he then applied to the starting level of groups. A survey of the whole of Benado's more comprehensive mathematical creation is provided by Rudeanu and Vaida [in press].

6.5 Formal Concept Analysis

As stated in the monograph by Ganter and Wille [1999], "formal concept analysis is a field of applied mathematics based on the mathematization of *concept* and *conceptual hierarchy*. It thereby activates mathematical thinking for conceptual data analysis and knowledge processing.". In this section we present the beginning of the theory.

Definition 6.5.1 A *formal context* (G, M, I) consists of two sets G, M and a relation $I \subseteq G \times M$. The elements of G are called *objects* (*Gebiete*), those of M are named *attributes* (*Merkmalen*), while I is the *incidence relation*.

Definition 6.5.2 For any $A \subseteq G$, put

(6.5.1) $$A' = \{m \in M \mid g\,I\,m \text{ for all } g \in A\}$$

(the set of attributes common to the objects of A), and for any $B \subseteq M$, put

(6.5.2) $$B' = \{g \in G \mid g\,I\,m \text{ for all } m \in B\}$$

(the set of objects which have all the attributes in B).

Definition 6.5.3 A *formal concept* of the context (G, M, I) is a pair (A, B), where

(6.5.3) $$A \subseteq G,\ B \subseteq M, \text{ and } A' = B,\ B' = A\,.$$

The sets A and B are called the *extent* and the *intent* of the concept (A, B). We denote by $\mathcal{C}(G, M, I)$ the set of all formal concepts.

Definition 6.5.3 reflects the fact, pointed out in general logic, that each concept has two aspects, which determine each other: the intent and the extent. Thus, e.g., the intent of the concept "square" consists of the properties common to all squares, while the extent comprises all the squares.

Proposition 6.5.1 *The maps $A \mapsto A'$ and $B \mapsto B'$ determine a Galois connection $A \subseteq B' \iff B \subseteq A'$ between 2^G and 2^M.*

PROOF: By Example 4.5.1, since the sets (6.5.1) and (6.5.2) are a rewriting of the sets (4.5.2) and (4.5.3), respectively. □

In fact, $A' = A^+$ and $B' = B^-$. According to Theorem 4.5.1, the map $A \mapsto A''$ is a closure operator on 2^G and the map $B \mapsto B''$ is a closure operator on 2^M. In the sequel the term *closed set* will refer to these closure operators.

Proposition 6.5.2 *A subset of G (of M) is the intent (the extent) of a concept if and only if it is closed.*

PROOF: If (A, B) is a concept, then $A = B' = A''$ and $B = A' = B''$. If $A = A''$ then (A, A') is a concept, and if $B = B''$ then (B', b) is a concept. □

To go further, we need:

Lemma 6.5.1 *Let P, Q be complete lattices and $^+ : P \longrightarrow Q$ and $^- : Q \longrightarrow P$ a Galois connection $x \leq y^- \iff y \leq x^+$. Then for every family $(x_j)_{j \in J} \subseteq P$ we have*

$$(\bigvee_j x_j)^+ = \bigwedge_j x_j^+ \, .$$

COMMENT The dual of this property does not hold.
PROOF: We check the inf property of $(\bigvee_j x_j)^+$.

For every $j \in J$ we have $x_j \leq \bigvee_j x_j$, hence $(\bigvee_j x_j)^+ \leq x_j^+$. If $y \leq x_j^+$ for all j, then $x_j \leq y^-$ for all j, hence $\bigvee_j x_j \leq y^-$, therefore $y \leq (\bigvee_j x_j)^+$. □

Now we make the set of all concepts into a complete lattice.

Theorem 6.5.1 *The set $\mathcal{C}(G, M, I)$ is a complete lattice with respect to the partial order*

$$(A, B) \leq (A_1, B_1) \iff A \subseteq A_1 \iff B_1 \subseteq B \, , \tag{6.5.4}$$

the meet and join operations being given by

$$\bigwedge_j (A_j, B_j) = (\bigcap_j A_j, (\bigcap_j A_j)') \, , \tag{6.5.5}$$

$$\bigvee_j (A_j, B_j) = ((\bigcap_j B_j)', \bigcap_j B_j) \, . \tag{6.5.6}$$

PROOF: If (A, B) and $A_1, B_1)$ are concepts, then

$$A \subseteq A_1 \iff A \subseteq B_1' \iff B_1 \subseteq A' \iff B_1 \subseteq B \, ,$$

therefore formula (6.5.4) is consistent and the relation $\leq$ is a partial order.

If (A_j, B_j), $j \in J$, are concepts, then the sets $\bigcap_j A_j = \bigcap_j B_j'$ and $\bigcap_j B_j = \bigcap_j A_j'$ are closed by Proposition 6.5.2, therefore the right sides of formulas (6.5.5) and (6.5.6) are indeed concepts, which shows that these formulas actually determine the meet and the join in $\mathcal{C}(G, M, I)$. □

The lattice $\mathcal{C}(G, M, I)$ can be characterized, up to an isomorphism, within complete lattices. We need one more definition, which we state for complete lattices, although it works for arbitrary posets.

Definition 6.5.4 A subset S of a complete lattice L is called *sup-dense* (*inf-dense*) if for every $x \in L$ there exists $S_x \subseteq S$ such that $x = \bigvee S_x$ ($x = \bigwedge S_X$).

Thus a complete lattice is by definition algebraic if the set of compact elements is sup-dense. As in that case, we have:

Remark 6.5.1 S is sup-dense (inf-dense) if and only if, for every $x \in L$,

$$x = \bigvee\{s \in S \mid s \leq x\} \ \ (x = \bigwedge\{s \in S \mid x \leq s\}) .$$

Theorem 6.5.2 *The following conditions are equivalent for a complete lattice L and a context $\mathcal{C}(G, M, I)$:*

(i) *L is isomorphic to $\mathcal{C}(G, M, I)$;*

(ii) *there exist $\gamma : G \longrightarrow L$ and $\mu : M \longrightarrow L$ such that $\gamma(G)$ is sup-dense in L, $\mu(M)$ is inf-dense in L, and for every $g \in G$ and $m \in M$,*

$$g\, I\, m \iff \gamma(g) \leq \mu(m) . \tag{6.5.7}$$

PROOF: (i)$\Longrightarrow$(ii) : Let $\varphi : \mathcal{C}(G, M, I) \longrightarrow L$ be an isomorphism. Writing simply x' for $\{x\}'$ and recalling that $x''' = x'$, we define γ and μ by

$$\gamma(g) = \varphi(g'', g'), \ \ \mu(m) = \varphi(m', m'') .$$

Let us prove that for every $(A, B) \in \mathcal{C}(G, M, I)$ we have

$$\varphi(A, B) = \bigvee\{\gamma(g) \mid g \in A\} = \bigwedge\{\mu(m) \mid m \in B\} .$$

Indeed,

$$\bigvee\{\gamma(g) \mid g \in A\} = \bigvee_{g \in A} \varphi(g'', g') = \varphi(\bigvee_{g \in A} (g'', g')) ;$$

using (6.5.6) we obtain

$$\bigvee_{g \in A} (g'', g') = ((\bigcap_{g \in A} g')', \bigcap_{g \in A} g') ,$$

while Lemma 6.5.1 implies

$$A' = (\bigcup_{g \in A} \{g\})' = \bigcap_{g \in A} g' ,$$

therefore

$$\bigvee\{\gamma(g) \mid g \in A\} = \varphi((\bigcap_{g \in A} g')', \bigcap_{g \in A} g') = \varphi(A'', A') = \varphi(A, B) .$$

One proves similarly that $\bigwedge\{\mu(m) \mid m \in B\} = \varphi(A, B)$.

To prove (6.5.7), note that (6.5.1) implies $g' = \{m \in M \mid g\, I\, m\}$, therefore

$$\gamma(g) \leq \mu(m) \iff \varphi(g'', g') \leq \varphi(m', m'') \iff (g'', g') \leq (m', m'')$$

$$\Longleftrightarrow g'' \subseteq m' \Longleftrightarrow \{m\} \subseteq g''' \Longleftrightarrow m \in g' \Longleftrightarrow g\,I\,m\,.$$

(ii)$\Longrightarrow$(i) : The proof is based on Proposition 4.3.7, which states that isomorphisms coincide with order isomorphisms. We will provide two isotone mappings $\varphi : \mathcal{C}(G,M,I) \longrightarrow L$ and $\psi : L \longrightarrow \mathcal{C}(G,M,I)$ such that $\varphi\psi = 1_L$ and $\psi\varphi = 1_{\mathcal{C}(G,M,I)}$.

Define $\varphi(A,B) = \bigvee(A)$. Then φ is isotone:

$$(A,B) \le (A_1,B_1) \Longleftrightarrow A \subseteq A_1 \Longrightarrow \gamma(A) \subseteq \gamma(A_1) \Longrightarrow \bigvee \gamma(A) \le \bigvee \gamma(A_1)\,.$$

For $x \in L$, define $\psi(x) = (A,B)$ with $A = \{g \in G \mid \gamma(g) \le x\}$ and $B = \{m \in M \mid x \le \mu(m)\}$.

First we prove that $\psi(x) \in \mathcal{C}(G,M,I)$. This will follow from

$$\bigvee \gamma(A) = x = \bigwedge \mu(B)\,, \tag{6.5.8}$$

because (6.5.1) and (6.5.8) imply $A' = B$:

$$m \in A' \Longleftrightarrow \forall g \in A \;\; g\,I\,m \Longleftrightarrow \forall g \in A \;\; \gamma(g) \le \mu(m)$$

$$\Longleftrightarrow \bigvee \gamma(A) \le \mu(m) \Longleftrightarrow x \le \mu(m) \Longleftrightarrow m \in B\,.$$

To prove (6.5.8), note first that $g \in A \Longrightarrow \gamma(g) \le x$, hence $\bigvee \gamma(A) \le x$. On the other hand,

$$\gamma(g) \le x \Longleftrightarrow g \in A \Longrightarrow \gamma(g) \in \gamma(A) \Longrightarrow \gamma(g) \le \bigvee \gamma(A)\,,$$

therefore using Remark 6.5.1 we get $x = \bigvee\{\gamma(g) \mid \gamma(g) \le x\} \le \bigvee \gamma(A)$. So $x = \bigvee \gamma(A)$ and similarly $x = \bigwedge \mu(B)$.

Furthermore, ψ is isotone because if $x \le x_1$ and $\psi(x_1) = (A_1,B_1)$, then $A \subseteq A_1$, that is, $(A,B) \le (A_1,B_1)$.

The next step is to prove that $\varphi\psi = 1_L$. Indeed, for every $x \in L$ we have $\varphi(\psi(x)) = \varphi(A,B) = \bigvee \gamma(A) = x$.

Finally, to prove the property $\psi\varphi(A,B) = (A,B)$ we start with $\psi(\varphi(A,B)) = \psi(\bigvee \gamma(A)) = (A_1,B_1)$, where

$$m \in B_1 \Longleftrightarrow \bigvee \gamma(A) \le \mu(m) \Longleftrightarrow \forall g \in A \;\; \gamma(g) \le \mu(m)$$

$$\Longleftrightarrow \forall g \in A \;\; g\,I\,m \Longleftrightarrow m \in A' \Longleftrightarrow m \in B\,,$$

therefore $\psi(\varphi(A,B)) = (A_1,B_1) = (A,B)$. □

Corollary 6.5.1 *In every context (G,M,I), the functions $\gamma : G \longrightarrow \mathcal{C}(G,M,I)$,*
$\gamma(g) = (g'',g')$ and $\mu : M \longrightarrow \mathcal{C}(G,M,I)$, $\mu(m) = (m',m'')$ satisfy (6.5.7) and the properties: $\gamma(G)$ is sup-dense in $\mathcal{C}(G,M,I)$ and $\mu(M)$ is inf-dense in $\mathcal{C}(G,M,I)$.

PROOF: By Theorem 6.5.2 for $L := \mathcal{C}(G,M,I)$ and $\varphi := 1_{\mathcal{C}(G,M,I)}$. □

Remark 6.5.2 The meaning of $g\,I\,m$ is "object g has property m", therefore g' is the set of all properties of g, g'' is the set of all objects which share the properties of g, m' is the set of all objects with property m, and m'' is the set of all properties of the objects with property m.

Remark 6.5.3 From the point of view of Galois connections, the shorthand $A^+ = A'$ and $B^- = B'$ is unambiguous, with the exception of the empty set: we have $\varnothing^+ = M$ (because $g \in \varnothing \Longrightarrow g\,I\,m$) and $\varnothing^- = G$ (because $m \in \varnothing \Longrightarrow g\,I\,m$).

Corollary 6.5.2 *We have* $G'' = G$ *and* $M'' = M$.

PROOF: $G = \varnothing^- = \varnothing^{-+-} = G^{-+} = G''$ and similarly $M = M''$. □

Corollary 6.5.3 *In* $\mathcal{C}(G, M, I)$ *we have* $(M', M) = 0$ *and* $(G, G') = 1$.

PROOF: $(M', M), (G, G') \in \mathcal{C}(G, M, I)$ by Corollary 6.5.2, and for every $(A, B) \in \mathcal{C}(G, M, I)$ we have $B \subseteq M$ and $A \subseteq G$, therefore $(M', M) \le (A, B) \le (G, G')$. □

Corollary 6.5.4 *In every context* (G, M, I), *the possible atoms (dual atoms) of* $\mathcal{C}(G, M, I)$ *are among the non-zero elements of the form* $\gamma(g)$ *(non-one elements of the form* $\mu(m)$*);* cf. Corollary 6.5.1.

PROOF: We prove that an element $(A, B) \in \mathcal{C}(G, M, I)$ which is not of the form $\gamma(g)$ cannot be an atom.

If $A = \varnothing$ then $(A, B) = (\varnothing, \varnothing') = 0$.

If $A \ne \varnothing$, take $g \in A$. Then $\{g\} \subseteq A$, hence $g'' \subseteq A'' = A$, therefore $\gamma(g) = (g'', g') \le (A, B)$, while $(A, B) \ne \gamma(g)$ by hypothesis. Besides, from $\{g\} \subseteq g''$ we infer $g'' \ne \varnothing$. So $0 = (\varnothing, \varnothing') < (g'', g') < (A, B)$, therefore (A, B) is not an atom. The other claim follows similarly. □

Corollary 6.5.5 *The atoms (dual atoms) of a finite concept lattice* $\mathcal{C}(G, M, I)$ *are the minimal non-zero elements of the form* $\gamma(g)$ *(the maximal non-one elements of the form* $\mu(m)$*).*

PROOF: Atoms do exist by the finiteness of $\mathcal{C}(G, M, I)$, they are among the non-zero elements $\gamma(g)$ by Corollary 6.5.4, and clearly the non-minimal elements $\gamma(g) \ne 0$ are not atoms. The other claim follows similarly. □

The above properties can be used in order to actually construct finite concept lattices.

The first idea is to work with the M-components, ordered by the dual of set inclusion.

The first step consists in detecting (the M-components of) the atoms of $\mathcal{C}(G, M, I)$ among (the M-components of) the elements $\gamma(g)$.

In the following steps we successively determine (the M-components of) the joins of 2 atoms, the joins of 3 atoms, etc., joining also with the non-atom elements $\gamma(g)$, until we reach (the M-component G' of) 1.

Finally we compute the G-components corresponding to the M-components which we have found.

We carry out this construction on two examples. The first one illustrates also the fact, mentioned in the beginning of this section, that formal concept analysis is not limited to objects and attributes, but is also applied in data analysis and knowledge processing.

Example 6.5.1 Consider the following context, which represents the service offers of an office supplies business: $G = \{1, 2, 3, 4, 5, 6, 7, 8\}$ and $M = \{f, c, x, t, s\}$ where

1 = Consulting, 2 = Planning, 3 = Assembly and installation, 4 = Instruction, 5 = Training workshops, 6 = Original spare parts and accessories, 7 = Repairs, 8 = Service contracts, and

f = Furniture, c = Computers, x = Copy-machines, t = Typewriters, s = Specialized machines, and

the relation I is given in Table 6.1 below.

	1	2	3	4	5	6	7	8
f	×	×	×			×	×	
c	×	×	×	×	×	×	×	×
x	×		×	×		×	×	×
t	×		×	×		×	×	×
s	×		×	×		×	×	

Table 6.1

We write $x \ldots z$ instead of $\{x, \ldots, z\}$.

Table 6.1. shows that

$f' = 12367$, $c' = 12345678$, $x' = 134678$, $t' = 134678$, $s' = 13467$,

$1' = fcxts$, $2' = fc$, $3' = fcxts$, $4' = cxts$, $5' = c$, $6' = fcxts$, $7' = fcxts$, $8' = cxt$.

We identify (A, B) with B. So, instead of $(A, B) \leq (A_1, B_1)$ we write $B \leq B_1$ ($\Longleftrightarrow B_1 \subseteq B$), and (6.5.6) in the particular case $j \in \{1, 2\}$ will be written $B_1 \vee B_2 = B_1 \cap B_2$.

With these conventions, the elements $\gamma(g)$ are f', c', x', t', s'. We have $c' = M = 0$ by Corollary 6.5.3. The atoms are 12367 and 134678 by Corollary 6.5.5. Besides, $134678 < 13467$.

The join of the two atoms is $12367 \vee 134678 = 12367 \cap 134678 = 1367$.

The single non-atom element $\gamma(g) \neq 0$ is $s' = 13467$. We have $134678 < 13467$ and we compute $12367 \vee 13467 = 12367 \cap 13467 = 1367$.

It follows from Table 6.1, Lemma 6.5.1 and Corollary 6.5.3 that

$1367 = f' \cap c' \cap x' \cap t' \cap s' = (fcxts)' = G' = 1$.

Therefore the construction of $\mathcal{C}(G, M, I)$ in terms of M-components is over. We have obtained the well-known non-modular lattice known as *the pentagon*, cf. Fig.4.3, with the following "dictionary": $0 := 12345678$, $a := 134678$, $b := 13467$, $c := 12367$, $1 := 1367$.

It remains to compute the corresponding G-components. From $g \in B' \Longleftrightarrow \{g\} \subseteq B' \Longleftrightarrow B \subseteq g'$ we infer that $B' = \{g \in G \mid B \subseteq g'\}$, therefore

$(12345678)' = c$, $(134678)' = cxt$, $(13467)' = cxts$, $(12367)' = fc$, $(1367)' = fcxts$.

The final result is depicted in Fig. 6.1.

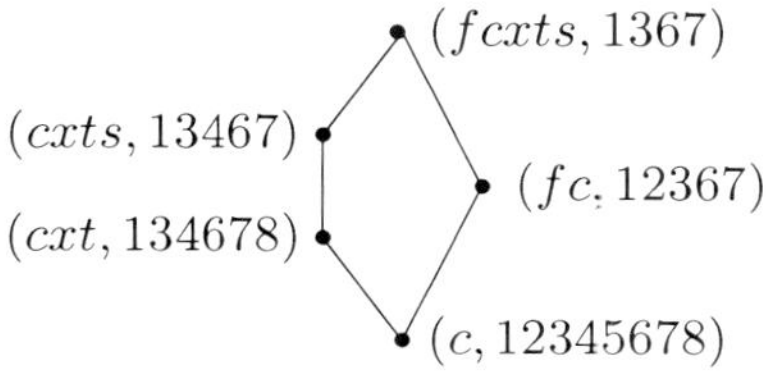

Fig. 6.1

Example 6.5.2 Consider the following context for the planets of our solar system:

$G = \{Me, V, E, Ma, J, S, U, N, P\}$, that is : Mercury, Venus, Earth, Mars, Jupiter, Saturn, Uranus, Neptune, Pluto,

and the following properties relative to size, distance from sun, and natural satellites:

1 = small, 2 = medium, 3 = large, 4 = near, 5 = far, 6 = yes, 7 = no.

Then the relation I is given in Table 6.2 below.

	1	2	3	4	5	6	7
Me	×			×			×
V	×			×			×
E	×			×		×	
Ma	×			×		×	
J			×		×	×	
S			×		×	×	
U		×			×	×	
N		×			×	×	
P	×				×	×	

Table 6.2

Table 6.2 shows that

Me′ = V′ = E′ = 147, Ma′ = 146, J′ = S′ = 356, U′ = N′ = 256, P′ = 156,

1′ = MeVEMaP, 2′ = UN, 3′ = JS, 4′ = MeVEMa, 5′ = JSUNP, 6′ = EMaJSUNP, 7′ = MeV.

The joins of 2 atoms are

$147 \vee 146 = 14, 147 \vee 356 = 147 \vee 256 = \varnothing, 147 \vee 156 = 1, 146 \vee 356 = 146 \vee 256 = 6, 146 \vee 156 = 16, 356 \vee 256 = 356 \vee 156 = 256 \vee 156 = 56$.

The last step is $1 \vee 6 = \varnothing$.

The M-labelled concept lattice is depicted in Fig. 6.2.

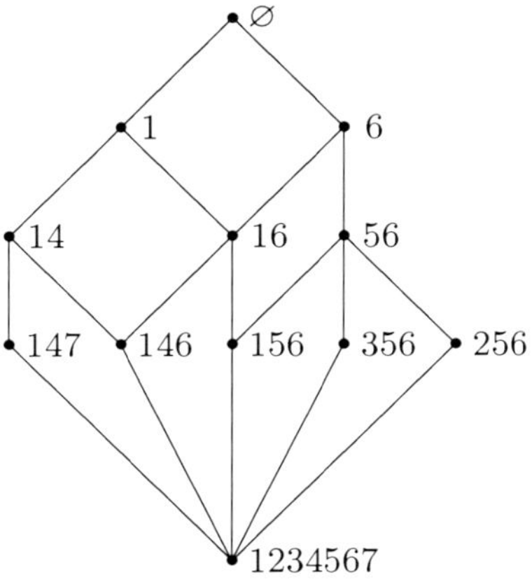

Fig. 6.2

We leave to the reader the task of computing the G-components. It will turn out that the elements (A, B) of this concept lattice are the following:

$(\varnothing, 1234567)$,
(MeV, 147), (EMa, 146), (P, 156), (JS, 356), (UN, 256),
(MeVEMa, 14), (EMaP, 16), (JSUNP, 56),
(MeVEMaP, 1), (EMaJSUNP, 6),
(MeVEMaJSUNP, $\varnothing$).

6.6 There is Much Algebra Behind Logic

The aim of this section is to illustrate the rôle of algebra in mathematical logic. We focus on the formal system of classical propositional calculus, which is the starting point in mathematical logic. We will conclude with a few references to several monographs dealing with algebraic methods in logic.

The formal systems dealt with in mathematical logic have two components: the *language* and the *logic*. The language consists of the set of strict rules which must be obeyed when we write formulas in the system. The logic has two components: *syntax* and *semantics*. The syntax provides an abstract model of mathematical *proofs*, while the semantics studies the way in which the *truth* of certain sentences can be inferred from the truth of other sentences.

The title of this section expresses a thesis which we wish to illustrate on the *formal system of the classical propositional calculus*. Of course, we cannot transform this section into a chapter about the classical propositional calculus, so we will only sketch it, emphasizing the algebraic aspects and omitting tedious computations. The point of our approach is to reproduce those short proofs which enlighten ideas, and to give without proofs those important results which follow from tedious calculations. We introduce (recall) propositional calculus without recalling the formal construction of compound propositions and without specifying the sets of axioms (nor an axiom system for Boolean algebras): the axioms are used only in the omitted proofs.

The reader is warned that there are numerous variations of this calculus as concerns the signature and the axioms, but all of them are equivalent.

We will work with algebras of signature $(\wedge, \vee, ', \rightarrow)$ of type (2,2,1,2), which corresponds to the fact that on the one hand we are familiar with compound propositions of the form $p \,\&\, q$, $p \vee q$ and $\neg p$, and on the other hand the implication $\rightarrow$, not so intuitive as conjunction, disjunction or negation, is essential for the theory.

The language of propositional calculus is provided by the following purely algebraic construction. Given a set V of elements called *propositional variables*, the *term algebra* is the free algebra $\mathcal{E} = (E, \&, \vee, \neg, \rightarrow)$ freely generated by V in the class of all algebras of the same signature. The elements of the support set E are called *terms*, or *well-formed expressions*, or simply *propositions*. The name "free algebra" means that 1) $< V >= E$, i.e., the subalgebra of E generated by V is E itself, and 2) for every algebra $\mathcal{A} = (A, \wedge, \vee, ', \rightarrow)$ of the same signature, every function $v : V \longrightarrow A$ has a unique homomorphic extension $v_{\mathcal{A}} \in \mathrm{Hom}(\mathcal{E}, \mathcal{A})$. This homomorphism is called an *$\mathcal{A}$-valuation*. When there is no danger of confusion, we may write simply v instead of $v_{\mathcal{A}}$.

This *universality property* has a crucial consequence for the semantics of propositional calculus. The usual coding for the two possible truth values of a sentence is "true" $= 1$, and "false" $= 0$. So the Boolean algebra $2 = (\{0,1\}, \wedge, \vee, ', \rightarrow)$, where $x \rightarrow y = x' \vee y$, becomes the algebra of truth values. Note that in the case of a *bivalent valuation*, meaning a 2-valuation v, the homomorphism conditions

$$v(\alpha \,\&\, \beta) = v(\alpha) \wedge v(\beta),\ v(\alpha \vee \beta) = v(\alpha) \vee v(\beta),\ v(\neg\alpha) = (v(\alpha))' ,$$

are nothing but the well-known *truth tables*.[3] That is why a bivalent valuation may be also called a *system of truth values*.

Summarizing, the fact that $\mathcal{E}$ is a free algebra ensures the property that every assignment of truth values 0 and 1 to the propositional variables is uniquely extended to a system of truth values for all the propositions.

Before continuing with propositional calculus, let us explore a bit more the relationship between $\mathcal{E}$ and the algebras $\mathcal{A}$ of the same signature. Using the homomorphism $v_{\mathcal{A}} : \mathcal{E} \longrightarrow \mathcal{A}$, we associate with each term $\alpha \in E$ the function

(6.6.1) $$\alpha_{\mathcal{A}} : A^V \longrightarrow A, \alpha_{\mathcal{A}}(v) = v_{\mathcal{A}}(\alpha) \text{ for all } v : V \longrightarrow A .$$

Then the following properties hold:

[3]The truth tables of $\&$, $\vee$, $\neg$ correspond to our intuition about the propositions $p \,\&\, q$, $p \vee q$ and $\neg p$, while the truth table of the implication $\rightarrow$ requires a more delicate discussion.

(6.6.2.1) $$p_{\mathcal{A}}(v) = v_{\mathcal{A}}(p) = v(p) \;\; for \text{ every } p \in V \ ,$$

(6.6.2.2) $$(\neg\alpha)_{\mathcal{A}}(v) = v_{\mathcal{A}}(\neg\alpha) = (v_{\mathcal{A}}(\alpha))' = (\alpha_{\mathcal{A}}(v))' \ ,$$

and for each binary function $\circ$ in the signature,

(6.6.2.3) $$(\alpha \circ \beta)_{\mathcal{A}}(v) = v_{\mathcal{A}}(\alpha \circ \beta) = v_{\mathcal{A}}(\alpha) \circ v_{\mathcal{A}}(\beta) = \alpha_{\mathcal{A}}(v) \circ \beta_{\mathcal{A}}(v) \ .$$

The meaning of the *function* $\alpha_{\mathcal{A}}$ *generated by* α is that $\alpha_{\mathcal{A}}$ is the function obtained from α if the symbols $'$ and $\circ$ of $\mathcal{E}$ are interpreted as operations of $\mathcal{A}$, and the propositional variables occurring in α are interpreted as variables in everyday mathematics. For instance, if $p, q \in V$, then $\alpha = p \vee q \in E$ is just a string of 3 symbols with no associated meaning. On the other hand, $\alpha_{\mathcal{A}}(v) = p_{\mathcal{A}}(v) \vee q_{\mathcal{A}}(v) = v(p) \vee v(q)$ is an element of A, and so are $x = v(p)$ and $y = v(q)$. Therefore $\alpha_{\mathcal{A}}(v) = x \vee y$ and when v runs over A^V, the variables x, y run over A.

The *identities* of algebra $\mathcal{A}$ are defined as those pairs $(\alpha,\beta) \in E^2$ for which $\alpha_{\mathcal{A}} = \beta_{\mathcal{A}}$, or equivalently, $v_{\mathcal{A}}(\alpha) = v_{\mathcal{A}}(\beta)$ for every $v \in A^V$. For instance, if $\alpha = p \vee q$ and $\beta = q \vee p$, where $p, q \in V$, then (α,β) being an identity of $\mathcal{A}$ means $x \vee y = y \vee x$ for every $x, y \in A$, that is, the commutativity of the operation $\vee$ of $\mathcal{A}$.

The name "free algebra" is probably explained by the fact that no identity in the everyday sense is imposed on $\mathcal{E}$. Therefore $\mathrm{Hom}(\mathcal{E},\mathcal{E}) = E^E$, whence an easy argument shows that the identities of $\mathcal{E}$ in the above technical sense are just the pairs (α,α).

A well-known theorem says that the identities common to all Boolean algebras are the same as the identities of the *two-element Boolean algebra* $2 = \{0, 1\}$. This can be paraphrased as follows: for every $\alpha,\beta \in E$,

(6.6.3) $$\alpha_2 = \beta_2 \Longleftrightarrow \alpha_{\mathcal{B}} = \beta_{\mathcal{B}} \text{ for every Boolean algebra } \mathcal{B} \ .$$

Now we come back to the logic of propositional calculus.

The *axioms* of propositional calculus are chosen in such a way that, if the symbols occurring in the axioms are interpreted as in everyday mathematics, one obtains true sentences. This property is shared by the larger set of *theses* of propositional calculus. The theses are obtained from the axioms by applying the so-called *deduction rules*. Now two approaches are possible: either one adopts *substitution* as a deduction rule, meaning that for every axiom or thesis, the replacement of the letters occurring in it by any terms results in a new thesis, or the axioms are interpreted as *schemes of axioms*, meaning that the letters $p, q, r, \ldots$ occurring in the axioms represent arbitrary propositions from $\mathcal{E}$. We prefer the latter approach, which exempts us from using the substitution rule.

In our sketch we do not specify the set Ax of axioms. As we have just explained, we don't need the substitution rule. We adopt *modus ponens*, that is,

$$\frac{\alpha,\alpha \to \beta}{\beta} \ ,$$

as the unique deduction rule. We are going to express this syntax in algebraic form.

Let us equip the set E with the partially defined operation *modus ponens*,

$$MP : E \times E \overset{\circ}{\longrightarrow} E, \quad MP(\alpha,\alpha \to b) = \beta, \text{ else undefined .}$$

If $\mathcal{A}$ is an algebra of support A and $X \subseteq A$, let $< X >$ denote the subalgebra of $\mathcal{A}$ generated by X.

With this notation the set of *theses* is defined as $< \mathrm{Ax} >$, and if H is an arbitrary subset of E, we introduce the set of propositions *deducible from H*, defined by $< \mathrm{Ax} \cup H >$. Setting $< \mathrm{Ax} \cup H >= \mathrm{Ded}H$, it is immediately seen that $\mathrm{Ded} : 2^E \longrightarrow 2^E$ is a *closure operator*. The meaning behind this definition is that the set $\mathrm{Ded}H$ represents the propositions which can be proved if we assume the truth of the propositions in H, where H models the axioms of a mathematical theory. The set $\mathrm{Ded}\varnothing$ of theses is a model for the propositions like $p \vee \neg p$, which we can prove because they are true by themselves. The notation $H \vdash \alpha$ means $a \in \mathrm{Ded}H$, while the fact that α is a thesis is denoted simply by $\vdash \alpha$. Two important tools are the *deduction theorem*, which says that $H \cup \{\alpha\} \vdash \beta \Longleftrightarrow H \vdash \alpha \to \beta$, and the *syllogistic rule*, saying that if $H \vdash \alpha$ and $\alpha \vdash \beta$ then $H \vdash \beta$.

The following result may be called a *compactness* theorem.

Proposition 6.6.1 *If $H \vdash \alpha$ then $H_0 \vdash \alpha$ for some finite subset $H_0 \subseteq H$.*

PROOF: We have $\alpha \in \mathrm{Ded}H \ =< \mathrm{Ax} \cup H >$. According to the so-called compactness theorem for partial algebras, there is a finite subset $S \subset \mathrm{Ax} \cup H$ such that $\alpha \in< S >$. Set $S \cap H = \{h_1, \ldots, h_n\}$ unless $S \cap H = \varnothing$, in which case take $h_1, \ldots, h_n \in H$ arbitrarily. Then $S = (S \cap \mathrm{Ax}) \cup (S \cap H) \subseteq \mathrm{Ax} \cup \{h_1, \ldots, h_n\}$, hence $\alpha \in< \mathrm{Ax} \cup \{h_1, \ldots, h_n\} >= \mathrm{Ded}\{h_1, \ldots, h_n\}$. □

An important topic is consistency. Like any mathematical theory, mathematical logic should be consistent.

The *consistency* of the propositional calculus means that we cannot have both $\vdash \alpha$ and $\vdash \neg\alpha$.

The concept of consistency is extended to the subsets of E as follows. A subset $H \subseteq E$ is called *consistent* if we cannot have $H \vdash \alpha$ and $H \vdash \neg\alpha$; in other words, we cannot deduce a falsity from H. The existence of consistent sets (see below) implies the consistency of propositional calculus.

A consistent set $H \subset E$ is said to be *maximal* if there is no consistent set properly including H.

Proposition 6.6.2 *Every consistent set is included in a maximal consistent set.*

PROOF: In view of the Kuratowski-Zorn lemma, it suffices to prove that if $\{H_t \mid t \in T\}$ is a chain of consistent sets, then $H = \bigcup_{t\in T} H_t$ is a consistent set.

Indeed, suppose there exists $\alpha \in H$ such that $H \vdash \alpha$ and $H \vdash \neg\alpha$. According to Proposition 6.6.1, there exist two finite subsets $H_1, H_2 \subseteq H$ such that $H_1 \vdash \alpha$

and $H_2 \vdash \neg\alpha$. Since Ded is a closure operator, it follows that both α and $\neg\alpha$ belong to $\mathrm{Ded}(H_1 \cup H_2)$. Set $H_1 \cup H_2 = \{h_1, \ldots, h_n\}$, where $h_i \in H_{t_i}$ for $i = 1, \ldots, n$. But $\{H_t \mid t \in T\}$ is a chain, therefore there is $j \in \{1, \ldots, n\}$ such that $H_{t_i} \subseteq H_{t_j}$ $(i = 1, \ldots, n)$. So $H_1 \cup H_2 \subseteq H_{t_j}$, hence $\mathrm{Ded}(H_1 \cup H_2) \subseteq \mathrm{Ded}H_{t_j}$, therefore both α and $\neg a$ belong to $\mathrm{Ded}H_{t_j}$, which contradicts the consistency of H_{t_j}. □

It is therefore natural to look for a characterization of maximal consistent sets.

Lemma 6.6.1 *If H is a maximal consistent set, then $H = \mathrm{Ded}H$.*

PROOF: If $H \neq \mathrm{Ded}H$, take an element $\alpha \in \mathrm{Ded}H \setminus H$ and compute

$$\mathrm{Ded}(H \cup \{\alpha\}) \subseteq \mathrm{Ded}(H \cup \mathrm{Ded}H) = \mathrm{Ded}\mathrm{Ded}H = \mathrm{Ded}H \subseteq \mathrm{Ded}(H \cup \{\alpha\}) ,$$

therefore $\mathrm{Ded}(H \cup \{a\}) = \mathrm{Ded}H \neq E$, which shows that H is not a maximal consistent set. □

Theorem 6.6.1 *A subset $H \subseteq E$ is a maximal consistent set if and only if it satisfies*

(i) $H = \mathrm{Ded}H$, *and*

(ii) *for every $\alpha \in E$, exactly one of the elements a and $\neg\alpha$ belongs to H.*

PROOF: If H is a maximal consistent set, we have just proved (i). Further, by the definition of consistency, we cannot have both α and $\neg\alpha$ in H.

If $\alpha \notin H$, then $H \subset H \cup \{\alpha\}$, therefore $\mathrm{Ded}(H \cup \{\alpha\}) = E$ by the maximality of H. In particular $H \cup \{\alpha\} \vdash \neg\alpha$, therefore $H \vdash \alpha \to \neg\alpha$ by the deduction theorem. But it is known that $\alpha \to \neg\alpha \vdash \neg\alpha$, hence $H \vdash \neg\alpha$ by the syllogistic rule, consequently $\neg\alpha \in \mathrm{Ded}H = H$.

If $\neg\alpha \notin H$, it follows that $H \vdash \neg\neg a$ by the previous property; but it is known that $\neg\neg\alpha \vdash \alpha$, therefore $\alpha \in \mathrm{Ded}H = H$.

Conversely, suppose H satisfies (i) and (ii). Then $\mathrm{Ded}H = H \neq E$. If $H \subset G$, take $a \in G \setminus H$. Then $\alpha \notin H$, therefore $\neg a \in H$, hence $\alpha, \neg\alpha \in G$, therefore the set G is inconsistent, which shows that H is a maximal consistent set. □

At this point we recall without (not even a sketch of) proof the theorem saying that the relation

$$\alpha \sim \beta \text{ iff } \vdash \alpha \to \beta \text{ and } \vdash \beta \to \alpha \text{ iff } \alpha \vdash \beta \text{ and } \beta \vdash \alpha \tag{6.6.4}$$

is a congruence on $\mathcal{E}$, and the quotient algebra $\mathcal{E}/\sim$, endowed with the partial order

$$|\alpha| \leq |\beta| \text{ iff } \vdash \alpha \to \beta \text{ iff } \alpha \vdash \beta , \tag{6.6.5}$$

where $|\gamma|$ stands for the class modulo $\sim$ of an element $\gamma \in E$, is a Boolean algebra. The proof relies on much calculation and depends on the choice of the axiom systems for propositional calculus and for Boolean algebras. We only mention that the second "iff" in (6.6.4) and (6.6.5) follows from the deduction theorem for $H = \varnothing$.

The algebra $\mathcal{E}/\sim$, known as the *Lindenbaum-Tarski algebra* of propositional calculus, enables transfers of properties and proofs from rather difficult calculations in $\mathcal{E}$ to the much more convenient framework of Boolean algebras : we are in a position to go back and forth between $\mathcal{E}$ and $\mathcal{E}/\sim$.

Proposition 6.6.3 *The element* 1 *of* $\mathcal{E}/\sim$ *is the class of theses,* that is, for every $\alpha \in E$,

$$(6.6.6) \qquad \vdash \alpha \Longleftrightarrow \mid \alpha \mid = 1 \, .$$

PROOF: Suppose $\vdash \alpha$. Then for every $\beta \in E$ we have $\alpha \in \mathrm{Ded}\varnothing \subseteq \mathrm{Ded}\{\beta\}$, that is $\beta \vdash \alpha$, therefore $\mid \beta \mid \leq \mid \alpha \mid$. Hence $\mid \alpha \mid = 1$.

Conversely, suppose $\not\vdash \alpha$. Then $\vdash \beta$ implies $\not\vdash \beta \to \alpha$ (otherwise $\vdash \alpha$), hence $1 = \mid \beta \mid \not\leq \mid \alpha \mid$, therefore $\mid \alpha \mid \neq 1$. □

The next result is a more concrete variant of the compactness property provided by Proposition 6.6.1.

Proposition 6.6.4 $H \vdash \alpha$ *if and only if there exists a finite subset* $\{h_1, \ldots, h_n\} \subseteq H$ *such that* $\mid h_1 \mid' \vee \cdots \vee \mid h_n \mid' \vee \mid \alpha \mid = 1$.

PROOF: We have $H \vdash \alpha$ if and only if $\{h_1, \ldots, h_n\} \vdash \alpha$ for some $h_1, \ldots, h_n \in H$ by Proposition 6.6.1, and for these elements $h_1, \ldots, h_n$ we have

$$\{h_1, \ldots, h_n\} \vdash \alpha \Longleftrightarrow \vdash h_1 \to (h_2 \to \ldots \to (h_n \to \alpha) \ldots) \text{ (cf. deduction theorem)}$$

$$\Longleftrightarrow \mid h_1 \mid \to (\mid h_2 \mid \to \ldots \to (\mid h_n \mid \to \mid \alpha \mid) \ldots) = 1 \text{ (cf. Proposition 6.6.3)}$$

$$\Longleftrightarrow \mid h_1 \mid' \vee \mid h_2 \mid' \vee \cdots \vee \mid h_n \mid' \vee \mid \alpha \mid = 1 \text{ (Boolean computations)} \, .$$

□

There is a parallelism between the properties of consistent sets and the properties of proper filters in the Lindenbaum-Tarski algebra.

To see this, let us introduce the notation $X/\sim = \{\mid x \mid \mid x \in X\}$ for every subset $X \subseteq E$. Note that when computing with cosets $\mid \alpha \mid$ we may suppose, without loss of generality, that

$$(6.6.7) \qquad \mid \alpha \mid \in X/\sim \Longleftrightarrow \alpha \in X \, .$$

Exercise 6.6.1 Using Proposition 6.6.4 and (6.6.7), prove the following properties, for every subset $H \subseteq E$:

(i) $(\mathrm{Ded}H)/\sim$ is a filter of $\mathcal{E}/\sim$;

(ii) H is consistent $\Longleftrightarrow$ the filter $(\mathrm{Ded}H)/\sim$ is proper ;

(iii) H is maximal consistent $\Longleftrightarrow$ $(\mathrm{Ded}H)/\sim$ is a maximal filter.

Now we pass to semantics.

We say that α is a *tautology* and we write $\models \alpha$, if $v(\alpha) = 1$ for every bivalent valuation $v : \mathcal{E} \longrightarrow 2$.

For every bivalent valuation v and every $H \subseteq E$, we denote by $v(H) = 1$ the statement that $v(\alpha) = 1$ for every $\alpha \in H$, in which case we say that v is a *bivalent model* for H.

We say that H *implies semantically* α and we write $H \models \alpha$, if $v(H) = 1 \Longrightarrow v(\alpha) = 1$ for every bivalent valuation v.

Proposition 6.6.5 *Every consistent set has a bivalent model.*

PROOF: Let H be a consistent set. According to Proposition 6.6.2, there is a maximal consistent set G such that $H \subseteq G$. Then $(\text{Ded}G)/\sim$ is a maximal filter by Exercise 6.6.1, hence its characteristic function $\varphi : \mathcal{E}/\sim \longrightarrow 2$ is a homomorphism. So, $\varphi((\text{Ded}G)/\sim) = 1$ and if $\pi : \mathcal{E} \longrightarrow \mathcal{E}/\sim$ is the canonical surjection, then the function $f = \varphi \circ \pi : \mathcal{E} \longrightarrow 2$ is a homomorphism, that is, a bivalent valuation, and $f(G) = 1$, therefore $f(H) = 1$. □

The major problem is that of *completeness*, meaning *soundness* and *adequacy*: what we prove is true, and what is true can be proved.

Proposition 6.6.6 *For every $\alpha \in E$, the following conditions are equivalent:*

(i) $\vdash \alpha$;

(ii) $\alpha_{\mathcal{B}} = 1$ for every Boolean algebra $\mathcal{B}$;

(iii) $\alpha_2 = 1$;

(iv) $| \alpha | = 1$ in $\mathcal{E}/2$.

Comment For a generalization of the equivalences (i)$\Longleftrightarrow$(ii)$\Longleftrightarrow$(iv) see Rasiowa [1974], 7.7.2.

PROOF: (i)$\Longleftrightarrow$(iv): By Proposition 6.6.3.

(i)$\Longrightarrow$(ii): By induction on the definition of theses. The axioms $\alpha \in \text{Ax}$ are chosen in such a way that it is well known that $\alpha_{\mathcal{B}}(v) = 1$. Further, if $\vdash \alpha$ and $\vdash \alpha \to \beta$ and $\alpha_{\mathcal{B}}(v) = (\alpha \to \beta)_{\mathcal{B}}(v) = 1$ for all v, then using the Boolean identity $1 \to x = x$ and (6.6.2), we obtain

$$\beta_{\mathcal{B}}(v) = 1 \to \beta_{\mathcal{B}}(v) = \alpha_{\mathcal{B}}(v) \to \beta_{\mathcal{B}}(v) = (\alpha \to \beta)_{\mathcal{B}}(v) = 1 \ .$$

(ii)$\Longrightarrow$(iv): The canonical surjection $\pi : \mathcal{E} \longrightarrow \mathcal{E}/\sim$, $\pi(\alpha) = | \alpha |$, is a homomorphism, that is an $(\mathcal{E}/\sim)$-valuation, therefore (1) implies $| \alpha | = \pi(\alpha) = \alpha_{\mathcal{E}/\sim}(\pi) = 1$.

(ii)$\Longrightarrow$(iii): Trivial.

(iii)$\Longrightarrow$(ii): Take a thesis β; then $\beta_{\mathcal{B}} = 1$ for every Boolean algebra $\mathcal{B}$, because (i)$\Longrightarrow$(ii). So $\alpha_2 = 1 = \beta_2$, therefore $\alpha_{\mathcal{B}} = \beta_{\mathcal{B}} = 1$ by (6.6.3). □

Corollary 6.6.1 $\vdash \alpha \Longleftrightarrow \models \alpha$.

PROOF: In view of (6.6.1), property (iii) is $v(\alpha) = 1$ for all v, that is, $\models \alpha$. □

This corollary is known as *the weak completeness theorem*: theses coincide with tautologies.

Corollary 6.6.2 *For every* $\alpha,\beta \in E$*, the following conditions are equivalent:*

(i) $\alpha \sim \beta$;

(ii) $\alpha_{\mathcal{B}} = \beta_{\mathcal{B}}$ *for every Boolean algebra* $\mathcal{B}$;

(iii) $\alpha_2 = \beta_2$.

PROOF: The equivalence (ii)$\Longleftrightarrow$(iii) is (6.6.3).

Further, it follows by Proposition 6.6.6 and (6.6.2), using the fact that in every Boolean algebra $x \to y = y \to x = 1 \Longleftrightarrow x \leq y \;\&\; y \leq x \Longleftrightarrow x = y$, that

$$\text{(i)} \Longleftrightarrow (\alpha \to \beta)_2 = (\beta \to \alpha)_2 = 1 \Longleftrightarrow (\alpha \to \beta)_2(v) = (\beta \to \alpha)_2(v) = 1 \text{ for all } v$$

$$\Longleftrightarrow \alpha_2(v) \to \beta_2(v) = \beta_2(v) \to \alpha_2(v) = 1 \text{ for all } v$$

$$\Longleftrightarrow \alpha_2(v) = \beta_2(v) \text{ for all } v \Longleftrightarrow (iii) \ .$$

□

Let Id$\mathcal{B}$ denote the set of all *identities* $\alpha_{\mathcal{B}} = \beta_{\mathcal{B}}$ valid in a Boolean algebra $\mathcal{B}$, and Id**Bol** denote the set of identities valid in all Boolean algebras.

Corollary 6.6.3 *We have* $\sim =$ Id**Bol**.

PROOF: This is the equivalence (i)$\Longleftrightarrow$(ii). □

Proposition 6.6.7 *The Lindenbaum-Tarski algebra* $\mathcal{E}/\sim$ *is the free Boolean algebra freely generated by* V.

PROOF: From $\mathcal{E}/\sim = \mathcal{E}/$ Id**Bol** and a well-known theorem in universal algebra. □

Thus every function $v : V \longrightarrow 2$ can be uniquely extended to a homomorphism from $\mathcal{E}/\sim$ to 2. This is the algebraic counterpart of the fact that every $v \in 2^V$ can be uniquely extended to a system of truth values.

Lemma 6.6.2 $H \vdash \alpha \Longrightarrow H \models \alpha$.

PROOF: by induction on Ded$H$. Let v be a bivalent valuation. If $\alpha \in H$ and $v(H) = 1$ then trivially $v(\alpha) = 1$. Now suppose $H \models \alpha$ and $H \models \alpha \to \beta$. If $v(H) = 1$ then (as in the proof of Proposition 6.6.6)

$$v(\beta) = 1 \to v(\beta) = v(\alpha) \to v(b) = v(\alpha \to \beta) = 1 \ ,$$

therefore $H \models \beta$. □

Lemma 6.6.2 is the easy part of the *strong completeness theorem*, which says that

(6.6.8) $$H \vdash \alpha \Longleftrightarrow H \models \alpha \ .$$

The proof of the implication $\Longleftarrow$ uses specific properties of propositional calculus.

*

Numerous non-classical logics have similar approaches. For instance, *Lewis' modal logic* introduces the modal operator "it is possible that". The Lindenbaum-Tarski algebra of Lewis' propositional calculus is a closure algebra, i.e., a Boolean algebra endowed with a closure operator (see e.g. §6.1). The *intuitionistic logic* accepts the tautology $\alpha \to \neg\neg\alpha$, but rejects $\neg\neg\alpha \to \alpha$. The corresponding Lindenbaum-Tarski algebra is a Heyting algebra, i.e., a (distributive) lattice equipped with a binary operation $\to$ which satisfies $a \wedge x \leq b \Longleftrightarrow x \leq a \to \beta$.

Boolean algebras, Heyting algebras and closure algebras have in common their relationship to logic and topology, besides their intrinsic algebraic interest. Nowadays the study of algebras related to logic is a huge field, in continuous development. The interested reader is referred e.g. to the following monographs, which we list in chronological order: Rasiowa and Sikorski [1970], Rasiowa [1974], Boicescu et al. [1991], Buşneag [2006], Galatos et al. [2007], Piciu [2007], Iorgulescu [2008], Di Nola, Georgics and Leuştean [in preparation].

Bibliography

I. Abian
1970 : *On the similarity of partially ordered sets.* Amer. Math. Monthly 77(1970), 1092-1094.

M.A. Arbib, E.G. Manes
1975: *The Categorical Imperative: Arrows, Structures and Functors.* Academic Press, New York 1975.

H. Bachmann
1967 : *Transfinite Zahlen.* Springer-Verlag, Berlin 1967.

D.W. Barnes, J.M. Mack
1975 : *An Algebraic Introduction to Mathematical Logic.* Springer-Verlag, Berlin 1975.

C. Berge
1970 : *Graphes et hypergraphes.* Dunod, Paris 1970.

Bernard
1969 : *Préfermetures sur un ensemble ordonné.* Thèse. Univ. Clermont-Ferrand, 1969.

G. Birkhoff
1967 : *Lattice Theory.* Third edition. Amer. Math. Soc. Coll. Publ. vol.25. Providence, RI, 1967.
1970 : *What can lattices do for you?* In: J.C. Abbott, ed., Trends in Lattice Theory. Van Nostrand, New York 1970, 1-38.

T.S. Blyth
2005 : *Lattices and Ordered Structures.* Springer-Verlag, London 2005.

V. Boicescu, A. Filipoiu, G. Georgescu, S. Rudeanu
1991 : *Lukasiewicz-Moisil Algebras.* North-Holland, Amsterdam 1991.

N. Bourbaki
1963 : *Eléments de Mathématique. Théorie des Ensembles. Chap.3.* Hermann, Paris 1963.

S. Burris, H.P. Sankappanavar
1981 : *A Course in Universal Algebra.* Springer-Verlag, New York 1981.

D. Buşneag
2006: *Categories of Algebraic Logic.* Ed. Academiei Române, Bucureşti 2006.

P. CORSINI, V. LEOREANU
2003 : *Applications of Hyperstructure Theory.* Advances in Math., vol. 3 . Kluwer Academic Publ. 2003.

P. CRAWLEY, R.P.DILWORTH
1973 : *Algebraic Theory of Lattices.* Prentice Hall, Englewood Cliffs 1973.

B.A. DAVEY, H.A. PRIESTLEY
1990 : *Introduction to Lattices and Order.* Cambridge Univ. Press, New York 1990.

A.C. DAVIS
1955: *A characterization of complete lattices.* Pacific J. Math. 5(1955), 311-319.

B. DAVVAZ, V. LEOREANU-FOTEA
2008 : *Hyperring Theory and Applications.* Hadronic Press, 2008.

R. DEDEKIND
1900: *Über die von drei Moduln erzeugte Dualgruppe.* Math. Annalen 53, 371-403. = Ges. Werke, vol.2, 236-271.

A. DI NOLA, G. GEORGESCU, I. LEUŞTEAN
???? : *An Introduction to MV-algebras.* (in preparation).

M.L. DUBREIL-JACOTIN, L. LESIEUR, R. CROISOT
1953 : *Leçons sur la Théorie des Treillis, des Structures Algébriques Ordonnées et des Treillis Géométriques.* Gauthier-Villars, Paris 1953.

PH. DWINGER
1961 : *Introduction to Boolean Algebras.* Physica Verlag, Würzburg 1961.

M. ERNÉ
1982 : *Einführung in die Ordnungstheorie.* Bibliographisches Institut Mannheim
/Wien/Zürich 1982.

A. FRAENKEL, Y. BAR-HILLEL
1958 : *Foundations of Set Theory.* Second revised edition. North-Holland, Amsterdam 1958.

P. FREYD
1968: *Abelian Categories.* Harper & Row, New York; John Weatherwille, Tokyo 1968.

N. GALATOS, P. JIPSEN, T. KOWALSKI, H. ONO
2007: *Residuated Lattices. An Algebraic Glimpse at Substructural Logics.* Elsevier, Amsterdam 2007.

B. GANTER, R. WILLE
1999 : *Formal Concept Analysis.* Springer-Verlag, Berlin 1999.

N. GASPARD, B. LECLERC, B. MONTJARDET
2007 : *Ensembles Ordonnés Finis: Concepts, Résultats et Usages.* Springer, Berlin 2007.

G. GIERZ, K.H. HOFMANN, K. KEIMEL, J.D. LAWSON, M.MISLOWE, D.S. SCOTT
1980 : *A compendium of Continuous Lattices.* Springer-Verlag, Berlin 1980.

J.S. GOLAN
1992 : *The Theory of Semirings with Applications in Mathematics and Theoretical Computer Science.* Pitman Monographs Surveys Pure Appl. Math., vol. 54. Longman Sci. & Tech. 1992.

G. GRÄTZER
1978a : *General Lattice Theory.* Academic Press, New York 1978.
1078b : *Universal Algebra.* Springer-Verlag, New York 1978.

P.R. HALMOS
1950 : *Measure Theory.* Van Nostrand, Princeton 1950.
1960 : *Naive Set Theory.* Van Nostrand, Princeton 1960 = Springer-Verlag, Berlin 1974.
1962 : *Algebraic Logic.* Chelsea, New York 1962.
1963 : *Lectures on Boolean Algebras.* Van Nostrand, Princeton 1963.

A. IORGULESCU
2008 : *Algebras of Logic as BCK Algebras.* Editura ASE, Bucharest 2008.

D.A. KAPPOS
1969 : *Probability Algebras and Stochastic Spaces.* Academic Press, New York 1969.

J.L. KELLEY
1957 : *General Topology.* D. Van Nostrand, Princeton, New Jersey 1957.

F. LAPSCHER
1972 : *Théorie des fermetures.* Cours, Univ. Montpellier II, 1972.

S. MAC LANE
1971 : *Categories for the Working Mathematician.* Springer-Verlag, New York 1971.

J.C.C. MCKINSEY, A. TARSKI
1944 : *The algebra of topology.* Ann. Math. 45, 141-191.

J.D. MONK, R. BONNET, eds.
1989 : *Handbook of Boolean Algebras.* Vol.1–3. North-Holland, Amsterdam 1989.

A. MONTEIRO
1947 : *Sur l'arithmétique des filtres premiers.* C.R. Acad. Sci Paris 255 (1947), 846-848.

E.H. MOORE
1910. *Introduction to a Form of General Analysis.* New Haven Coll. 1906, Yale Univ. Press, New Haven, 1910.

L. NACHBIN
1947 : *Une propriété caractéristique des algèbres booléennes.* Portugaliae Math. 6(1947), 115-118.

C. NĂSTĂSESCU
1973 : *Introducere in Teoria Mulţimilor (Introduction to Set Theory).* Univ. Bucurşti (lito), 1973.

G. NÖBELING

1954 : *Grundlagen der Analytischen Topologie.* Springer-Verlag, Berlin 1954.

J. von Neumann
1923 : *Zur Einführung der transfiniten Zahlen.* Acta Szeged 1(1923), 199-208 = Collected Works, Pergamon Press, New York 1961, vol.1, 24-33.

O. Onicescu, I. Cuculescu
1976 : *Probability Theory on Boolean Algebras of Events.* Editura Academiei, Bucureşti 1976.

R. Padmanabhan, S. Rudeanu
2008 : *Axioms for Lattices and Boolean Algebras.* World Scientific, Singapore 2008.

D. Piciu
2007 : *Algebras of Fuzzy Logics.* Ed. Universitaria, Craiova 2007.

R.S. Pierce
1968 : *Introduction to the Theory of Abstract Algebras.* Holt, Reinehart and Winston, New York 1968.

D. Ponasse, J.-C. Carrega
1979 : *Algèbre et Topologie Booléennes.* Masson, Paris 1979.

N. Popescu
1971 : *Categorii Abeliene (Abelian Categories).* Ed. Academiei R.S.România, Bucureşti 1971.

N. Popescu, A. Radu
1971 : *Teoria Categoriilor şi a Fascicolelor (Category and Sheave Theory).* Ed. Ştiinţifică, Bucureşti 1971.

G. Preuss
1972 : *Allgemeine Topologie.* Springer-Verlag, Berlin 1972 .

H.A. Priestley
1970a : *Representation of distributive lattices by means of ordered Stone spaces.* Bull. London Math. Soc. 2(1970), 186-190.
1970b : *Ordered topological spaces and the representation of distributive lattices.* Proc. London Math. Soc. (3) 24(1970), 507-530.

Gh. Radu
1988 : *Algebra Categoriilor şi a Functorilor (Algebra of Categories and Functors).* Ed. Junimea, Iaşi 1988.

H. Rasiowa
1974 : *An Algebraic Approach to Non-Clssical Logics.* North-Holland, Amsterdam 1974.

H. Rasiowa, R. Sikorski
1970 : *The Mathematics of Metamathematics.* Panstwowe Wydawnictwo Naukowe, Warszawa 1963. 3rd ed. 1970.

S. Roman
2008 : *Lattices and Ordered Sets.* Springer-Verlag, New York 2008.

S. Rudeanu

1963 : *Elemente de Teoria Mulţimilor (Elements of Set Theory).* Univ. Bucureşti (lito), 1963.

1974 : *Boolean Functions and Equations.* North-Holland, Amsterdam 1974.

2001 : *Lattice Functions and Equations.* Springer-Verlag, London 2001.

S. RUDEANU, D. VAIDA

Revisiting the works of M. Benado. J. Multiple-Val. Logic Soft Comput.(in press)

E.T. SCHMIDT

1969 : *Kongruenzrelationen Algebraischer Strukturen.* VEB Deutscher Verlag Wiss., Berlin 1969.

1970 : *Eine Verallgemeinerung des Satzes von Schmidt-Ore.* Publ. Math. Debrecen 12(1970), 283-287.

W. SIERPIŃSKI

1965 : *Cardinal and Ordinal Numbers.* PWN, Warsaw, 1965.

R. SIKORSKI

1964 : *Boolean Algebras.* Springer-Verlag, Berlin 1964.

M.H. STONE

1937 : *Applications of the theory of Boolean rings to general topology.* Trans. Amer. Math. Soc. 41(1937), 375-481.

M. SUZUKI

1956 : *Structure of a Group and the Structure of its Lattice of Subgroups.* Springer-Verlag, Berlin 1956.

G. SZÁSZ

1963 : *Introduction to Lattice Theory.* Academic Press, New York & Akadémiai Kiado, Budapest 1963.

T. SZELE

1950 : *On Zorn's lemma.* Publ. Math. (Debrecen) 17(1950), 254-256.

R. VAIDYANATHASWAMY

1960 : *Set Topology.* Chelsea Publ. Co., New York 1960.

D.A. VLADIMIROV

1969 : *Bulevy Algebry.* Isdatel'stvo Nauka, Moskva 1969.

Index

Made in United States
North Haven, CT
04 November 2026

21719580R00146